Systems Analysis and Policy Planning

Systems Analysis and Policy Planning

APPLICATIONS IN DEFENSE

Edited by
E. S. QUADE
and
W. I. BOUCHER
The RAND Corporation, Santa Monica, California

American Elsevier Publishing Company, Inc.
New York

AMERICAN ELSEVIER PUBLISHING COMPANY, INC.
52 Vanderbilt Avenue, New York, N.Y. 10017

ELSEVIER PUBLISHING COMPANY
335 Jan Van Galenstraat, P.O. Box 211, Amsterdam, The Netherlands

Standard Book Number 444-00033-X
Library of Congress Catalog Card Number 68-22241

Seventh Printing, 1975

PRINTED IN THE UNITED STATES OF AMERICA

PREFACE

"One of the key problems of contemporary national security policy," as Henry Kissinger has said, "is the ever-widening gap that has opened up between the sophistication of technical studies and the capacity of an already overworked leadership group to absorb their intricacy." This book, a survey of the nature, aims, and limitations of systems analysis in current defense planning, is an attempt to help close that gap. We focus on systems analysis because it is unquestionably the most powerful and widely influential approach to systematic inquiry that decisionmakers and policy-oriented analysts have at their disposal today – and are likely to have in the foreseeable future. We focus on its applications to problems of national security because here, in its traditional domain, there is a continuing and probably growing need, as Kissinger suggests, to understand the concepts and procedures of analysis. Almost daily, of course, there are new indications that systems analysis is beginning to discover a role in policy planning outside the area of defense, on all levels of government, in industry and commerce, and elsewhere. For this effort to be successful, however, we feel that it is essential to understand how systems analysis has worked in defense, and why. Here, too, in the search for new applications, this book may make a contribution, perhaps in helping to keep a possible gap from opening.

As a pioneer in the development of systems analysis and the techniques on which it relies, The RAND Corporation has long recognized the need to clarify the nature of analysis in defense planning. To this end, some twelve years ago RAND first began to offer short courses on systems analysis to senior military officers and civilians associated with the Armed Forces. To bring this material to a wider audience, the lectures given in 1959, much revised and amplified, were declassified and published in the open literature.* The most recent course, which was held in 1965, provided the basis for the present volume. Needless to say, those lectures given in 1965 that are retained in this volume have been thoroughly updated for publication at this time; indeed, most of them have also been enlarged or extensively revised. In addition, several new chapters have been written especially for this book. The resulting collection, we believe, extends the discussion of systems analysis in certain fundamental ways:

It makes an effort to account for the development of analysis in the last decade;

It discusses at length certain methods of analysis that either are receiving great emphasis today (such as a Monte Carlo computer

*E. S. Quade (ed.), *Analysis for Military Decisions*, Rand McNally & Company, Chicago, 1964.

routine that simulates the operations and maintenance requirements of complex equipment at one or several locations) or are relatively new (such as the analytic scenario and the Delphi technique);

It attempts to anticipate the problems that analysts will have to overcome in the future, and to explore the reasons why these problems will arise;

And it reexamines earlier conclusions, particularly in the areas of establishing criteria for decisionmaking, weighing the utility of accepted standards for measuring the effectiveness of alternatives, defining the principles of suboptimization, handling the problem of uncertainty, and responding to the interaction between the pace of technological advance and the proper role and character of analysis itself.

This volume, like its predecessor, attempts to demonstrate that systems analysis can and does provide knowledge that decisionmakers need; that it can serve to sharpen intuition; that its usefulness is not limited solely to questions of policy and planning that can be quantified; and, most important, that whatever its weaknesses, it produces more fruitful results, of far greater consequence and reliability, than any of its alternatives. We should emphasize, however, that this book, while concerned with such essentials, is not simply introductory. It is intended more as a sophisticated guide to users of analysis than as a manual for those who prepare such material.

Its basic aim is to provide detailed answers – both practical and theoretical – to the questions that will be important to those responsible for sponsoring, evaluating, or implementing the analyses of others. What is systems analysis? Why is it necessary? When and where is it appropriate? How does one approach, and carry out, a systems analysis? What methods can be used? How can a good analysis be recognized? What can one expect from a systems analysis? How has analysis changed over the years? Why? What changes can be expected in the future?

The organization and contents of the whole reflect this emphasis on satisfying the needs of the *users* of systems analysis. The first six chapters explore the basic concepts of systems analysis. Included are an introductory example of analysis; a discussion of the problem of selecting operationally useful objectives, measures of their attainment, and criteria; a somewhat mathematical discussion of uncertainty; and an examination of the place and function of technological considerations in analyzing the merits of proposed systems. The next three chapters discuss the character and importance of costs in systems analysis. Resource analysis and cost-sensitivity analysis are illustrated in depth. The following nine chapters concern models – what they are, how they are constructed and used, what their limitations are, and what place they have in analysis as a whole. It is here that game theory, simulation, scenarios, war gaming, and political analysis are considered, as well as some newer techniques that attempt to provide a framework for obtaining the judgments of experts on problems that are not amenable to any of the methods of quantitative analysis.

Of the remaining four chapters, the first discusses a large variety of flaws, in both analysis and analysts, that can seriously affect the conduct or evaluation of systems studies. The second examines the character of analysis in the recent past, compares it to that of the present, and looks briefly ahead. The third returns to the opening example of analysis and shows, in light of what has gone between, how the problem might better have been approached and solved. The final chapter attempts to draw together the threads essential to the earlier discussions and, again, to take a look ahead.

Nowhere does this book presume an advanced knowledge of such specific tools of analysis as linear programming or probability theory, or their special applications to military problems.

In assembling this volume, we have made no attempt to eliminate the informality of the chapters that originally were lectures; nor have we attempted to impose a common viewpoint on the book as a whole. The authors remain their own agents, and each – it will be noted – takes a critical approach to his own topic, and that of the others. As editors, what we have tried to do is to give the book a unity, and at the same time discourage more than that minimum of repetition that would allow each chapter to stand by itself should the reader choose to revise our ordering. If we have succeeded, our greatest debt is to the authors themselves. But we also owe a good share of the credit to the assistance of our colleagues – in particular, R. L. Belzer, G. H. Clement, R. J. Lew, E. T. Lowe, and Col. J. W. Shinners (USAF), who helped us in selecting the topics and lecturers for the original seminar and provided critical comment.

This book and the course from which it derives were undertaken by The RAND Corporation as a part of its research program for the United States Air Force.

E.S.Q.
W.I.B.

Santa Monica, California
December, 1967

CONTENTS

PREFACE . . . v
FIGURES . . . xix
TABLES . . . xxiii

Chapter

1 INTRODUCTION, by E. S. Quade . . . 1
Introductory Remarks . . . 1
Definition of Systems Analysis . . . 1
Origins . . . 2
Relation to Other Types of Analysis . . . 3
The Need for Analysis . . . 5
Types of Applications . . . 7
The Essence of Systems Analysis . . . 11
The Elements of Analysis . . . 12
Development of Systems Analysis . . . 13
The Current Status of Systems Analysis . . . 15
Objectives of This Book . . . 18

2 THE TRADE-OFF BETWEEN GROUNDPOWER AND AIR SUPPORT: AN INTRODUCTORY EXAMPLE OF SYSTEMS ANALYSIS, by M. G. Weiner . . . 20
Introduction . . . 20
Casualty Production by Airpower . . . 22
Casualty Production by Groundpower . . . 23
Some Additional Assumptions . . . 23
Effect of Blue Groundpower . . . 24
Effect of Blue Airpower . . . 25
Combined Effects of Airpower and Groundpower . . . 25
Effect of Blue Force Attrition . . . 25
Other Air-Ground Force Mixes . . . 26
Costs . . . 27
Trade-off Conclusions . . . 27
Some Larger Considerations . . . 28

3 PRINCIPLES AND PROCEDURES OF SYSTEMS ANALYSIS, by E. S. Quade . . . 30
Introduction . . . 30
The Essence of Systems Analysis . . . 30

The Alternatives 31
The Process of Analysis 33
Formulation 35
Search 41
Evaluation 42
Interpretation 50

4 CRITERIA AND THE MEASUREMENT OF EFFECTIVENESS, by L. D. Attaway 54
Introduction 54
Elements of Analysis 54
An Example: Selection of a New Aircraft Engine . . 55
Overspecification of Criteria 57
Maximizing Effectiveness/Cost 57
Dominance 58
A Second Example: Choice of Operational Mode for Interceptors 59
Selection of a Scale of Effectiveness 59
Measurement of Effectiveness 61
Selection of an Alternative 62
Suboptimization 63
Aggregation 63
A Third Example: Allocation of a Strategic Budget . . 64
The Definition of Goals 64
Limitations of Effectiveness Scales 66
Alternatives in an Uncertain Future 68
Measurement of Effectiveness 72
Criteria 75
Summary 78

5 UNCERTAINTY, by Albert Madansky 81
Probabilities–Objective and Subjective 81
Utilities–The Theory of Criterion Selection . . . 85
Examples of the Use of Utility Theory 89
Models of Uncertainty in Systems Analysis . . . 91
Treatment of Uncertainty in Systems Analysis . . . 94
Buy Time 95
Buy Information 95
Buy Flexibility as a Hedge 95
Use *A Fortiori* Analysis 95
Use Sensitivity Analysis 96
Postscript 96

6 TECHNOLOGICAL CONSIDERATIONS, by H. Rosenzweig 97
Introduction 97
Types of Decisions 98
Concept Formulation 98
Contract Definition 101
Full Development 101
The Number of Units Deployed 102
Modifications 102
Replacement or Phase-out 103
Summary 104
Technical Feasibility 104
Selection of Best Technical Approach 108
Scope of the Problem 108
Formulation of the Problem 110
Design Parameters and Parametric Analysis . . . 110
Trade-off Analyses 113
Technological Uncertainty 115
Comparison with Competing Systems 118
Nonrelative Comparisons 120
External Constraints 121

7 RESOURCE ANALYSIS, by G. H. Fisher . . . 124
Introduction 124
Some Definitions and Concepts 125
Context 127
Time Horizon 127
Type of Decision 129
Scope of the Problem 129
Impact of Context on Concepts and Techniques Used in Resource Analysis 130
Summary Remarks 137

8 COST-SENSITIVITY ANALYSIS: AN EXAMPLE, by R. L. Petruschell and A. J. Tenzer 138
The Context of Cost Analysis 138
Definition of Cost-Sensitivity Analysis 138
Assumptions of the Example 139
Tactics of SLBM Interception 139
Construction of the Model 141
Effects of System Variations on Aircraft Cycle Time . . 142
Number of Aircraft vs. Endurance of the Aircraft . . 143
Endurance vs. State-of-the-Art 144

Maintenance Policy vs. Number of Aircraft . . . 145
Loading Time vs. Endurance of the Planes Airborne . 147
System Costs 147
Usefulness of Sensitivity Analysis 151

9 THE B-x: A HYPOTHETICAL BOMBER COST STUDY, by W. E. Mooz 153
Introduction 153
The B-x System and Its Estimated Cost 155
Initial Assumptions 155
Calculation of Personnel Requirements 156
Operating Personnel 158
Maintenance Personnel 159
Administrative Personnel 162
Support Personnel 163
Summary 164
Major Cost Categories 165
Calculation of RDT&E Costs for the B-x System . . 167
Design and Development 167
System Test 169
Major Cost Categories for Initial Investment . . . 173
Calculation of Initial Investment Costs 173
Facilities 174
Primary Mission Equipment 175
Unit Support Aircraft 178
Aerospace Ground Equipment 178
Miscellaneous Equipment 178
Stocks 178
Spares 179
Personnel Training 179
Initial Travel 179
Initial Transportation 180
Major Cost Categories for Annual Operation . . . 180
Calculation of Annual Operating Costs 180
Facilities Replacement and Maintenance . . . 180
PME Replacement 181
PME Maintenance 183
PME POL 184
Unit Support Aircraft POL and Maintenance . . 184
AGE R&M 184
Personnel Pay and Allowances 185
Personnel Replacement Training 185

Annual Travel 186
Annual Transportation 187
Annual Services 187
Summary and Discussion of Results 188
Sensitivity Analysis of the B-x System 192
Possible Errors in Estimating 193
PME Maintenance, PME POL, and AGE R&M . . 194
Maintenance Personnel 194
PME Replacement 195
Summary 197
Possible Variations in Operating Parameters . . . 199
Aircraft Constraints 199
Crew Constraints 202
The B-x Operating Envelope 207
The B-x Cost Envelope 209

10 THE NATURE OF MODELS, by R. D. Specht . . 211
What Is a Model? 211
Design and Use of Models: An Example 214
Omissions in This Example 217
The Problem of Selecting Criteria 218
The Problem of Deciding What Is Relevant . . . 218
The Necessity of Being Explicit 219
The Treatment of Nonquantitative Considerations . . 219
The Static Character of the Model 219
Other Observations 220
Types of Models 221
Analytical Models 222
Computer Models 224
People Models 224
People and Computer Models 224
Verbal Models 225
Conclusion 226

11 MATHEMATICAL MODELS OF CONFLICT, by Melvin Dresher 228
Introduction 228
An Air Defense Game 229
Description of Payoff 230
Optimal Strategies 231
Generalization of the Air Defense Model . . . 231
A Tactical Air War Game 233

Formulation of the Tactical Air Game . . . 234
Description of Payoff 235
Further Simplifying Assumptions 236
Strategies of the Tactical Air Game 236
Optimal Tactics 236
Generalization of the Tactical Air Game Model . . 238
Applications of Game Theory 239

12 SIMULATION, by Norman C. Dalkey 241
Introduction 241
Simulation Basics 243
Choice of Simulation Techniques 246
Examples of the Use of Simulation 248
Pros and Cons 250
New Developments 252

13 SAMSOM: A LOGISTICS SIMULATION, by
Chauncey F. Bell 255
Introduction 255
Description of SAMSOM 255
Estimating Requirements for Manpower and Equipment . 260
Conclusions 262

14 GAMING, by M. G. Weiner 265
Introduction 265
What War Gaming Is Not 265
What War Gaming Is 266
Types of War Games 266
Varying Requirements 266
Essential Characteristics 267
Some War Games at RAND 267
Special Conditions for Limited War Game Play . . 268
Phases of War Gaming 269
Preparation Phase 269
Play Phase 271
Analysis Phase 272
Some Examples 273
A Tactical Game for Comparing Different Weapons . 274
A Tactical Game for Examining Force Requirements . 274
A Tactical Game for Comparing Programs of Military
Assistance 275
A Lesson from Prince Ladislaus 277

15 THE ANALYSIS OF FORCE POLICY AND POSTURE INTERACTIONS, by Roger Levien 279
Interaction of Policy and Posture 279
Other Determinants of Posture 280
Role of Analysis 281
What Is Needed? 281
Traditional Tools of Analysis 282
Can the Traditional Methods Meet the Requirements? . 283
The SAFE Game 285
The Menu of Options in SAFE 286
Routine of Play 290
A Set of SAFE Plays 291
Types of Analysis of SAFE Plays 293
Limitations of SAFE as an Analytic Tool . . . 295
Benefits of the SAFE Game 296

16 SCENARIOS IN SYSTEMS ANALYSIS, by Seyom Brown . 298
Introduction 298
Definition of Scenario 299
Importance of Discussing Scenarios Systematically . . 300
Form and Content Determined by Research Task . . 301
Levels of Analysis 302
The Problem of Political Credibility 305
Scenarios and Specialists 309

17 U.S. SPACE POLICY: AN EXAMPLE OF POLITICAL ANALYSIS, by Alton Frye 311
Some Analytical Distinctions 311
Political-Strategic Analysis: A Perpetual Process . . 312
Typical Challenge: Military Role in Space . . . 314
U.S. Policy of Restraint 315
Appraising Soviet Intentions in Space 317
Outstanding Issues for U.S. Space Policy . . . 321
Rational Decisionmaking in a Technical-Political Environment 322

18 WHEN QUANTITATIVE MODELS ARE INADEQUATE, by E. S. Quade 324
Introduction 324
The Basic Role of Judgment 324
Identifying Experts 325
Utilizing Individual Experts 327
Utilizing Groups of Experts 328

The Contextual Map 329
Operational Gaming 329
The Delphi Technique 333
Summary 343

19 PITFALLS AND LIMITATIONS, by E. S. Quade . . . 345
Ordinary Errors and Expert Blunders 345
Some Common Fallacies in Analysis 348
Underemphasis on Problem Formulation . . . 348
Inflexibility in the Face of Evidence 349
Adherence to Cherished Beliefs 350
Parochialism 352
Communication Failure 353
Overconcentration on the Model 353
Excessive Attention to Detail 353
Neglect of the Question 354
Incorrect Use of the Model 354
Disregard of the Limitations 355
Concentration on Statistical Uncertainty . . . 355
Inattention to Uncertainties 356
Use of Side Issues as Criteria 357
Substitution of a Model for the Decisionmaker . . 357
Neglect of the Subjective Elements 359
Failure To Reappraise the Work 360
Limitations of Analysis 360
Analysis Is Necessarily Incomplete 361
Measures of Effectiveness Are Inevitably Approximate . 361
No Satisfactory Way Exists To Predict the Future . . 363
Systems Analysis Falls Short of Scientific Research . . 363

20 THE CHANGING ENVIRONMENT FOR SYSTEMS ANALYSIS, by James R. Schlesinger 364
Introduction 364
Major Environmental Changes 365
Increased Perception of Political Fluidity . . . 365
Greater Sophistication Regarding the Character of Nuclear War 366
Increased Emphasis on Highly Specialized Weapon Systems . 369
Rising Costs of R&D 371
Two Contemporary Analytical Problems 372
Criteria for Resource Allocation for Damage Limitation . 373

Constraining Future Strategic Options by an Early-on Force-Structure Decision 380
Some Final Inferences 384
Uncertainties: State of the World, Objectives, and Strategies 384
Disparate Approaches to Analysis 385
Developing and Selecting U.S. Military Capabilities . . . 385
The Problem of Cherished Beliefs 386
Complementarities and Mission Trade-offs . . . 387
Quantitative Precision 387

21 THE TRADE-OFF STUDY REVISITED, by L. H. Wegner and M. G. Weiner 388
Introduction 388
Establishing a Point of View 389
Levels of Analysis and Criteria 390
Level 1 – Trade-offs Between Different Forces to Accomplish the Same Specific Task 390
Level 2 – Trade-offs Between Different Forces in the Same Situation 393
Level 3 – Trade-offs Between Different Forces to Implement National Policy 397
An Illustration 400
The Force Mixes and Their Cost 400
The Threat 402
The Model 403
Examples of Output 407
Discussion of Results 409
A Final Note 416

22 BY WAY OF SUMMARY, by E. S. Quade 418
Precepts for the Systems Analyst 419
Principles of Good Analysis 422
Nature of the Decisionmaker and His Responsibility . . 425
Some Dangers of Analysis 426
The Future of Systems Analysis 427

Selected Bibliography 430

Bibliographic Note 440

Index 441

FIGURES

1.1 The Range of Problems in Which Analysis Can Be Useful . 8
1.2 The Structure of Analysis 14
1.3 The Key to Successful Analysis 14
2.1 Effect of Blue's Groundpower 24
2.2 Effect of Blue's Airpower 25
2.3 Combined Effects of Airpower and Groundpower . . 26
2.4 Effect of Blue's Force Attrition 27
2.5 Effect of Various Air-Ground Mixes 27
3.1 Activities in Analysis 35
4.1 Cost and Effectiveness 56
4.2 Effectiveness/Cost Ratio 58
4.3 Dominance 58
4.4 Strategic Choices 68
4.5 Effectiveness/Cost for a Series of Alternatives . . . 69
4.6 Percentage of Plays with Effectiveness, E . . . 74
4.7 Fixed Contingency Effectiveness Results . . . 75
4.8 Typical Fixed Contingency Results (Fixed Cost) . . 76
4.9 Cost and Effectiveness 79
4.10 Cost, Effectiveness, and Contingency 79
5.1 Subjective Probability that Reliability Is at Least ρ . . 83
5.2 Worth as a Function of Risk 86
5.3 Percentage of Plays with Effectiveness, E . . . 90
5.4 Worth as a Function of Effectiveness (Case 1) . . . 90
5.5 Worth as a Function of Effectiveness (Case 2) . . . 91
6.1 Parametric Analysis Output (Constant Budget) . . . 112
6.2 Range-Payload Trade-off 114
6.3 Cruise Versus Dash Range Trade-off 115
6.4 Effect of Uncertainty 120
6.5 Nonrelative Comparison 122
6.6 Effect of External Constraints 123
7.1 System "Life Cycle" Identification Plotted Against Time (Idealized Curves) 131
7.2 Missile System Cost Versus Payload for Various Types of Propellants and Ground Environments (Fixed Number of Ready Missiles) 135
8.1 Defense Zone (1000 Miles Out from Both Coasts) . . 140
8.2 Boost-Intercept 141
8.3 Mid-course Intercept 141
8.4 Counter-battery 141
8.5 Aircraft Cycle Time 142

8.6 Number of Aircraft Versus Endurance Hours (Case A) . 143
8.7 Number of Aircraft Versus Endurance Hours (Case B) . 144
8.8 Number of Aircraft Versus Endurance Hours (Case C) . 144
8.9 Number of Aircraft Versus Endurance Hours (Case D) . 145
8.10 Number of Aircraft Versus Endurance Hours (Case E) . 146
8.11 Number of Aircraft Versus Endurance Hours (Case F) . 146
8.12 Percentage Airborne Versus Endurance Hours . . . 147
8.13 Cost Versus Endurance 148
8.14 Cost Versus Defense Zone Extent 149
8.15 Number of Air Stations Versus Defense Zone Extent . . 150
8.16 Cost Per Pound Versus Payload Weight 150
8.17 Time-Phased Resource Impact 151
9.1 Organization Chart for B-x Wing 157
9.2 Estimating Relationship for Obtaining Bomber Maintenance Personnel Factors 161
9.3 Cost of Installed Underground Fuel Storage . . . 175
9.4 Cost-Quantity Relation 177
9.5 Aircraft Destroyed Versus System Flying Hours . . 182
9.6 Static Analysis of B-x System Costs (5-Year Operation) . 190
9.7 Time-Phased Costs for B-x System 191
9.8 Effect of Estimating Errors Upon System Operating Costs . 198
9.9 Effect of Estimating Errors Upon Total System Cost Estimate 199
9.10 Monthly Time Distribution Between Flying Activities and Ground Duty Activities 201
9.11 Effect of Maintenance Requirements upon Aircraft Activities 202
9.12 Maintenance and Flying Hours as a Function of Ground Alert Duty 203
9.13 Duty Schedule for Crews, at 152 Hours per Month Flying and Ground Alert Duty 204
9.14 Effect of Flying and Training Constraints upon Crew Duty . 205
9.15 Possible Ground Alert and Flying Duty in Terms of Aircraft and Personnel Constraints 206
9.16 B-x Operating Envelope 209
10.1 Categories of Model Forms 223
13.1 Model Logic Schematic 256
13.2 Simulated Daily Flying Program (Typical Manning) . . 258
13.3 Simulated Daily Flying Program (Manpower as Needed) . 259
13.4 Possible Daily Flying Program – Typical Manning Reallocated 260
13.5 Important Considerations in Establishing Resource Requirements 261
13.6 Time-Oriented Data 261

13.7 Manpower Utilization Analysis (Adjusted) . . . 263
14.1 Play Phase (Two-Sided Games) 271
14.2 A Princely Failure 278
15.1 Can the Traditional Methods Do the Job? . . . 284
15.2 Blue Menu of Options in SAFE (Circa 1962) . . . 287
15.3 Two Blue Menu Items 288
15.4 Six SAFE Plays 292
20.1 Trade-off Curve for Damage Limiting 374
20.2 Trade-off Curves for Damage Limiting 375
20.3 Measuring War Outcomes 376
20.4 Variation in Optimal Resource Allocation Depending on Assumed Character of War 379
21.1 Basic Weapon-Target Matrix 391
21.2 Weapon-Target Matrix with a Distance Measure Added . 392
21.3 Trade-off Criteria on Levels 1 and 2 395
21.4 Trade-off Criteria on All Three Levels 399
21.5 Structure of the TAGS-II Model 405
21.6 Case 2: Movement of the FEBA (West) . . . 411
21.7 Case 2: Blue Air Sorties 412
21.8 Case 2: Blue Division Losses 412
21.9 Case 2: Red Aircraft Losses 413
21.10 Case 2: Red Division Losses 413
21.11 Case 2: Planned Versus Actual Red Ground Force Commitments 414
21.12 Case 2: Planned Red Ground Force Commitments, and Divisions Surviving at the Front 414
21.13 Case 2: Blue Aircraft Losses 415
22.1 Precepts for the Systems Analyst 419
22.2 Principles of Good Analysis 422

TABLES

3.1 Hypothetical War Outcomes for Three Alternatives . . 44
3.2 A Framework for Evaluating Alternatives . . . 46
4.1 The Performance of Two Possible Strategic Systems . . 66
4.2 Hypothetical Performance of Two Other Strategic Systems . 67
7.1 Resource Analysis Categories for Individual Systems . . 132
7.2 Component Structure for a Ballistic Missile (Or Similar Aerospace Vehicle) 134
9.1 Aircraft Characteristics – The B-x 156
9.2 Operating Personnel – B-x Bomber Squadron . . . 159
9.3a Armament Maintenance Personnel – B-x System . . 160
9.3b Total Maintenance Personnel – B-x System . . . 161
9.4 Direct Personnel – B-x System 163
9.5 Support Personnel – B-x System 164
9.6 Total Personnel Requirements – B-x System . . . 164
9.7 Cost Categories for Aircraft 166
9.8 Cumulative Average Production Costs – B-x System . . 173
9.9 Summary of Costs – B-x System 188
9.10 Those Costs Comprising 85 Per Cent of the B-x System Cost 189
9.11 Effect of Varying Attrition Rate 196
9.12 Investment Cost Comparison with Different Attrition Rates Based Upon Ten Wings 197
9.13 Operating Cost Comparison with Different Attrition Rates Based Upon Ten Wings 198
20.1 Effects of Different Allocations of Strategic Forces . . 378
21.1 Force Mix Comparisons against 15 Index Values (Days) . 410

Chapter 1

INTRODUCTION

E. S. QUADE

This chapter provides a preliminary view of the most general features of systems analysis. It describes the origins and scope of systems analysis; gives an idea of its nature; suggests where, why, and how it is helpful; and contrasts it with other forms of analysis. It also discusses the objectives of the other chapters in this book and how they are intended to contribute to the whole.

INTRODUCTORY REMARKS

Since 1961, the United States has introduced a new philosophy, technique, and style of defense management. To some extent this was inevitable; military planning today presents a new problem, different from earlier military planning, not in any deep logical or philosophical sense, but in a practical sense. The radical change in weapons, with their almost exponential increase in complexity, and the concomitant need for research and development, forced a new emphasis on science and engineering and rendered past military experience a far less certain guide to future conflict. Central to this new concept of defense management is the acceptance by decisionmakers of policy advice provided by systematic analytic studies. Such studies by engineers and physical and social scientists working as part of, or in collaboration with, the military services and the Department of Defense have thus become an essential part of the policymaking process.

In scope, these studies range from attempts to increase the efficiency of the routine peacetime housekeeping operations of the Armed Forces to advising decisionmakers on the broadest issues of national security. In its research for the United States Air Force, The RAND Corporation has played a leading role in developing an approach to the full range of these problems and in bringing the methods used to national attention. This approach, which we call "systems analysis," is the subject of this book.

DEFINITION OF SYSTEMS ANALYSIS

What is systems analysis? Most of the defense community interprets the term narrowly, restricting it to the application of quantitative economic analysis and scientific methods to such matters as weapon design and the determination of force composition and deployment. But systems analysis is not a method or technique; nor is it a fixed set of techniques. Because systems analyses take their character largely from the problems they address, they often seem to bear little resemblance to each other. The techniques used differ from study to study, and there is but the thinnest thread of method that ties these studies together. Similarly, the problem addressed and the questions asked about the problem will induce a wide variation in the specific form of the results. It would, therefore, also be a mistake to define systems analysis in terms of the reports or briefings to which it leads, as if to say that this or that document, all starched and fresh, was "a systems analysis." This is, of course, a useful shorthand, and the authors of this book are not alone in depending on it, but it tends to blur the fact that systems analysis is actually what goes on before such documents can be prepared – and this includes all the false starts. If, then, systems analysis is not a method, a set of techniques, or a type of report, what is it?

We would suggest that, properly speaking, it is a research strategy, a perspective on the proper use of the available tools, a practical philosophy of how best to aid a decisionmaker with complex problems of choice under uncertainty. In the absence of a good brief definition, systems analysis, as the term is intended to be understood in this book, can be characterized as *a systematic approach to helping a decisionmaker choose a course of action by investigating his full problem, searching out objectives and alternatives, and comparing them in the light of their consequences, using an appropriate framework – in so far as possible analytic – to bring expert judgment and intuition to bear on the problem.*

ORIGINS

The idea that analytic techniques might be applied to policy and strategy formulation in the military establishment was suggested by the success of operations analysis in dealing with military operations in World War II. Operations analysis became a more or less formal and distinct occupation early in that war, although much the same type of analysis was done on occasion in earlier wars[1] and even in very ancient times. The major impetus to this activity was provided by the introduction of new weapons based on, and requiring for their operation, technical know-how foreign to past military experience. These weapons and weapon systems (radar is the outstanding example) were so novel in concept and design that their exploitation could not be planned purely on the basis of traditional military experience. The questions addressed were largely tactical: how first

[1]For instance, some of the work done by the Statistical Branch of the General Staff, U.S. War Department, during the first World War, under the direction of Colonel Leonard P. Ayres.

to use "window" or "chaff" as a radar countermeasure; how to determine more effective bombing patterns; how to determine better antisubmarine search procedures; or how to deploy destroyers to best protect a convoy. New methods of analysis had to be developed. These formed the beginnings of a body of knowledge called at that time "operations analysis" and later, in its various extensions, "operations research," "systems engineering," "management science," "cost-effectiveness analysis," and, by RAND, "systems analysis." The term "systems analysis" came into use because the first postwar efforts were concerned with the selection and evaluation of weapon systems for development. Since development requires several years, these studies no longer dealt exclusively with those operations for which the inputs were known, the objectives clear, and the uncertainties limited.

Later, around 1950, weapon system analysts (particularly at The RAND Corporation) began an attempt to include issues of national security policy and strategy in their research and to make these issues the subject of studies in themselves. The initial reaction of experienced "military analysts" in the Pentagon was cool indeed. They argued that because military policy and other large national security problems were so different from the questions of weapon systems optimization and selection that RAND and others had been reasonably successful in answering, there was little chance that the techniques and concepts of the original systems studies would carry over. Strategy and policy planning were arts, and would remain so.

Fortunately, these skeptics were only partially right. It is true that additional concepts and methodologies, significantly different from those of earlier analysis, have had to be developed. But there has been a large transfer and substantial progress. In fact, recent years have seen a dramatic increase in the extent to which analysis, in this broader sense, has influenced decisionmakers on even the most critical issues of national security.

Relation to other types of analysis

Systems analysis is sometimes described generally as the application of the scientific method to problems of economic choice. In no case, military or nonmilitary, is it scientific research, however. Its objective, in contrast to that of pure science, is primarily to recommend – or at least to suggest – policy, rather than merely to understand and predict. Thus, it is more nearly engineering than science. For the purposes of making a distinction here, one might say that science seeks to find things out, while engineering uses the results of science to do things well and cheaply. Yet *military* systems analysis differs from ordinary engineering in its enormous responsibility, in sometimes being forced by the nature or urgency of a problem to substitute intuition for verifiable knowledge, in the unusual difficulty of appraising – or even discovering – a value system applicable to its problems, and in the absence of ways to test its validity.

The difference between the various extensions of World War II operations analysis is largely a matter of terminology or emphasis. There are no differences in principle, and hence no clear lines of demarcation can be drawn.

The analyst who practises operations research is usually trying to use mathematics, or logical analysis, to help a client improve his efficiency in a situation in which everyone has a fairly good idea of what "more efficient" means. He rarely has to concern himself with discovering the purpose of the operation or how to tell whether it is successful or not. A major aim is to develop common structures (or "models") relevant to a wide variety of situations.

Someone has remarked that systems analysis is to operations research as strategy is to tactics. At the national policy level, this is certainly the case.

> ... Systems Analysis ... differs in scope from Operations Research in the conventional sense, and it is not performed exclusively or even primarily by people who might be identified as operational researchers ... It is a discipline with a logic of its own, similar in many respects to that of Operations Research, but also different in some fundamental aspects.
>
> Like operations research, this kind of analysis can and must be honest, in the sense that the quantitative factors are selected without bias, that the calculations are accurate, that alternatives are not arbitrarily suppressed, and the like. But it cannot be 'objective' in the sense of being independent of values. Value judgments are an integral part of the analysis; and it is the role of the analyst to bring to light for the policy-maker exactly how and where value judgments enter so that the latter can make his own value judgments in the light of as much relevant information as possible.
>
> Again, analysis at this level cannot prove the optimality of any national security policy. I don't doubt for a moment that, given a specified set of ships and aircraft and equipment, and a particular task such as tracking down and killing submarines in a given area, operations analysis can indicate the optimal way to go about doing it. There, only one value judgment enters in. That is, that it is desirable to kill enemy submarines. You cannot do that at the national policy level. Rather, at that level, analysis can only trace out implications of alternative policies.[2]

These more comprehensive studies also involve, at one point or another, a comparison of the possible alternative courses of action in terms of their effectiveness and costs. This comparison often requires major attention and, by a natural substitution of the part for the whole, the entire study is sometimes called a cost-effectiveness analysis. Such an analysis typically stresses the selection, from among the available alternatives, of a "least-

[2] Alain C. Enthoven, Assistant Secretary of Defense (Systems Analysis), "Operations Research and the Design of the Defense Program," in *Proceedings of the 3rd International Conference on Operational Research*, Dunod, Paris, 1964, pp. 530, 534.

cost" scheme for carrying out some specified task. Because the analyst usually accepts as inputs someone else's statement of the objectives of the system and the possible alternatives, his results may not represent a complete systems analysis. In contrast, the systems analyst is the fellow who is likely to be forced to deal with problems in which the difficulty lies precisely in deciding what ought to be done, not simply in how to do it. He thus puts greater attention on the suitability of the task and the augmentation of alternatives. The staff study, by the way, lies in here somewhere: it may be a systems analysis, although frequently time allows little chance to make it as quantitative or complete as the ideal systems analysis would be.

This distinction between systems analysis and cost-effectiveness analysis can perhaps be clarified by a homely example.

Suppose T. C. Mits has decided to buy a washing machine for his wife. His objective is fairly clear and the alternatives are probably well-defined. If so, the situation is one for a cost-effectiveness analysis. The available machines have differences in both performance and cost. With a little care, making due allowance for uncertainty about maintenance, water, and electrical costs, he can then estimate, say, the five-year procurement and operating cost of any particular machine, and do so with a feeling that he is well inside the ball park. He will discover, of course, that finding a standard for measuring the effectiveness of the various machines is somewhat more difficult. For one thing, the problem is multidimensional – Mr. Mits must consider convenience, length of cycle, load capacity, residual water in the clothes, and so forth. But ordinarily one consideration – perhaps capacity – dominates. On this basis, he can go look at some machines, compare costs against capacity, and finally determine a best buy.

Now suppose Mr. Mits has simply decided to spend more money and thus increase his family's standard of living – a decision similar to one to strengthen the U.S. defense posture by increasing the military budget. How can he decide how to allocate the money among various possibilities? This is a situation for systems analysis, and he should probably call in his wife. Together, they first would need to investigate their goals or objectives, and then establish criteria, determine measures of effectiveness, look into the full range of alternatives – a new car, a piano, a trip to Europe. Here, because the alternatives are so dissimilar, determining what they want to do is the major problem; determining what it costs and how to attain it may become a comparatively minor one.

The need for analysis

The acceptance of systems analysis at the national policy level has been due in part to the success of its forerunners in World War II and the Korean

War, and to the impressive record of its extensions since that time in helping to solve complex problems in the military and in business and industry. A more important reason, however, has simply been the recognition that in the present state of the world a need exists for an analytic approach to national security problems. The radical changes in the weapons of war that began in 1945 and are still in process today strongly imply that military experience relevant to large-scale war may no longer provide adequate guidance. Nations – particularly the United States and its major potential enemies – are vulnerable in totally new ways, and military preparations can never again be put off until after hostilities have started. The capability of traditional methods of decisionmaking, based largely on policies of making incremental changes to permit the steady gaining of experience, has thus declined. Both the military professional and his civilian co-worker have been driven to the use of analytic methods to devise reasonably adequate and meaningful substitutes for experience: without calculation there is no way to discover how many missiles may be needed to destroy a target system, or how arms control may affect security.[3]

In addition, the magnitude of the defense effort is now so great that it invites critical scrutiny. As long as the total national defense requirements in peacetime were not a significant part of the national product, the country managed to survive very well with the "requirements" approach to national security planning. This approach made little use of analysis as we know it today. It proceeded in several steps. First, in the light of "national objectives" and "sound military doctrine," each service determined the kinds of military capabilities it needed. Next, by considering the technological possibilities and the operating constraints – for example, certain missions were prohibited to the Air Force – each service determined how it would like these capabilities to be obtained. The services then appraised enemy capabilities and prepared estimates of the number of items they would require. Finally, they submitted these estimates to budget authorities, who weighed them with little more than intuition as a guide and usually ended up cutting them by appeals to "national fiscal limitations." We could afford a large measure of inefficiency then. But today national security requires a more efficient utilization of resources.

To obtain it, defense decisions now depend heavily on systems analysis, applied within the context of a modern management system, known as the Planning-Programming-Budgeting System (PPBS). Program budgeting, as it is often called for short, is designed as a tool for the formulation and

[3] This is not to say that every aspect of such problems can be analyzed, much less quantified, or that analysis is without its defects, but only that it is not sensible to formulate national defense policy without careful consideration of whatever relevant alternatives can be discovered.

continuous review of defense programs. Its distinguishing characteristics fall into three categories: (1) a budget format that indicates planned expenditures over an extended period in terms of the national security objectives these expenditures are expected to attain; (2) a management information system to keep track of expenditures and the progress of programs and to provide data for analysis; and (3) systems analyses, at all levels of activity, to search out, examine, and evaluate possible courses of action.[4] In their modern form, the ideas for the PPBS were first proposed by David Novick of RAND's Cost Analysis Department and were brought to the Defense Department in 1961 by Charles Hitch when he left RAND to become Assistant Secretary of Defense, Comptroller. In implementing this change in management practices in something less than two years, Mr. Robert McNamara, the Secretary of Defense, established himself as the foremost military administrator of our time. The implications of this system – President Johnson called it revolutionary and directed that it be implemented in all Federal departments and agencies[5] – are vast; the methods are being studied with interest throughout the world, and, at least in the United States, are now being adopted by civil administrators at all levels of government.

TYPES OF APPLICATIONS

Analytic techniques can be applied to military problems which range from routine day-by-day operations of the services to critical onetime decisions of national security. This spectrum may be divided into the following categories:

1. Management of operations;
2. Choice of tactical alternatives;
3. Design and development of weapon systems;
4. Determination of major policy alternatives.

Roughly speaking, the order here increases with respect to the policy level involved and decreases with respect to the ability of analysis to produce firm and actionable recommendations. While the division is fairly arbitrary, and there is considerable overlapping, it will allow us to make certain broad distinctions that will help to set later discussions in place. With these qualifications in mind, let us consider Fig. 1.1, which illustrates typical problems attempted in each category.

In the first category, analysis takes its most mathematical – and, in a certain sense, its most fruitful – role. Except for the context, much of the

[4] See David Novick (ed.), *Program Budgeting: Program Analysis and the Federal Budget*, 2nd ed., Harvard University Press, Cambridge, 1967.

[5] "Transcript of the President's News Conference on Foreign and Domestic Matters," *The New York Times*, August 26, 1965.

Problem Areas	*Examples*
MANAGEMENT OF OPERATIONS	Determining the inventory at a parts depot.
	Establishing maintenance procedures for a fighter base.
CHOICE OF TACTICAL ALTERNATIVES	Determining the most effective armament for an interdiction mission.
	Selecting a fire control system for a new fighter.
DESIGN AND DEVELOPMENT OF WEAPON SYSTEMS	Selecting a preferred set of space boosters.
	Determining the need for Army airlift.
DETERMINATION OF MAJOR POLICY ALTERNATIVES	Determining the role of space systems in national defense.
	Deciding between a policy of military superiority and one of parity with the Soviet Union.

Fig. 1.1 – The range of problems in which analysis can be useful

analysis is essentially no different from the analysis done to support decisionmaking and resource allocation in commerce and industry – stock control, personnel assignment, reliability checkout, transportation routing, and so forth. It is management science or operations research in the strict sense – an attempt to increase the efficiency of a man-machine system in a context where it is fairly clear that "more efficient" means something like the military equivalent of maximizing profits. A characteristic of problems in this category is that they are so well-structured that management can be helped simply by applying systematic computational routines to a generic "model," which can be made relevant to a wide variety of operations merely by modifying its parameters. (Queuing theory, for example, is relevant to many aspects of the operations of military – and nonmilitary – communication systems, airfields, service facilities, ground traffic systems, and so on.) Notwithstanding the importance of this type of analysis, it is not discussed in this book except incidentally.

The analyses in the remaining three categories almost always involve the element of *conflict* and require that it be explicitly considered; it is not ordinarily present in the first. In true conflict situations – those that concern military analysts – it is the interaction with the enemy, and not the interaction of one's own alternatives and their costs, that is likely to be the main problem. The analysis is bound to be more difficult, of course, for conflict introduces an additional set of uncertainties and complexities.

The simplest type of conflict analysis is the analysis of tactical alternatives. Such problems as the determination of an efficient search pattern for enemy submarines, or the allocation of a missile payload between warhead, decoys, and protection, belong in this category. For these problems, the objective of the operation is usually clear and some satisfactory measure of its effectiveness exists. Sometimes the situation can be represented by means of a mathematical model with such realism that the theory of games proves useful.[6] But analysis of problems of this type is again not a principal subject of this book.

Analysis in the next category is the original source of the name "systems analysis"; it involves the planning and design of new systems to perform existing operations better or to implement operations never before performed.[7] While the bulk of the man-hours applied to such problems of research and development goes into technological research to satisfy performance requirements (a feature that helps to set this category apart), this work is done by engineers and scientists, and most of the *analyst's* time may be spent to determine just what these requirements are and how they can best be achieved. Conflict, of course, has to be considered. Thus, a military communications system must be designed to survive a physical attack and to operate in the face of countermeasures. To do this the weight and nature of possible attacks and the load on the system must be forecast. What represents the most distinctive characteristic of analyses of this sort, however, is that they are always concerned with systems that are years from actual operational use. The uncertainties of the future – including those associated with the behavior of an enemy – must somehow be taken into account. Typical problems might be to design a national communication system for wartime command and control or to determine the configuration and armament of a long-endurance antisubmarine aircraft to patrol and destroy missile-firing submarines. Analyses of this type, which clearly are more demanding than those in the previous categories, are the central topic of this book.

Major attention, however, is also devoted in the following chapters to the next category – the systematic analysis of major policy alternatives. A typical analysis in this category might attempt to aid in determining the objectives of the future strategic force and in planning its composition, or to help in deciding whether or not forces based in the United States and backed by the airlift necessary for a rapid response could replace forces

---

[6] See Chapter 11.

[7] Various aspects of these tasks are sometimes called "systems research," "systems design," or "systems engineering" – and, unfortunately, these terms are all too often used to designate the entire approach to problems in this third category.

based overseas without weakening our military capabilities, prestige, or alliances.

It should be emphasized that systems analysis is by no means exclusively military, but is used extensively by managers and engineers in large industrial enterprises, such as telephone companies and electric power utilities. In two respects, however, the normal business systems analysis is conceptually simpler. For one thing, in such analyses there is usually a single over-all objective – the maximization of profits – which can be measured and expressed in the same terms as the costs. For another – as we have already seen – conflict plays only a minor role.

It should also be emphasized that systems analysis has other non-military applications. As the central tool of the program budgeting effort, the potential applications of systems analysis are being explored throughout the Federal government for every possible social, technological, and governmental purpose. Hailed as the most valuable by-product of the national defense and space effort, it is being touted as the vehicle to convey recent scientific and technological advances directly into the life of the ordinary citizen. Bills have been introduced in both houses of Congress "to mobilize and utilize the scientific and engineering manpower of the Nation to employ systems analysis and systems engineering to help fully employ the Nation's manpower resources to solve national problems."[8] The uses of systems analysis are also being explored on other levels of government. In California, for example, major problem areas of concern to the state have been subjected to systematic analysis by engineers of a number of aerospace firms: transportation systems (North American Aviation), criminal justice and the prevention of delinquency (Space-General), the flow of information needed for the state's operation (Lockheed), the control and management of wastes (Aerojet-General), regional land-use information systems (TRW Systems), and the state's social welfare operations (Space-General).[9]

One further point. Analyses in the later categories frequently include studies in the earlier categories as components. Often it is the completion of these component studies which absorbs most of the man-hours and makes the broader analysis possible. Nevertheless, the solution of broad military problems depends only in slight part on the narrowly technical and traditional disciplines of the natural or social sciences or engineering – and still less on the knowledge that can be found stored away in textbooks. There are no experts in this field in the sense that there are experts in navigation or in thermodynamics. Any advice that is given must come as the

[8] H. R. 14076, 1st session, 89th Congress (1966), 213.
[9] "Aerospace 'Think Tanks' Getting Earthy Look," Los Angeles *Times*, August 14, 1966, p. I–1.

result of study applied to the particular situation, not as a deduction from some well-established theory.

A broad systems study usually makes use of an interdisciplinary team. This is not merely because a broad study is complex. Even more important is that the questions it raises will look different to an economist, a mathematician, a lawyer, a political scientist, an engineer, or a military professional – and different ways of looking at a problem are of first importance in finding a solution.

THE ESSENCE OF SYSTEMS ANALYSIS[10]

The idea of an analysis to provide advice is not new and, in concept, what needs to be done is simple and rather obvious. One strives to look at the entire problem, as a whole, in context, and to compare alternative choices in the light of their possible outcomes. Three sorts of inquiry are required, any of which can modify the others as the work proceeds. There is a need, first of all, for a systematic investigation of the decisionmaker's objectives and of the relevant criteria for deciding among the alternatives that promise to achieve these objectives. Next, the alternatives need to be identified, examined for feasibility, and then compared in terms of their effectiveness and cost, taking time and risk into account. Finally, an attempt must be made to design better alternatives and select other goals if those previously examined are found wanting.

Even though the concept is simple in practice, the actual conditions of the analysis pose many problems, some of which we have already mentioned. At bottom, these difficulties arise because systems analysis itself and the entire process of policy planning lack an accepted theoretical foundation. Since analysts and decisionmakers alike are thus faced with serious problems of choice that yield only partially to quantitative reasoning, they are forced sooner or later to rely on the judgment, largely intuitive, of specialists with experience in the field. *The approach that makes this possible – and, hence, the very essence of systems analysis – is to construct and operate within a "model"* – an idealization of the situation appropriate to the problem. Such a model, which may range from an elaborate computer program to a war game played on a sand table, introduces a precise structure and terminology that serve primarily as an effective means of communication, enabling the participants in the study to make their judgments in a concrete context. Moreover, through feedback – the results of computation or the countermoves in the war game – the model helps the decisionmaker, the analysts, and the experts on whom they depend to arrive at a clearer understanding of the subject matter and the problem.

---

[10] The notions discussed in this section and the one following are considered at length in subsequent Chapters, especially 3, 4, 7, and 10.

THE ELEMENTS OF ANALYSIS

The central importance of the model can be seen most readily, perhaps, by looking at its relation to the other elements of analysis. There are five altogether. Each of them is present in every analysis of choice, although they may not always be explicitly identified.

1. *The objective* (*or objectives*). Systems analysis is undertaken primarily to help choose a policy or course of action. The first and one of the most important tasks of the systems analyst is to discover what objectives the decisionmaker is, or should be, trying to attain through the options open to him, and how to measure the extent to which they are, in fact, attained. This done, strategies, forces, or equipment are examined, compared, and chosen on the basis of how well and how cheaply they can accomplish these objectives.

2. *The alternatives*. The alternatives are the means by which it is hoped the objectives can be attained. They need not be obvious substitutes or perform the same specific function. Thus, to protect civilians against air attack, shelters, "shooting" defenses, counterforce attack, and retaliatory striking power are all alternatives.

3. *The costs*. The choice of a particular alternative for accomplishing the objectives implies that certain specific resources can no longer be used for other purposes. These are the costs. In analyses for a future time period, most costs can be measured in money, but their true measure is in terms of the opportunities that they preclude. Thus, if we are comparing ways to eliminate guerrillas, the injury or death of nonparticipating civilians caused by the various alternatives must be considered a cost, for such casualties may recruit more guerrillas.

4. *A model* (*or models*). A model is a representation of reality which abstracts the features of the situation relevant to the question being studied. The means of representation may vary from a set of mathematical equations or a computer program to a purely verbal description of the situation, in which judgment alone is used to assess the consequences of various choices. In systems analysis, or any analysis of choice, the role of the model (or models, for it may be inappropriate or absurd to attempt to incorporate all the aspects of a problem in a single formulation) is to estimate the consequences of the choice; that is, the costs that each alternative will incur and the extent to which each alternative will attain the objective.

5. *A criterion*. A criterion is a rule or standard for ranking the alternatives in order of desirability and indicating the most promising. It provides a means for weighing cost against performance.

The process of analysis takes place in five overlapping stages. In the first, the formulation stage, the issues are clarified, the extent of the inquiry limited, and the elements identified. In the second, the search stage, information is gathered and alternatives generated. The third stage is evaluation; the fourth, interpretation; and the fifth, verification.[11]

To start the process of evaluation or comparison (see Fig. 1.2), the various *alternatives* (which may have to be discovered or invented as part of the analysis) are examined by means of the *models*. The models tell us what consequences or outcomes can be expected to follow from each alternative; that is, what the *costs* are and the extent to which each *objective* is attained. A *criterion* can then be used to weigh the costs against performance, and thus the alternatives can be arranged in the order of preference.

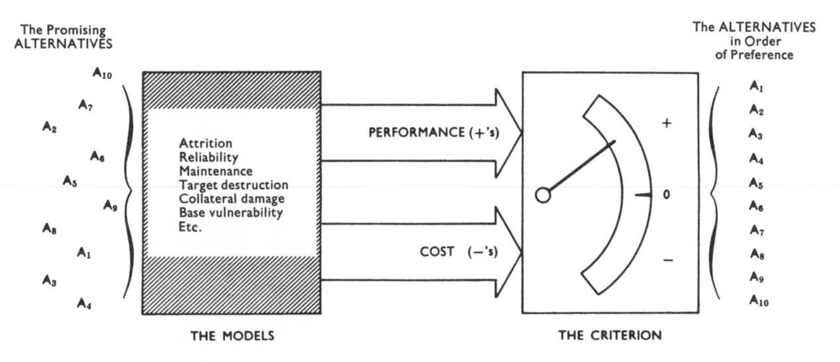

Fig. 1.2 – The structure of analysis

Unfortunately, things are seldom tidy: Too often the objectives are multiple, conflicting, and obscure; alternatives are not adequate to attain the objectives; the measures of effectiveness do not really measure the extent to which the objectives are attained; the predictions from the model are full of uncertainties; and other criteria that look almost as plausible as the one chosen may lead to a different order of preference. When this happens, we must take another approach. A single attempt or pass at a problem is seldom enough. As indicated in Fig. 1.3, the key to successful analysis is a continuous cycle of formulating the problem, selecting objectives, designing alternatives, collecting data, building models, weighing cost against performance, testing for sensitivity, questioning assumptions and data, re-examining the objectives, opening new alternatives, building better models, and so on, until satisfaction is obtained or time or money force a cutoff.

Note that there is nothing really new about these procedures. They have been followed more or less successfully since ancient times. The need for considering cost relative to performance must have occurred to the earliest planner. Systems analysis is thus not a catchword to suggest we are doing

[11] We shall discuss these five stages in detail in Chapter 3.

something new; at most, we are doing something better – "better" in deliberately attempting to be systematic, analytic, and comprehensive, in making use of mathematical and computer techniques, and in paying careful attention to sensitivity.

Development of systems analysis

Although systems analysis has, in fact, contributed substantively to long-range defense planning, the contribution has not been uniform over the range of problems to which it has been applied. In the early days of systems analysis, the studies were highly preoccupied with the analytic techniques of operations research – linear programming and game theory, for example. Complicated mathematical models, featuring an astronomically large number of machine computations designed to pick out the optimum system,

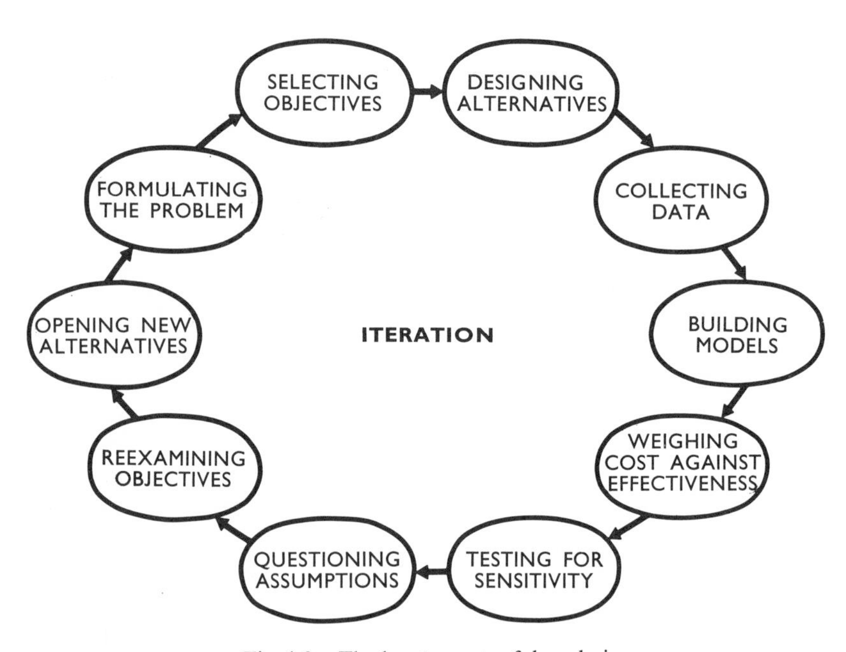

Fig. 1.3 – The key to successful analysis

were popular. We soon realized, however, that real-world defense planning was too complicated for such a purely quantitative approach. Nevertheless, at the start, problems concerning strategic bombing or air defense received a good deal more attention than those, say, of limited war. This was not surprising; limited war is obviously a tri-service problem and full of factors – political, social, and economic – that cannot be easily quantified.

When the first systems analysis was done, central war problems looked relatively free of such clutter. It was not obvious until much later that this appearance was deceiving, and that attention must be given to such aspects of strategic war as initiation, intra-war deterrence and negotiation, termination, and recovery.

Today, analyses no longer look as "analytical" as they did in the past. To an increasing extent they deal with strategy as well as with tactics – with the ability to achieve general foreign policy objectives, rather than merely with the ability of weapon systems to influence the character of a single military clash. As advisors, our objective is *insensitivity* as frequently as it is optimization: we seek to define systems that will work well under many widely divergent contingencies and even give some sort of reasonably satisfactory performance under a major misestimate of the future. Thus, RAND's first defense study focused its attention on the tactics for shooting down enemy bombers. Today, the corresponding study would look for the less obvious values of defense, relate active defense to other military missions, consider the use of warning systems for surveillance and related "nonspasm" uses, investigate the different kinds of contingencies in which defense might be useful, and so on. We now realize that the impact of subjective considerations – such as the system's flexibility, its compatibility with other systems (some yet unborn), its contributions to national prestige abroad, and its effect on domestic political constraints – can play as important a role in the choice of alternatives as any calculation of war outcomes. In addition, we realize that such intangibles as the extent to which superiority in residual forces can be effectively used to coerce the enemy to discontinue the conflict, or the perception each side has of its own or its enemy's strengths, must be taken into account. Thus, it should be no surprise that many of the component studies, and even a major part of the over-all analysis, are verbal rather than quantitative in nature.

THE CURRENT STATUS OF SYSTEMS ANALYSIS

Where they can be obtained, quantitative estimates of costs and effectiveness are clearly helpful to any intelligent discussion of national security. In current Department of Defense practice, these quantitative estimates are now obtained as part of the planning-programming-budgeting process. The analytic part of this process is systems analysis as we have ideally defined it. But many people – some of them, perhaps, readers of this book – are vaguely uneasy about the particular way these estimates are made and their increasingly important role in military planning. This is especially so when cost-effectiveness or the use of computers is mentioned.

For example, an Air Force officer writes that computer-oriented planning techniques are dangerous; that mathematical models of future wars are inadequate for defense planning; and that scientific methods cannot handle those acts of will which determine the conduct of war.[12] A Congressman says, "We should not allow cost-effectiveness to cost us our effectiveness in national security matters."[13] A Senator remarks, "Our potential enemies may not use the same cost-effectiveness criteria and thus oppose us with the best weapons their technology can provide. This would create an intolerable peril to the national security."[14]

The cost-effectiveness aspect of these analyses is most often criticized when it deals with engineered systems like the TFX and the nuclear-powered carrier. Just over a year ago, for instance, the House Armed Services Committee spoke of the Defense Department in these words:

> . . . the almost obsessional dedication to cost-effectiveness raises the specter of a decisionmaker who . . . knows the price of everything and the value of nothing.[15]

That Oscar Wilde's famous definition of a cynic should be quoted in such a context is remarkable. Later, the chairman, Representative L. Mendel Rivers, complaining about the inadequacy of current antisubmarine warfare capability and a program to build logistics ships in a single shipyard, remarked:

> All of this is being rationalized on the basis of cost/effectiveness studies. Do you know that the M-14 rifle costs more than a bow and arrows? From a cost/effectiveness standpoint we obviously would be better off if we went back to bows and arrows. A beer bottle filled with gasoline and stuffed with a rag wick is a fairly effective weapon at close quarters, and it is cheaper to produce than a land mine or a hand grenade. From a cost/effectiveness viewpoint, we should be collecting beer bottles and old rags.[16]

Lt. General Ira C. Eaker, USAF (Ret.), in a newspaper column entitled "Most Expensive Thing a Nation Can Buy Is a Cheaper Weapon," gave his opinion:

> One of the prime obstacles to adequate defense weapons and measures in recent years has been a hurdle called cost-effectiveness. This test applied by scientists and theorists has killed off many new weapons, urgently requested by military leaders.
>
> If Hitch applies cost-effectiveness to the curriculum at California, philosophy will have to go. It does not give the financial return to graduates which they can get from medicine, engineering or law. The department of education no doubt will be eliminated also. Teaching does not pay as well as dentistry.[17]

---

[12] Col. F. X. Kane, USAF, "Security Is Too Important To Be Left To Computers," *Fortune*, Vol. 69, No. 4, April 1964, pp. 146, 231+.

[13] Representative Melvin Laird of Wisconsin, quoted in *Missile/Space Daily*, April 7, 1964, p. 161.

[14] Senator John O. Pastore of Rhode Island, quoted in *U.S. News and World Report*, January 6, 1964, p. 6.

[15] House Report 1536, May 16, 1966.

[16] *Congressional Record*, October 3, 1966, pp. A–5088–5089.

[17] Los Angeles *Times*, August 22, 1965, p. G–7. Charles J. Hitch had recently resigned as an Assistant Secretary of Defense to become Vice President for financial affairs of the University of California.

One might wonder why an approach that seems so logical is so violently opposed. Cost-effectiveness analysis seeks to increase value received (effectiveness) for the resources expended (cost). It is something we always practise when buying an automobile, or planning a vacation, or building a house. Hence, as a practical matter, it is not the method that should be under attack. The deficiencies of cost-effectiveness (or systems analysis, for that matter) exist only when the work is not competently done or when the results are used without their limitations in mind. And in this connection it is worth noting that those who have used cost-effectiveness extensively have a high opinion of its value. Charles J. Hitch, for example, in speaking of the critics, remarks:

> In a way, it is quite ironic that the very people who are so insistent that they want "the best and most modern" in Defense hardware are opposed to the "best and most modern" in Defense analysis and decisionmaking techniques.[18]

We might add that those who hold that national defense decisions are being made solely by considering calculations and numbers are not only premature in their belief (to say the least), but probably have a basic misunderstanding of how such decisions must, in fact, be made. Even a nodding familiarity with the process reveals that it is today rampant with dogma, service rivalries, special interests, and horse-trading – so much so that, in the opinion of some analysts, a computerized solution that paid no attention to these human constraints might lead to something better. This book will attempt to show, however, that even in the narrowest of military contexts, considerations not subject to rigorous, quantitative, computer-based analysis are always present – and that this situation is not likely to change. At best, calculations by themselves give us, for each set of specific assumptions – about the political and economic state of the world, the actions of the enemy, the outcome of various technological investigations, and so on – a somewhat less than objective appraisal of, say, the effectiveness, for a fixed cost, of proposed forces or weapons for attaining given goals. Such appraisals are not good enough; they must be supplemented by informed and considered judgment, and there are many sets of assumptions that might be chosen. As Charles Hitch once noted, "there is nothing inherent in . . . systems analysis that calls for ignoring military judgment or for relying on computers for anything other than computation . . .."[19]

---

[18] Charles J. Hitch, "Cost/Effectiveness," address before the 13th Military Operations Research Symposium, Washington, D.C., April 29, 1964.

[19] Charles J. Hitch, "Programming's Role in Defense," address before the first International Meeting of the Western Section of the Operations Research Society of America, Honolulu, Hawaii, September 1964; quoted, in part, in *Aviation Week and Space Technology*, Vol. 81, No. 15, October 12, 1964, p. 17. Incidentally, interservice rivalries and bargaining have at least the virtue of insuring that some nonquantitative considerations are not neglected.

Objectives of this book

It is not easy to tell someone how to carry out a systems analysis. We lack an adequate theory to guide us. This must be expected, for systems analysis is a fairly new discipline, and history teaches us that good theory usually comes late in the development of any field and after many false starts. Where the attention of systems analysis has turned to methods, it has focused mainly on the development of mathematical techniques for handling certain specialized problems, common in the practice of operations research – rather than on building a basic theory for the treatment of the broad questions typical in defense planning. This attention to technique *has* met with great success. Models have become easier to manipulate, even with many more variables represented, and the computational obstacles in systems analysis now cause comparatively little difficulty. The more important philosophical problems, however, such as occur in providing assurance that the model is meaningful, in devising schemes to compensate for uncertainty, or in choosing appropriate measures of effectiveness, still remain troublesome. Of the matters we could discuss in the following pages, it is these fundamental problems that most deserve a critical examination. Consequently, the many important and useful operations research techniques essential to systems analysis are treated only very cursorily. Many university and college courses, and books in profusion, handle these subjects adequately. We propose here to emphasize concepts and understanding instead – areas where the analyst as well as the user of analysis is more likely to err.

The intended reader of this book has four roles to consider with respect to analysis, for he *sponsors*, *produces*, *evaluates*, or *implements* it. The objective of this book is to provide help in each of these roles.

For the sponsor, we attempt to point out the role of analysis in the military context, what kind of analysis is appropriate to what sort of problem, and what to expect from it. At the same time, we also attempt to indicate how overspecification of the problem by the sponsor and an arbitrary determination of assumptions and methods can lead to poor results.

We also aim to help the producer of analysis. But, in viewing the producer, we are taking the approach we just mentioned, that instruction in such well-established techniques of operations research as linear programming, queuing theory, and decision theory should not be our first order of business. We feel that the real pitfalls in analyzing the broad and complex problems faced by military and government decisionmakers lie elsewhere. They concern the design and definition of the problems, the selection and understanding of rules of choice, and the interpretation of the results of analysis in the light of considerations not taken into account in the analysis. Hence, we put our emphasis on those subjects.

To assist those who must evaluate analyses, we hope to describe fully the major characteristics of good analysis and to suggest the proper questions to ask to uncover weaknesses and errors. We hope to make perfectly obvious the virtues, as well as the drastic limitations, of an analytic approach to military problems.

Finally, for those who must implement analysis, we hope to produce the kind of understanding that will provide appropriate confidence in its results.

Chapter 2

THE TRADE-OFF BETWEEN GROUNDPOWER AND AIR SUPPORT: AN INTRODUCTORY EXAMPLE OF SYSTEMS ANALYSIS

M. G. WEINER

This chapter presents a highly simplified cost-effectiveness analysis of the trade-offs between groundpower and air support. Its purpose is to highlight the virtues and deficiencies of such analyses, and provide an insight into some of the principles and methods discussed in subsequent chapters.

INTRODUCTION

One of the most common forms of systems analysis to which the decision-maker is exposed is the "trade-off" study, part of which usually includes a cost-effectiveness analysis. A critical examination of the trade-off study should provide us, therefore, with some insight into systems analysis as a whole – how the analyst works, what he works with, what his analysis looks like, and why it looks just that way. While we could approach these questions philosophically, and talk in broad terms about the importance of assumptions, the significance of the variables or parameters chosen for a study, or how the validity of such an analysis is determined, we will instead actually go through a specific instance of a trade-off study. To keep matters within bounds, we have simplified our example considerably. Its purpose is merely to provide a vehicle for discussion, and it should not be taken too seriously. As a vehicle, it is a Model T Ford compared with some of the impressive Cadillacs that are currently in use. Nevertheless, it may get us to our destination, which is a critical appraisal of the important ingredients of trade-off studies.

First, a word or two about the differences between cost-effectiveness and trade-off analysis. A cost-effectiveness analysis is, in the simplest terms, an attempt to determine whether or not a system that might be purchased is worth the cost. To find out, it is necessary, in the first place, to formulate some notion of "worth," some "measure of effectiveness."[1] How well does the system do the job it is designed to do? Does its effectiveness warrant the cost? In many cases, these turn out to be difficult questions to answer in the military field. Take, for example, a tactical weapon system like a

[1] A problem discussed at length by L. D. Attaway in Chapter 4.

fighter-bomber. An analyst can indeed determine *some* aspects of its effectiveness – such as the number of weapons it can deliver, its delivery accuracy, its survivability – even though he often finds that the task of expressing them quantitatively is a complex one. But his problem of measuring effectiveness becomes still more difficult when he has to add to his account of the fighter-bomber a description of characteristics like its deterrent value or political demonstration value. Moreover, the analyst must formulate estimates of costs, and here he runs into the same difficulties. He can, of course, make some fairly accurate estimates of dollar costs, but what about some of the other costs, like the political or psychological?[2] As those who have some familiarity with cost-effectiveness studies know well enough, it is easier to say that these costs should be analyzed than it is to analyze them.

If a cost-effectiveness study is a way of determining whether or not a system is worth its cost, a trade-off study is a way of determining whether or not one system is better than another. Here the difficulties are much the same as they are in cost-effectiveness analyses.

We are all familiar with cost-effectiveness and trade-off studies in our everyday life. Put in overly simple terms, any time we compare several TV sets, listing the good and bad points about each (including the cost), and then decide which one we would like "on balance," we do a cost-effectiveness analysis. And every time we compare a TV set with the new dishwasher our wife wants, we do a trade-off analysis – again going through the pros and cons and costs of each, but with somewhat greater difficulty because dishwashers and TV sets are not easily compared, and other aspects of our relations with our wife may play an important role in the final decision.

The trade-off study which we will present is a good example of the problem of trying to compare quite different things. It is a study of the trade-off between airpower and groundpower.

Now what does this mean? What exactly are we going to trade off? Can the trade-off between airpower and groundpower be reasonably expressed as a trade-off between divisions of ground forces and wings of airpower? Should it be considered for the entire spectrum of limited war, including nuclear operations? Is it dominated by one potential conflict area, such as the Far East or Europe? To what extent, if any, should some of the non-quantifiable military and non-military factors – such as the value of these forces in showing the flag – be included in the analysis? Can pure trade-offs of air and ground forces be considered, or only different "mixes" of

---

[2]The nature of resource analysis and cost-sensitivity analysis is explored in Chapters 7–9. That of the more or less nonquantitative costs is considered throughout this book, especially in Chapters 3, 16, 17, 18, and 20.

the two? These and other questions must be answered, however tentatively, before analysis of force trade-offs can begin.

By the time the analyst has thought about these complex possibilities for a while he is ready to ask for reassignment. However, if this fails, he pushes on. He tries to narrow the problem down a bit. And, as every analyst knows, the minute he tries this, he provides critics with ammunition. "What about all those things which you haven't included?" they ask with a disarming smile, while the blood – usually the analyst's – drips.

But narrow down he must. So, for our purpose, we will say that the analysis will consider only the military trade-off of various mixes of divisions and wings in a non-nuclear war.

Unfortunately, this simplification has not made the task much less complicated. True, we can now state the problem more specifically, but, as analysts, we still have other major questions to face. Among the most significant is that of discovering a way of comparing the effectiveness of ground divisions and air wings.

Several bases for comparison suggest themselves: tonnage delivered, rapidity of response, ground-holding capability, casualties inflicted, casualties taken, and some others, singly or in combination. It might be objected, of course, that several of these measures are related: the tonnage delivered affects the casualties that are inflicted; the ability to hold ground is related to the ratio of casualties inflicted to casualties taken; and so on. With this variety of interrelations, we would probably want to try several different measures. But for this illustrative example, let us use "casualties inflicted" as a measure since it appears to be part of most of the others. We will therefore begin our trade-off study at that point, by comparing the casualty-producing capability of an air wing and a ground division.

CASUALTY PRODUCTION BY AIRPOWER

What is the casualty-producing capability of airpower? It depends on many things: the situation, the type of aircraft, the number of aircraft, the types of weapons and their effectiveness, and so on. For our purposes, let us start with an assumption about the casualty-producing capability of one jet aircraft equipped with non-nuclear weapons.

Although various estimates of this capability can be obtained from historical data and analyses, they will depend on the particular conditions appropriate to each source of data. In a fuller analysis than the one attempted here, we would, of course, want to use several different values and see how each one influences the outcome of the study. We would also want to make some side investigations to obtain as valid a set of values as possible.[3]

[3]Sensitivity analysis and the treatment of uncertainty are discussed throughout this volume. In particular, note Chapters 5, 8, 10, and 17.

But in the present example, where our aim is simply a broad sketch of trade-off analysis, we will *arbitrarily* select a number from the range of available values and say that one aircraft will produce 25 enemy casualties in each attack or sortie against organized ground units. For this illustration, we will consider these units to be enemy divisions of approximately 13,000 troops.[4]

Now, since each wing contains 75 aircraft,[5] and we can reasonably assume that each aircraft will fly one sortie a day,[6] we can calculate that one wing will produce 1875 casualties daily (75 acft $\times$ 1 sortie $\times$ 25 casualties per sortie).

Casualty production by groundpower

The casualty production of groundpower must be estimated in a somewhat different manner. Using historical data, we find that the attrition of a main enemy force on the offensive (assuming an enemy superiority between 2:1 and 4:1) is 11.2 per cent casualties a month.[7] Since we are using an enemy division size of 13,000 troops in our estimates, this amounts to approximately 1500 casualties in 30 days (0.112 $\times$ 13,000), or 50 casualties a day.

Some additional assumptions

Having made two sets of calculations of the casualty-producing effectiveness of airpower and groundpower, we now wish to relate these to some type of conflict situation. One method of doing this is to hypothesize a conflict environment, including the forces and objectives of the antagonists in the conflict. This activity, sometimes called "scenario" construction, can be done in various degrees of detail.[8] For our purposes, we will construct a hypothetical and highly simplified non-nuclear conflict situation in which Red commits 60 divisions and Blue has 30 divisions initially available to meet the assault.[9] In addition, we will assume that Blue has 15 divisions in reserve which he can commit to combat throughout the first 30 days of combat.

---

[4]Our choice of this figure of 13,000 troops as the size of an enemy division is arbitrary. The Army's *Handbook on Aggressor Military Forces*, FM 30–102 (Headquarters, Department of the Army, January 1963), cites a strength of 13,900 for a fictitious aggressor motorized rifle division.

[5]*Questions and Answers About the United States Air Force*, U.S. Government Printing Office Document 1965–0–764–115, Washington, D.C., April 1965.

[6]The sortie rate of aircraft depends on many factors. We use a rate of one sortie per aircraft per day because it approximates some actual combat experience, as indicated in *Air Force and Space Digest*, Vol. 49, No. 4 (April 1966), p. 46.

[7]Adapted from the *Staff Officers' Field Manual: Organization, Technical, and Logistical Data*, FM 101–10, Part I, Headquarters, Department of the Army, October 1961.

[8]A point that Seyom Brown will discuss more fully in Chapter 16.

[9]For convenience in what follows, we will use the nomenclature of war gaming to refer to the antagonists. Thus, "Red" denotes the aggressor's forces; "Blue" denotes the defender's forces.

For our hypothetical scenario, we further assume that both sides have considerable airpower. Blue uses the bulk of his airpower to oppose Red's air strength. He attacks Red's airfields and supply lines, and commits some aircraft to air defense. But he also commits two wings (150 aircraft) to attacking ground troops. That is, Blue flies 150 sorties per day for 30 days against enemy ground forces.

Finally, we will posit two rules. The first is that when a ground division has suffered 30 per cent casualties, it is ineffective for combat.[10] For Red, this means that about 3900 casualties (0.30 × 13,000 troops) will force the withdrawal of a division from combat. The second rule is that once a division has been withdrawn from combat, it cannot be replaced. In our example, this rule implies that Red cannot add new divisions to the original 60 he commits to the conflict, and this, in turn, assumes that Blue's air operations will be successful in preventing Red from bringing in reserves.

Effect of Blue groundpower

Let us now determine the effectiveness of Blue's groundpower in our example. Red commits 60 divisions. The attrition to each division, as calculated above, is 50 casualties a day from Blue's ground forces. This amounts to 3000 casualties for the 60 divisions each day, or 90,000 casualties in 30 days (60 divisions × 50 casualties per day per division × 30 days). Since, by our defeat criterion, each division that suffers 3900 casualties is withdrawn from combat,[11] the 90,000 casualties amount to defeating 22.5 Red divisions in 30 days. This result is portrayed graphically in Fig.

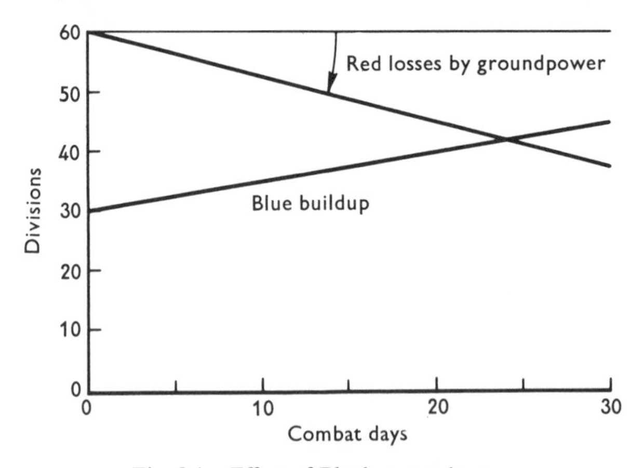

Fig. 2.1 – Effect of Blue's groundpower

[10]While various numbers might be used as "defeat criteria" we have chosen 30 per cent of the force since it represents the approximate magnitude of casualties suffered by the Germans in the Ardennes operation of 1944–1945.

[11]For convenience, we will use the figure of 4000 casualties in the following calculations.

2.1, which shows both Blue's buildup and the reduction in Red's ground force strength for the 30-day period.

Effect of Blue Airpower

The effectiveness of Blue's airpower is calculated in the same manner. Against a Red force of 60 divisions, the two wings of Blue aircraft in ground support will produce 3750 casualties per day (two wings × 1875 casualties per day, the casualty rate calculated earlier). For the 30-day period, this amounts to 112,500 Red casualties, or the defeat of approximately 28 divisions (112,500/4000). This result is shown graphically in Fig. 2.2.

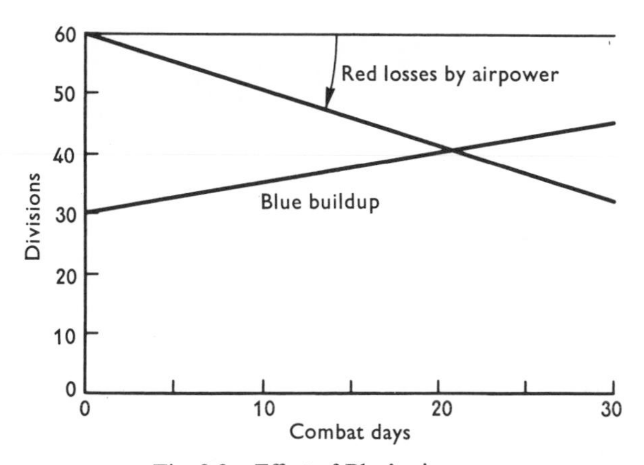

Fig. 2.2 – Effect of Blue's airpower

Combined Effects of Airpower and Groundpower

If, as is done in Fig. 2.3, we combine the effects of airpower and groundpower, we see that the total reduction in Red strength over the 30-day period will be 50.5 divisions: 22.5 divisions from groundpower, 28 divisions from airpower. Further, we find that the ground situation is stabilized on approximately the 14th day of combat; that is, on that day, both Red and Blue have the same number of ground divisions in combat.

Effect of Blue Force Attrition

Thus far, we have not estimated the attrition of Blue's ground forces. Because we have assumed that Blue's air operations are successful in preventing Red's aircraft from making any substantial contribution to the ground campaign, we need account only for the attrition produced by Red's ground forces. This can be calculated as it was for Red's ground forces, although corrections must be made to reflect the fact that Blue's ground forces are on the defensive and numerically smaller. In these terms, Blue's total attrition in the ground campaign amounts to the defeat of approximately 14 divisions in 30 days.

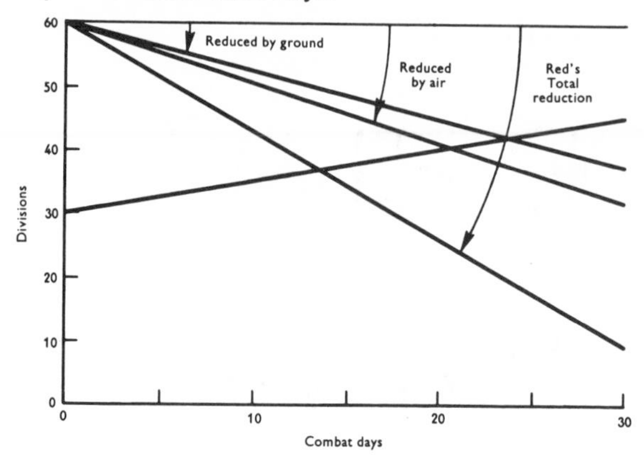

Fig. 2.3 – Combined effects of airpower and groundpower

If we compare our calculations of Red and Blue attrition, we find (Fig. 2.4) that the situation becomes stabilized in approximately 18 days; that is, the ground forces on each side are then of equal strength.

OTHER AIR-GROUND FORCE MIXES

Our calculations so far have assumed 60 Red divisions, 30 Blue divisions initially (plus 15 more from mobilization), and two wings of Blue aircraft in ground support. We can, of course, look at other combinations, among them these four:

1. Twenty-five Blue divisions plus 15 from mobilization.
2. Thirty-five Blue divisions plus 15 from mobilization.
3. Each of the above ground strengths with only one wing of Blue aircraft in ground support.
4. Each of the above ground strengths with three wings of Blue aircraft in ground support.

In considering each of these possibilities, we will leave unchanged our assumptions regarding Red's initial ground strength (60 divisions), the defeat criterion for a division (30 per cent casualties), and Red's inability to bring additional reserves into the ground campaign.

Calculating as before for each of these mixes, we can derive the results shown in Fig. 2.5. Of the various stabilization points that can be identified, we have indicated two on the Figure (the black dots). These points show the alternative mixes of airpower and groundpower that produce stabilization on the 16th day, and it is these two equal-effectiveness mixes we shall want to trade off.

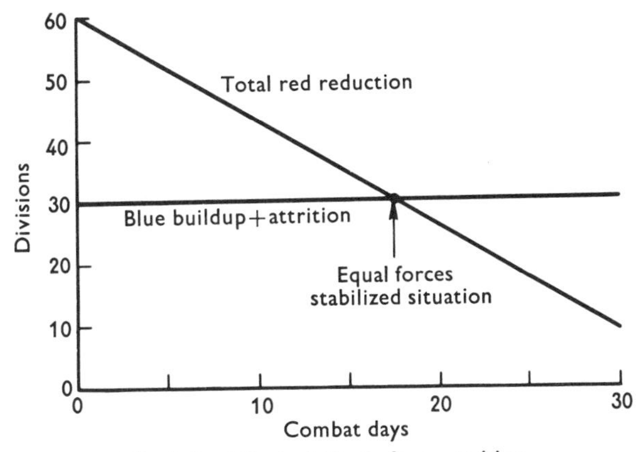

Fig. 2.4 – Effect of Blue's force attrition

Costs

At this point, we can introduce cost considerations. Expressing them as simply as possible, we can assume that one wing of the aircraft we are concerned with here will cost $400 million,[12] and one Blue division will cost just twice as much.[13] These figures are taken to include initial investment plus 5-year operating costs.

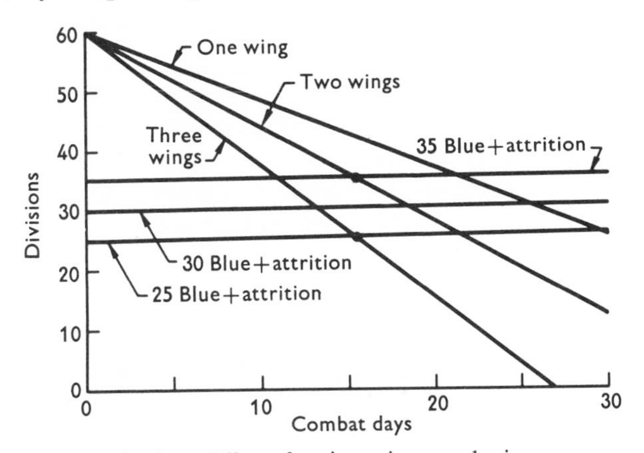

Fig. 2.5 – Effect of various air-ground mixes

Trade-off conclusions

Using these cost data and the earlier measures of effectiveness, we can

[12]This value is based on an arbitrary assumption of an initial investment cost of $2.0 million each for 75 aircraft plus $50 million a year for operating costs.

[13]This value is taken as the average of the costs of all types of U.S. divisions, as given in *Aviation Week & Space Technology*, Vol. 80, No. 8 (February 24, 1964), p. 65.

now calculate the trade-off between our two mixes. For the first mix, the total cost is $28.8 billion:

2 wings at $400 million	=	$0.8 billion
35 divisions at $800 million	=	$28.0 billion
Total	=	$28.8 billion

For the second mix, the total cost is $21.2 billion:

3 wings at $400 million	=	$1.2 billion
25 divisions at $800 million	=	$20.0 billion
Total	=	$21.2 billion

The difference in cost between the two is $7.6 billion. Thus, since both mixes lead to a stabilized conflict situation on the same day of combat, we can conclude that the second mix will produce *a saving of $7.6 billion over the first mix, without any loss of effectiveness.*

SOME LARGER CONSIDERATIONS

As far as systems analysis is concerned, it is appropriate to conclude this illustration by asking two major questions: First, "Is this study adequate?" Second, "If not, why not?"

Is there an aspect of this analysis that seems suspicious? If so, what is it? Is it the statement of the problem as a trade-off and cost-effectiveness analysis? Is it that the problem was narrowed down to a mix of wings and divisions? Is it the assumptions made in the scenario about the Red or Blue air or ground forces? Is it the criterion that we specified (the casualty-producing capability of our airpower and groundpower) or the way we calculated it? Is it the "pay-off" measure (length of the conflict) that we used? Is it the data we used on the effectiveness of airpower and groundpower? Is it our selection of alternative mixes of divisions and wings? Is it the cost estimates? Is it the conclusions we drew about trade-off savings and equal effectiveness? Or is it something else?

If the analysis is not credible, is it possible to pinpoint the reason why, or the remedies that might be taken?

The answer should be "yes." Despite the volumes of calculations, the detailed scenarios, the pages of analytic data, the computer programs and outputs, the war games, or the mathematical formulae that are part of many systems analyses, the basic structure of an analysis is usually simple. If the user or doer of an analysis cannot identify this structure and examine it critically, he may not understand what the analysis is all about, or whether or not it is valid. In short, it is the structure, rather than all the supporting data, calculations, games, computer programs, and so on, on which the analysis must stand or fall.

The remainder of this book, in various ways and from various points of view, elaborates on this theme and some others implied in our example. In Chapter 21, we will return specifically to the example and attempt to see, in the light of what the authors in between have had to say, how the analysis might have been designed and conducted differently.

Chapter 3

PRINCIPLES AND PROCEDURES OF SYSTEMS ANALYSIS

E. S. QUADE

This chapter surveys the "hows" of systems analysis and the "whys" behind them. Its basic purpose is to provide an overview of, and a context for, the individual questions of theory and technique discussed in detail by the following authors. Accordingly, the chapter limits itself to a general description of the steps common to all analysis; the principles governing each; their interaction; the character of the results they permit; the utility of these results for the decisionmaker; and, in the broadest terms, the nature of "successful" or "good" analysis.

INTRODUCTION

The RAND Corporation has produced analyses of national security problems for quite a number of years – in fact, since World War II. Although collectively we have learned a great deal that should be useful to anyone attempting to analyze such problems, we have not yet learned enough to supply a sequence of steps or rules that, if followed mechanically – by the numbers, so to speak – would automatically guarantee solutions that will stand the tests of time. In the main, this is so because military systems analysis is to some extent still an art – or at least a craft – rather than a form of engineering or an exact science. It is not, like statistics or physical chemistry, say, a body of knowledge and skills that can be acquired largely without becoming involved in particular applications.

Now, of course, some techniques of an art – even some of the most important ones – can be taught, but not by means of fixed rules which need only be followed exactly. Thus, in our analyses, we must sometimes do things that we think are right but cannot really justify or even check in the output of the work. We must accept many subjective judgments as inputs, and we must present answers based partly on judgment to be used as a basis for other judgments. Hence, a discussion of "how systems analysis is done" must content itself with indicating some guidelines, some principles, and some illustrative examples.

THE ESSENCE OF SYSTEMS ANALYSIS

If systems analysis is largely "art" and "judgment," what does the "analysis" contribute? Our answer to this question was expressed in the introductory chapter. There we stated our view that to a large extent systems analysis is successful in areas where there is no accepted theoretical foundation (defense planning is an example), precisely because it is able to make a more systematic and efficient use of expert judgment than can its alternatives. The essence of the method is to construct and operate within a model – an idealization of the situation appropriate to the problem. Such a model – in the example given by M. G. Weiner in Chapter 2, it is a series of rules or planning factors taken from official records – introduces a precise structure and terminology that serve primarily as a means of communication. As such, it enables the participants to make their judgments concretely, and, through feedback – which, in the previous example, would be the outcomes predicted by the planning factors – it helps the analysts, the experts, and the decisionmakers to arrive at a clearer understanding of both the problem and its context.

To keep the discussion from becoming too abstract, we will attempt to illustrate the points we intend to make by reference to the following hypothetical example:

Suppose a new, lighter-payload missile system is being advocated to replace or supplement the Minuteman. It would make use of the Minuteman silos and other ground facilities. Supporters claim that it will be more reliable and much more accurate and that these advantages far outweigh its somewhat higher cost and lower payload. Assume also that although development is advanced, several variants are possible, and that a decision should be made soon whether or not to freeze the design and plan procurement.

How can we proceed with an analysis to provide advice on this decision?

The alternatives

Before we answer this question, we might examine briefly the alternative sources of such advice. One of the most common, unfortunately, is pure intuition. It is in no sense analytic, since no effort is made to structure the problem or to establish cause and effect relationships and operate on them to arrive at a solution. The intuitive process is to learn everything possible about the problem, to "live with it," and to let the subconscious provide the solution. Someone using this method does not feel any obligation to show how he arrived at the solution.

Between pure intuition, on the one hand, and systems analysis, on the other, there are other sources of advice that can, in a sense, be considered analytic, although the analysis is ordinarily less systematic, explicit, and quantitative. One alternative is simply to ask an expert for his opinion.

What he says can, in fact, be very helpful, if it results from a reasonable and impartial examination of the facts, with due allowance for uncertainty, and if his assumptions and chain of logic are made *explicit*, so that others can use his information to form their own considered opinion. But an expert, particularly an unbiased expert, may be hard to find. National security problems – even those like our example, which is one of the simpler types – are complex and what should be done depends on many widely different disciplines. An expert's knowledge and opinions are likely to be more valuable if they can be formulated in direct association with other experts. This suggests systems analysis, for, as remarked above, that approach, with its models and feedback, is essentially a device for providing a framework for the systematic and efficient employment of the knowledge, judgment, and intuition of the available experts.

Another way of handling a problem is to turn it over to a committee. Now, although there is no reason why a committee cannot engage in systematic analysis, this is not likely to happen. Committees are much less likely than experts to make their reasoning explicit, since their findings are usually obtained by bargaining – by the effort to reach a consensus or an acceptable compromise. How this effort can affect originality, precision, and efficiency hardly need be mentioned. This is not to say that a look by a "blue ribbon" committee into our missile problem might not be useful, but its greatest utility is likely to be in the critique of work done by others.

Answers obtained from experts working individually or as a committee depend largely on subjective judgment. *So do the answers obtained from systems analysis.* As one writer has put it:

> Subjectivity is inherent because of the essential content of political values in public policy questions. Public policy by definition pertains to human conduct – the behavior and relations among men in political society. Because of its human impact public policy – and strategy in particular – cannot be free of questions of political value and hence cannot be decided except through the exercise of human judgment. The ingredient of human judgment – be it only the simplest kind of intuition – is therefore an essential part of any study of policy, no matter how analytical. Judgment can be aided and augmented by the techniques of scientific analysis, but it can never be supplanted.[1]

But the analytic approach, in contrast to its alternatives, provides its answers by processes that are accessible to critical examination and can be retraced by others, who can modify them more or less readily on the basis of their own judgments as errors appear or as new information becomes available.

However, no matter whether the advice is supplied by an expert, a committee, or a formal study group, an *analysis* of a problem of choice involves

[1] Col. Wesley W. Posvar, "The Realm of Obscurity," in *American Defense Policy*, prepared by Associates in Political Science, United States Air Force Academy, The Johns Hopkins Press, Baltimore, Md., 1965, p. 224.

the same five elements and basic structure we considered in Chapter 1:[2] the objectives; the alternatives for attaining them; the costs, or what we must give up; the models, which allow us to see the costs of the alternatives and the extent to which they attain the objectives; and finally, the criteria, which tell us what alternatives to choose.

We now turn to the process by which these elements are identified and the analysis carried out.

THE PROCESS OF ANALYSIS

The process of systems analysis represents a conscious attempt to extend the approach and methods – and, ideally, the standards – of the "hard" sciences into areas where controlled experimentation is seldom possible. Unfortunately, some people have exaggerated the significance or success of this attempt, and we find them saying such things as that systems analysis and operations research are really nothing more than the "scientific method" extended to problems outside the realm of pure science. Leaving aside the question whether there is anything that might be called *the* scientific method, what such statements must mean, in part, is that the analysis advances (by iteration or successive approximation) through something like the following stages:

FORMULATION (The Conceptual Phase)	Clarifying the objectives, defining the issues of concern, limiting the problem.
SEARCH (The Research Phase)	Looking for data and relationships, as well as alternative programs of action that have some chance of solving the problem.
EVALUATION (The Analytic Phase)	Building various models, using them to predict the consequences that are likely to follow from each choice of alternatives, and then comparing the alternatives in terms of these consequences.
INTERPRETATION (The Judgmental Phase)	Using the predictions obtained from the models and whatever other information or insight is relevant to compare the alternatives further, derive conclusions about them, and indicate a course of action.
VERIFICATION (The Scientific Phase)	Testing the conclusions by experiment.

[2] See pp. 16–18 and Fig. 1.2.

All analyses involve these five activities to some extent, but often the fourth is done largely by the policy-maker and the last must be done indirectly, if at all. There is a class of problems – our missile comparison is an example – in which verification may be possible in principle, but the costs of an actual test would certainly be too high. Thus, if we want to estimate what damage our missiles might do to the Soviet Union, the best we can do is use simulation to devise a vicarious experiment.

The process of analysis may be represented as in Fig. 3.1. Here the activities appear neatly separated. This is seldom the case, however, for to one degree or another they all occur simultaneously. In our missile comparison, for example, the prescription for carrying out the work might run as follows:

1. Define and limit the problem. Are we helping to make a force posture decision or is the decision really only one of whether or not to continue a promising development?

2. Classify the objectives or goals that one hopes to attain with the system being considered. Are we striving for deterrence to prevent a nuclear attack on the United States or are we striving for an even more comprehensive deterrent?

3. Forecast the political and military environment in which the systems are to operate. Do we need to consider scenarios in which the war starts as a result of the degeneration of a crisis situation, or deliberate escalation, or, as is often done, solely an attack "from out of the blue"?

4. Determine ways to measure the degree of attainment of the goals or objectives. This requires us to identify the mission to be assigned the missiles.

5. List and define the alternative systems that offer some reasonable hope of accomplishing the objectives, and select appropriate criteria for choosing among these systems.

6. Choose the approach. Shall we compare the systems for a fixed budget or shall we first fix the mission requirement? Do we start with a computer model or a manual war game?

7. Formulate a scheme for working out the dollar costs that takes account of changes in operating philosophy and development time. Explore the nonmonetary costs. Are there significant resource restraints or are there undesirable side effects that interfere with programs?

8. Examine the risks and timing in the development. Are we seeking to advance the state-of-the-art or merely to improve current capabilities?

9. Compare the systems. Do the important differences stem from unresolvable uncertainties about the future state of the world, or are they matters of engineering?

10. Perform sensitivity analyses by varying key parameters across a

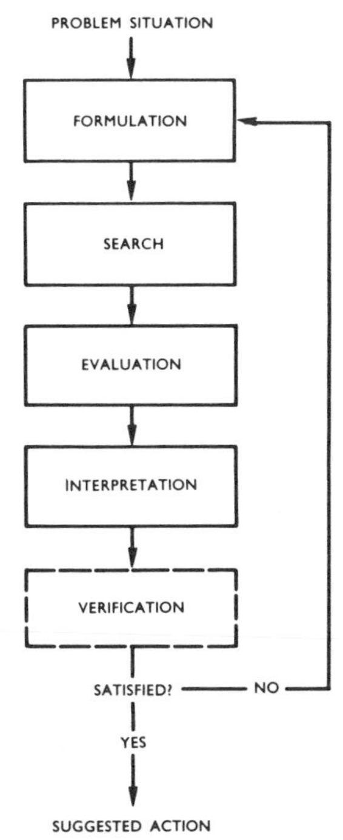

Fig. 3.1 – Activities in analysis

range of values, to see that major uncertainties are thoroughly explored.

11. Consider the factors that we have so far not taken into account, and test them against various assumptions where we think we have some knowledge of what the outcome should be.

12. Decide what we can really recommend on the basis of the analyses.

13. Document our work. This should include the rationale and assumptions, more than a mere summary.

Some of these steps are clearly part of problem formulation, others belong in the domain of the comparison, and still others might be classified as part of the interpretation stage. The search stage, as is typically the case, permeates the whole process and is especially difficult to isolate as a separate activity. If, however, we stand back from the clutter of the real world, there are several points we can usefully make about the process of analysis as it appears in each of the activities named in Fig. 3.1. Let us take them in turn, beginning with "Formulation."

Formulation

Formulation implies an attempt to isolate the questions or issues involved, to fix the context within which these issues are to be resolved, to clarify the objectives, to discover the variables that are operative, and to state relationships among them. These relationships may be extremely hypothetical if empirical knowledge is in short supply, but they will help make the logical structure of the analysis clear. In a sense, formulation is the most important stage, for the effort spent restating the problem in different ways or redefining it clarifies whether or not it is spurious or trivial and points the way to its solution.

The process of formulation is highly subjective. We must, for example, consider what evidence will be meaningful and significant to the decision-maker we are trying to help. Thus, in our missile comparison, will it be sufficient to compare the new missile with the Minuteman alone, or must other missiles or even manned bombers be considered? Will it be adequate to make the comparisons in a U.S. second strike situation? Greater reliability and accuracy may show to more advantage in wars initiating in other ways. Are we helping to make a force posture decision or should we really only be trying to demonstrate that we have a promising development that should be continued for its growth potential alone?

The tendency all too frequently is to accept the client's original statement of what is wanted, and then to set about building a model and gathering information, scarcely giving a thought to whether the problem is the right problem or how the answer will contribute to the decisions which it is meant to assist. In fact, because the concern is with the future, the major job may be to decide what the policy-maker should want to do. Since systems studies have resulted in rather important changes, not only in how the policy-maker carries out his activity, but in the objectives themselves, it would be self-defeating to accept without inquiry the customer's or sponsor's view of what the problem is.

But how is the analyst to know that his formulation of the problem is superior? *His only possible advantage lies in analysis.* That is, the process of problem formulation itself has to be the subject of analysis. What this means is that, using the few facts and relationships that are known at this early stage and assuming others, the analyst must simply make an attempt to solve the problem. It is this attempt that will give him a basis for better formulation. He always has some idea as to the possible solutions of the problem; otherwise, he probably should not be working on it, for his analysis might prove to be too formal and abstract.

Let us consider a classic example. For fiscal year 1952, Congress authorized approximately $3.5 billion for air base construction, about half to be spent overseas. RAND was asked to suggest ways to acquire, construct, and maintain air bases in foreign countries at minimum cost. The analyst who reluctantly took on this problem regarded it at first as essentially one of logistics. He spent a long time – several months, in fact–

thinking about it before he organized a study team. Although he had little of the information needed to make recommendations, he was able to see the problem in relation to the Air Force as a whole. He came to the conclusion that the real problem was *not* one of the logistics of foreign air bases, but the much broader one of *where and how to base the nation's strategic air forces and how to operate them in conjunction with the base system chosen.* He argued that base choice would critically affect the composition, destructive power, and cost of the entire strategic force and thus that it was not wise to rest a decision about base structure and location merely on economy in base cost alone. His views prevailed and he led the broader study, the results of which contributed to an Air Force decision to base SAC bombers in the continental United States and use overseas installations only for refueling and restaging.[3] An Air Force committee later estimated that the study recommendations saved over $1 billion in construction costs alone. In addition, it sparked a tremendous improvement in strategic capability, particularly with regard to survival, and stimulated a good deal of additional research on related questions.

In analysis, the problem never remains static. Interplay between a growing understanding of what it involves now and might involve in the future forces a constant redefinition. Thus, the study just mentioned, originally conceived as an exercise to reduce costs, became in the end a study of U.S. strategic deterrent policy. Its recommendations led to a major reduction in SAC vulnerability; that costs were also reduced was secondary.

Primarily as the result of discussion and intuition, the original effort to state a problem should suggest one or more possible solutions or hypotheses. As the study progresses, these original ideas are enriched and elaborated upon – or discarded – and new ideas are found. The process of analysis is an *iterative* one. Each hypothesis serves as a guide to later work – it tells us what we are looking for while we are looking. As a result, the final statement of the conclusions and recommendations usually rests on a knowledge of facts about the problem which the analyst did not have at the start. In the early stages it is not a mistake to hold an idea as to the solution; the error is to refuse to abandon such an idea in the face of mounting evidence.

It is important to recognize that anything going on in one part of an activity, organization, or weapon system is likely to affect what goes on in every other part. The natural inclination in problem-solving is to select a

[3] For the full report, see A. J. Wohlstetter, F. S. Hoffman, R. J. Lutz, and H. S. Rowen, *Selection and Use of Strategic Air Bases*, The RAND Corporation, R–266, April 1954. A nontechnical account of this study appears as Chapter VI in Bruce L. R. Smith's *The RAND Corporation: Case Study of a Nonprofit Advisory Corporation*, Harvard University Press, Cambridge, Mass., 1966.

part of the problem and analyze it separately, or to reduce the problem to one that looks manageable. "Many scientists owe their greatness not to their skill in solving problems but to their wisdom in choosing them."[4] Systems analysis, however, does not offer us this freedom, at least not at the outset. We have to solve the problem that exists. It calls for us to extend the boundaries of the problem as far as required, determine which interdependencies are significant, and then evaluate their combined impact.

But even for small-scale problems, the number of factors under consideration at any one time must be reduced until what is left is manageable. In systems analysis, the complexity of the full problem frequently far outruns analytic competence. To consider in detail anything like the complete range of possible alternative ways to deliver weapons on strategic targets may be impossible. The use of suitcases or automobiles as delivery systems does not belong in our missile comparison. Fortunately, the vast majority of alternatives will be obviously inferior, and can be left out without harm. The danger is that some alternative better than the one ultimately uncovered by the analysis might also be left out. Thus, although constraints must usually be imposed to reduce the number of alternatives to be examined, this should be done by preliminary analysis, not by arbitrary decree. Moreover, such constraints should be flexible, so that they may be weakened or removed if it appears in later cycles that their presence is a controlling factor. In analyzing our missile system, for example, we do not simultaneously seek to determine the ideal ground support weapon for the Tactical Air Command or the ratio of medical corpsmen to cooks in the base support battalion. We call such a restriction of the problem a "suboptimization."

The necessity for suboptimization compounds the difficulties in the selection of criteria and objectives. It is inevitable that not all decisions can be made at the highest level or by one individual or group; some must be delegated to others. Analysts and decisionmakers must thus always consider actions that pertain to only a part of the military problem. Other choices are set aside temporarily, possible decisions about some things being neglected and specific decisions about others being taken for granted. What is crucial is that the criteria and objectives for the suboptimization be consistent with those that would apply to the full problem.

The most troublesome problems in analysis – those of selecting criteria, objectives, and ways to measure effectiveness – are discussed in detail in the next chapter. We might note several general points beforehand, however.

---

[4] E. Bright Wilson, Jr., *An Introduction to Scientific Research*, McGraw-Hill Book Company, Inc., New York, 1952, p. 1.

It is commonly supposed that goals should, and can, be set independently of the plans to attain them. Yet there is considerable evidence that operationally significant objectives are, more often than not, the result of opportunities that possible alternatives offer rather than a source of such alternatives. For one thing, it is impossible to select satisfactory objectives without some idea of the cost and difficulty of attaining them. Such information can only come as part of the analysis itself. For another, only some of the possible consequences of different alternatives can be anticipated before the analysis. The newly discovered consequences may then become goals. Thus, for example, the invention of a near-perfect system for continuous peacetime strategic reconnaissance might, in some circumstances, make a first strike to disarm the enemy an objective worth considering.

In fact, a characteristic of systems analysis is that solutions are often found in a set of compromises which seek to balance and, where possible, to reconcile conflicting objectives and questions of value. It is more important to choose the "right" objective than it is to make the "right" choice between alternatives. The choice of the wrong alternative may merely mean that something less than the "best" system is being chosen. Since we must frequently be satisfied with at most a demonstration that a suggested action is "in the right direction," this may not be tragic. For, as we shall see, such a demonstration may be the best that can be done anyway. But the wrong objective means that the wrong problem is being solved.

The choice of the objective must be consistent with higher, or national, objectives. Since these are seldom operationally defined, however, the analyst has a great responsibility to exercise care and good judgment. In our missile comparison, for instance, if we choose as an objective one that puts a great premium on keeping collateral damage and civilian casualties low, we bias the analysis toward the more accurate missile. Since in the example we are attempting to determine whether or not to replace a current capability with a more accurate one, we should select deterrence as our objective, measuring it proximately by the total mortalities inflicted, for accuracy is not so significant here. We then can argue *a fortiori* that if the lighter payload shows up better given this objective, then the case for it is all the stronger.

At some stage we must decide on a specific approach to the problem. The essence of the questions with which systems analysis is concerned is uncertainty, not only about economic, technical, and operational parameters – which can be serious but are to a large extent under our control and somehow appear limited – but also about future environments or contingencies. It is almost impossible to forecast what these might be, let alone to predict what an enemy might do about them. Hence, except in very narrowly defined problems, we must look for an approach to the analysis which offers a hope of producing something constructive in spite of the great uncertainties.

A point of view that has evolved from bitter experience runs something like the following. An attempt to determine a sharp optimum or the unique best solution in a problem having largely indeterminate parameters, some subject to enemy influence, is probably doomed to failure. The goal instead should be to find and recommend a system that is close to the optimum for the expected circumstances, but, at the same time, is to a large degree insensitive to many uncertainties – specifically, a system that might work well under many widely divergent contingencies and could even give some sort of reasonably satisfactory performance under unexpected and thus possibly catastrophic circumstances, This characteristic is probably what the military man intends when he speaks of "flexibility."

It is helpful here to recall that the primary purpose of systems analysis is to advise a decisionmaker, help answer his questions, sharpen his intuition, and broaden his basis for judgment. In practically no case should we expect to "prove" to the decisionmaker that a particular course of action is uniquely best. The really significant problems are just too difficult and there are too many considerations that cannot be handled quantitatively. If we insist on a strictly quantitative treatment, we are likely to end with such a simplified model that the results will be almost meaningless, or arrive too late to be useful to the decisionmaker.

These observations suggest two rules of thumb:

1. Throughout the inquiry, it is well to look for *gross* differences in relative costs and effectiveness among the alternatives and, specifically, for differences of the sort that have a chance of surviving many likely resolutions of the various uncertainties and intangibles. Thus, in comparing future systems, the question to address is which systems have a *clear* advantage, rather than that of precisely how much better one is than another. Something of this attitude no doubt underlies a motto of the Systems Analysis Office in the Office of the Secretary of Defense: "It is better to be roughly right than exactly wrong."[5]

2. All comparisons should be made with the uncertainties in mind. A choice between missile systems, therefore, must depend on a careful investigation of a wide range of enemy offensive and defensive capabilities.

---

[5] But one RAND analyst notes that even the discovery that the quantitative differences among the alternatives are insignificant can have a considerable value to the decisionmaker. "This is especially true if sensitivity analyses have been made . . . [and] the final results are still within relatively narrow ranges. Given results of this kind, the decisionmaker can be less concerned about making a mistake regarding the quantitative aspects of the problems, and he may then feel somewhat more comfortable about focusing more of his attention on the qualitative . . . considerations." (G. H. Fisher, *The Analytical Bases of Systems Analysis*, The RAND Corporation, P-3363, May 1966, p. 14.)

But this is not enough. When, as in our example, a proximate criterion must be used – the capability of the force to inflict mortalities being supposed to substitute for deterrence – the effort must be made to test with criteria other than the one used for preliminary screening. Moreover, the performance of the alternatives must be considered under a variety of contingencies that involve major changes in the environment – those due, say, to political acts or actions that depend on human caprice.

Search

The search phase is concerned with finding both the alternatives and the data, or evidence, on which the analysis is to be based. It is as important to look for new alternatives (and evidence to support them) as it is to look for ways to compare them. Obviously, if we have no alternatives or no ideas about them, there is nothing to analyze or to choose between. If in the end we are to designate a preferred course of action, we must have discovered earlier that such a course exists.

In military analysis, many facts are hard to come by. The actual operational performance of future missiles in combat, for instance, cannot be predicted with any great degree of certainty by purely theoretical studies. Nor, for that matter, is it very likely that an individual systems analyst modeling a weapon system will be thoroughly familiar with all the aspects of the system and its environment. For these reasons, systems analysis must depend a good deal more on informed judgment than do most types of engineering. The analyst must be assisted by others, civilian as well as military, and depend on their judgment, not only for facts, but for opinions about the facts. Of course, it may not always be possible to recommend a course of action even when the facts are known, but it is a knowledge of them which makes a solution possible.

One aspect of the search stage that will be emphasized throughout this book is the role of component or supporting studies – scientific, engineering, and cost – in systems analysis. A sensible answer to broad questions obviously requires a great many facts at hand, and these supporting studies are designed to provide them. Thus, in our example, the feasibility of the alternative missiles, or the range of force postures in which they might be imbedded, may hinge primarily on discovering the trade-offs between such engineering parameters as range, payload, and accuracy and how these affect the cost, effectiveness, and availability of the missiles. It is here that most of the man-hours invested in a study must go. As a result, it is largely the technical competence found in his associates in engineering and science that may account for an analyst's successes. Certainly, this is the case at RAND. One might get the idea from much of what is discussed in this book that systems analysis is done mainly by men with their feet on a desk, men once described as "pipe-smoking, tree-full-of-owls types." And this may sometimes be the case; but typically it takes a lot of detailed individual research, conferences, and traveling by engineers, cost analysts, economists, political scientists, operations analysts, and other specialists to produce an analysis that makes a useful contribution.

The search for data can, of course, be endless, since in principle the uncertainties of most planning problems can never be completely eliminated. When should the theoretical analysis begin? What proportion of effort should be devoted to empirical research? D. M. Fort offers these suggestions:

> . . . the proper balance between theoretical analysis and fact-gathering depends on the problem. It is important, of course, to get the facts on the proper subject; a preliminary theoretical analysis can be very useful to this end, in pointing out what information is lacking and most needed. Much effort can be and often is wasted gathering the wrong data, for failure to do the necessary theoretical homework first. On the other hand, much effort is also wasted applying sophisticated analytical techniques to inadequate data, trying to make silk purses out of sows' ears. Physical experiments and data gathering in general are expensive; making plans and decisions in the face of uncertainty, even if aided by the best possible systems analysis, can also be very expensive. A proper balance may well call for much more emphasis on fact-gathering than has been customary.[6]

Expert opinion must be called upon when it is necessary to use numerical data or assumptions that cannot be based on theory or experience – when, say, we want to obtain something like an estimate of the guidance accuracy of our new missile in the presence of counter-measures that have been conceived in theory but have not yet been developed. Chapter 18 describes a method for doing this systematically.[7]

Evaluation

In order to choose among alternatives, a way to estimate or predict the various consequences of their selection must exist. This may be as elementary as calling on the intuition of a single expert, but the more formal process of using a model or a set of models usually leads to better results. The role of the model in systems analysis is to provide a way to obtain cost and performance estimates for each alternative. Sometimes these estimates are obtained from a single over-all model – say, an elaborate computer program which combines into a single computation all the various submodels for determining dollar cost, reliability, lives lost, targets destroyed, and so on. At other times, consequences of different types are obtained separately by a wide variety of processes – gaming, computation, or political analyses.

Later, we will devote separate chapters to models in general and to

---

[6] D. M. Fort, *Systems Analysis as an Aid in Air Transportation Planning*, The RAND Corporation, P–3293–1, March 1966, p. 10.

[7] The Delphi technique, pp. 435 ff.

three types common in systems analysis: simulations, war games, and scenarios.[8] Hence, the discussion here will confine itself to a few general remarks about the way models are used in systems analysis, especially as these models involve quantification and the use of judgment.

Consider our example of how to advise a decisionmaker on a substitute for the Minuteman. A typical military systems analysis such as this usually takes one of two forms. In the first, some level of military effectiveness (the objective) is fixed and an attempt is made to determine the alternative which will attain the desired effectiveness at minimum cost. In the second, the budget level is fixed and we seek to maximize effectiveness. Suppose we decide to take this latter approach.

To carry out the analysis, a specifically dated budget must be assumed and, using various models, the forces attainable with that budget must be worked out. This task, which is by no means simple, requires a cost model. In part, this model is constructed by measuring the purchase price not only of the various weapons and vehicles involved, but also of the whole materiel and manpower structure. The costs must take into account the entire system of utilization, extended over a period of time prolonged sufficiently to reflect the important factor of peacetime maintenance. It takes a great deal of research and sophisticated knowledge to cost a system that does not yet exist.[9]

Next, an environment and a mode of war initiation must be specified. Rather than base the analysis on a set of assumptions forced reluctantly from some consultant political scientist, an analytic scenario might be useful. Such a scenario starts with the present state of the world and shows, step by step, how one or more future situations might evolve out of the present one and how, in those situations, war might begin.

To carry on from here, a step-by-step procedure, called the campaign model, is used to work out what the war outcomes might be. Then, finally, some criterion or payoff function is used to weigh the various war outcomes and determine a preference ordering of the alternatives.

This process may break down at almost any stage. Some problems are so ill-structured and the cause and effect relationships so poorly understood that we cannot build a model with any feeling of confidence. When this is so, we cannot work out the consequences of adopting the various strategies or compare outcomes. The alternative is then to use a model which compares the salient characteristics of the possible strategies. This is the "consumers' research" approach, in which experts or "potential users" rate the alternatives. Again, of course, some way is needed to bring the various ratings together – a problem we have already looked at, but not

[8] See Chapters 10–18.

[9] An example of this sort of cost analysis is given in Chapter 9.

when value judgments were involved. We will consider this in greater detail in a moment, for the same type of difficulty arises even when we can, in one sense or another, compute the outcomes.

It should be emphasized that, for many important problems, we are in fact unable to build really quantitative or even formal models. The most obvious function of a model is "explanatory," to organize our thinking. What counts, therefore, is not whether the model was mathematical or was run on a computer, but rather whether an effort was made to compare alternatives systematically, in terms as quantitative as possible, using a logical sequence of steps that can be retraced and verified by others.

Usually, we can go beyond this bare minimum, and although we may not be able, at least initially, to abstract the situation to a mathematical model or series of equations, some way can generally be found to represent the consequences that follow from particular choices. Simulation, for example – the process of imitating, without using formal analytic techniques, the essential features of a system or organization and analyzing its behavior by experimenting with the model – can be used to tackle many seemingly unmanageable or previously untouched problems where a traditional analytic formulation is at least initially infeasible. Operational gaming – that is to say, simulation involving role-playing by the participants – is another particularly promising technique, especially when it is desirable to employ several experts with varying specialities for different aspects of the problem. Here the game structure – again a model – furnishes the participants with an artificial, simulated environment within which they can jointly and simultaneously experiment, acquiring through feedback the insights necessary to make successful predictions within the gaming context and thus indirectly about the real world.

Getting back to our example, suppose there is general agreement (highly unlikely!) that the model accurately reflects the real situation and that the calculations are valid. Suppose further that, for a particular set of assumptions (about such things as the way war begins, the strength and disposition of the enemy forces, and so on), the expected or average "war outcomes" as computed by the model are those shown in Table 3.1:

TABLE 3.1

Hypothetical war outcomes for three alternatives

Expected War Outcomes	Alternatives		
	A	B	C
Number of Enemy Targets Destroyed	80	100	150
Hours to Destroy 50 Enemy Targets	1	2	4
U.S. Lives Lost (millions)	20	25	50
Cost to the Enemy to Cut His Losses by 50 per cent ($ billions)	3	12	180

Of course, many other outcomes might have been computed or estimated from war gaming exercises that took other considerations into account – flexibility, contributions to our limited war capabilities, and so on. But given what we have, how does one decide which alternative to prefer? Fifteen years ago the rule was: Pick the system which destroyed the most targets for the given cost. Today we realize we must be interested in the other outcomes as well – some of which we cannot compute. One unrecommended way to determine a preference *a priori* is to use a payoff function which takes only the various numerical outcomes into account.[10]

A single decisionmaker would probably operate differently. He need only make up his mind, arguing with himself – thinking, for example:

> "C should be chosen. The potential threat it represents means that the probability of war will be reduced practically to zero and the cost to the enemy to counter it will collapse his economy."

Or, alternatively:

> "A is best. The primary purpose of these systems is to create a threat of unacceptable damage; 80 targets are as good as 150 for this purpose. C is too threatening; it leaves the enemy no choice but to attack."

In the usual case, there are a number of decisionmakers. The process changes accordingly, for what is needed is a collective judgment from them and the experts on whom they lean for advice.

Whenever possible, of course, this judgment should be "considered" judgment; that is, supplemented by inductive and numerical reasoning and made explicit. But it is judgment nonetheless.

How, then, might we apply group judgment to the problem of choosing one of the systems A, B, and C? We might seek a consensus by using one of several methods that allow us to pool the judgments of experts when faced with factual value uncertainty. The Delphi technique is a possibility.[11] Another is simply to ask each of our decisionmakers or experts to fill in an

---

[10] For example, confining ourselves to the four war outcomes we have listed, we might say: Pick the system for which the product of the number of targets destroyed and the logarithm of the cost to the enemy, divided by the product of the number of lives lost and the time to destroy 50 targets, is greatest. Using this payoff, the analyst would reach these results:

A: $(80 \log_e 3)/20 = 4.4$

B: $(100 \log_e 12)/50 = 5.0$

C: $(150 \log_e 180)/200 = 3.9$

This indicates that B should be the choice. But the use of such a function is extremely arbitrary; it might be just as absurd to use the square root instead of the logarithm of the cost to the enemy. The values for A, B, and C would then be 6.9, 6.9, 10.1, respectively. Either payoff function would give a lesser weight to the cost factor than to the other factors involved, but by what logic can we choose such a function? This approach is never satisfactory unless there is a logical argument or empirical evidence to determine the form of the payoff.

[11] See pp. 435 ff. for a description of this method.

array such as the one illustrated in Table 3.2. After the experts had estimated the military worth of the various considerations relevant to the decision – using, say, a number between 0 and 10 – we could then work out a numerical measure.[12]

TABLE 3.2

A framework for evaluating alternatives

Consideration	Rating of Alternatives		
	A	B	C
Targets Destroyed	—	—	—
Time to Destroy 50 Targets	—	—	—
U.S. Lives Lost	—	—	—
Cost to Enemy	—	—	—
Intra-war Deterrence Capability	—	—	—
False Alarm Security	—	—	—
Flexibility	—	—	—
Growth Potential	—	—	—

In addition to uncertainty as to the outcomes as listed in such a table, and moral or value uncertainty as to which combination of outcomes would be preferable, this problem also presents uncertainty as to the state of the world and the actions of the enemy. (This further complicates our problem, for we would have a display such as Table 3.1 for each contingency). But even if we went no further than to display systematically the opinions and judgment of a single decisionmaker for his own use, the exercise would be likely to help him. If the quantitative judgments of others are presented along with their arguments, they should be still more valuable, even though we might not make use of feedback to bring the various judgments more nearly to a consensus.

These, then, are some of the general notions the analyst cannot ignore. In a certain sense, specifically in their application to the problem of building models, they can be reduced to two heads: questions involving quantification or the treatment of uncertainty. Since almost every author in this book discusses uncertainty, we may content ourselves here with a simple example that points out what is meant by the explicit treatment of uncertainty. Its problems are, of course, intimately associated with those of quantification.

> A farmer must decide what crop or crops to plant without knowing whether the weather will be wet, moderate or dry. An analysis is performed to help him decide.

[12] A very similar approach is advocated by Everett J. Daniels and John B. Lathrop in "Strengthening the Cost-Effectiveness Criterion for Major System Decisions," a paper presented at the October 1964 meeting of the Operations Research Society of America.

A popular approach is employed in which the analysis is repeated in turn for each of the three distinguishable types of weather, in each case determining the best crop to plant for that type of weather. Considering all possible crops, it is found that for wet weather corn would be best, for moderate weather oats would be best, and for dry weather wheat would be best. The principal results presented to the farmer consist of the findings concerning the best crops in the three types of weather, the best yields achievable in each contingency (i.e., the wet-weather yield of corn, the moderate-weather yield of oats, and the dry-weather yield of wheat), and estimates of the probabilities of wet, moderate and dry weather. The implication is that the farmer ought to make his choice from the "preferred" crops, corn, oats or wheat, or perhaps a combination of these to provide some all-weather insurance.

The farmer is not satisfied with the analysis, however. He points out that the analysis tells him what crop he should plant if he knew for certain what the weather would be, but he doesn't see how this helps him to decide what to plant when he doesn't know what the weather will be, except for the weather probabilities. He would like to know, for example, what will happen if he plants corn and the weather turns out moderate or dry, and similarly for the other crops. The analyst therefore prepares a two-way contingency table, showing for each of the three "preferred" crops the yields in wet, moderate and dry weather. Yields for various mixtures of these crops are also shown in the various types of weather. It is found that each of the three crops is rather narrowly tailored for that type of weather in which it is best, and gives disastrously poor yields in other types of weather. Oats, for example, gives poor yields in wet or dry weather, but very good yields in moderate weather. The farmer can insure against disaster by planting a mixture of corn, oats and wheat, thereby obtaining a fair overall yield whatever the weather.

The farmer is still not satisfied, however. The contingency table does give him the information he wants on the three "preferred" crops, but he would like to see the same information for some other crops, even though they have been ruled out as "inferior" in the analytical optimization. The analyst obligingly expands the contingency table to show the yields of various other "inferior" crops in wet, moderate and dry weather. At this point it is noted that cane, which is inferior to corn in wet weather, inferior to oats in moderate weather, and inferior to wheat in dry weather, gives a "pretty good" yield in all types of weather, providing better all-weather insurance than can be achieved with any combination of the three "preferred" crops. This particular farmer, having a pronounced aversion to risk, decides that of all the crops he prefers the weather-yield pattern of cane over that of any other crop or combination of crops. Another farmer, looking at the same table, might prefer to take somewhat more of a chance on alfalfa, another "inferior" crop shown to give a rather good yield in wet or moderate weather but a poor yield in dry weather. Still another might prefer to take a greater chance on corn, but not necessarily because it was one of the "preferred" crops in the original analysis.[13]

The first approach described above, which determines which of the farmer's various options is "preferred" for each situation or specific set of assumptions about the uncertain factors, is far from uncommon in actual applications of systems analysis. It is useful in indicating some of the systems that merit consideration by the decisionmaker or planner. It can be worse than useless, however, if it leads him to limit his attention only to those "preferred" systems.

The approach that evolves toward the end of the example has the advantage of not ruling out systems that ought to be considered. It has

[13] D. M. Fort, *Systems Analysis as an Aid in Air Transportation Planning*, pp. 12–13.

the disadvantage, however, of not ruling out very many systems at all, for it eliminates only those systems which are "inefficient"; that is, systems which are inferior or at most equal to other systems in *all* situations or for *all* assumptions about uncertain factors. Some means must be found to narrow the list further. This may require going beyond the bounds of strictly quantitative analysis, by such expedients as eliminating systems or uncertainties by direct application of the analyst's judgment or that of experts on whom he might call.

> Whatever approach is used in narrowing down the list of systems to be presented to the customer, the approach should be described as explicitly as possible. The presentation should include, among other things, a contingency table, showing for each system its performance and cost in each of the various relevant situations and/or for each set of assumptions about the uncertain factors. Digesting this information and using it in making decisions or plans puts a heavy burden on the decision-maker or planner, but it can't be helped. Systems analysis does not relieve the customer of the responsibility for facing the uncertain consequences of his decisions or plans; it can, however, help him face uncertainty with a better appreciation of the relevant considerations than he might otherwise have had.[14]

Why is quantification desirable? Some aspects of problems of choice in national security require numbers; others do not. When a quantitative matter is being discussed, the greatest clarity of thought is achieved by using numbers instead of avoiding them, even when uncertainties are present. Only in rare cases is it possible to make a convincing comparison of alternatives without a quantitative analysis.

> What is at issue here really is not numbers or computers versus words or judgment. The real issue is one of clarity of understanding and expression. Take, for example, the statement "Nuclear power for surface ships offers a major increase in effectiveness."
>
> Precisely what does that mean? Does it mean 10 per cent better or 100 per cent better? When that sort of question is asked a frequent answer is, "It can't be expressed in numbers." But it has to be expressed with the help of numbers. Budgets are expressed in dollars, and nuclear power costs more than conventional power. If nuclear power costs, say 33 per cent more for some ship type, all factors considered, then, no matter what the budget level, the Navy and the Secretary of Defense have to face the choice of whether to put the nation's resources into four conventional or three nuclear ships, or for a larger budget, eight conventional or six nuclear ships, and therefore whether by "major increase" is meant more than 33 per cent, about 33 per cent, or less than 33 per cent. Because the Secretary of Defense has to make the decision in these terms, the statement "major increase" is not particularly helpful. It must be replaced by a quantitative analysis of the performance of various missions, leading to a conclusion such as, "Nuclear power for surface ships offers something between X and Y per cent more effectiveness per ship. Therefore, $1 billion spent on nuclear powered ships will provide a force somewhere between A and B per cent more or less effective than the same dollars spent on conventionally powered ships.[15]

---

[14] D. M. Fort, *Systems Analysis as an Aid in Air Transportation Planning*, p. 15.

[15] Alain C. Enthoven, Assistant Secretary of Defense (Systems Analysis), "Choosing Strategies and Selecting Weapon Systems," *United States Naval Institute Proceedings*, Vol. 90, No. 1, January 1964, p. 151.

Some variables are difficult to quantify, either because they are not calculable, like the probability of war, or because no satisfactory scale of measurement has yet been devised for them, like the effect on NATO solidarity of some unilateral U.S. action. This sometimes leads either to their neglect, for they tend to be ignored, or to their being recognized only by modifying a solution reached in fact by manipulating quantified variables. Thus, when the problem arises of using the model to recommend an action, the analyst may have trouble weighing these variables properly: the effect of the quantitative variables is built in, while that of the nonquantitative ones may be easily lost in the welter of qualitative considerations that must be taken into account.

As we have already seen, certain variables can be eliminated, either because they are irrelevant or trivial in their quantitative effects or because they have roughly the same effect on all the alternatives under consideration. The second explanation is the more important. Indeed, the fact that many variables fall into this category makes analysis possible. If the results were *not* insensitive to all but a relatively small number of variables, analysis would have to yield completely to guesses and intuition. *The point is that this insensitivity must be discovered.* Sometimes logical reconnoitering alone is sufficient, but usually analysis is required, possibly with arbitrary values assigned to the variables we are unable to calculate.

If nonquantitative variables are not to be neglected without mention or dismissed with some spurious argument, such as the one that they act in opposite direction and hence cancel out,[16] then how are they to be treated? The usual method is the one mentioned a moment ago – to attempt to take them into account through modification of the solution rather than to incorporate them into the model. But this in itself represents a particular method of quantification, for, by altering the solution to take account of the previously omitted variables, the analyst is implicitly valuing them. Since we nearly always have some insight into the range of values that a factor might take, we can, in many cases, assign it an arbitrary value and observe the effect on the solution.

In the general process of investigating a problem and gathering data about it, the analyst will have developed ideas of what considerations are likely to be most influential in determining the possible courses of action. To construct a model, he uses these insights – which actually represent crude preliminary models – and conducts pencil and paper experiments to illuminate their implications. Analysis, being iterative, is self-correcting; as the study goes on, early models are refined and then replaced, so that

[16] It is not enough to know that two variables act in opposite directions; their quantitative impact must also be estimated.

the behavior of the relationships being investigated is represented with greater accuracy.

For most phenomena, there are many possible representations; the appropriate model depends as much on the question being asked as on the phenomena about which it is asked. A town can be modeled by a map if the question being asked is how to walk from A to B; but if the question is how to speed up the flow of traffic between the same two points, a much more elaborate model may be needed. The point is that there are no "universal" models – that is to say, no one model that can handle all questions about a given activity.

"Working" the model, trying out various strategies and concepts of operation, is the closest systems analysis comes to scientific experimentation. Deductions based on operating with the model frequently suggest new directions of effort. That is to say, starting with the relatively few parameters that characterize a system in terms of the model, it is sometimes possible to show that changing them would improve the performance of the system as measured by the model, which, in turn, might suggest corresponding improvements that could be made in the real system as it performs in the real world. In this way, working the model contributes to system design.

It is also important to go outside the model, to contemplate changes that violate its assumptions, and thereby perhaps achieve a better model. But whether or not one model is better than another does not depend on its complexity or computability, but solely on whether it gives better predictions. Unfortunately for systems analysis, but possibly fortunately for the world, this test is not usually an operational one when military problems are being considered.

Interpretation

At this stage, not only does the analyst attempt to interpret his work, but so does the sponsor or the decisionmaker. Thus, the real world gets into the iterative cycle again, possibly to counteract its always imperfect mapping onto the model and, hopefully, to produce better answers.

As we remarked earlier, good criteria can only be found by working with the problem; that is, they cannot be developed *a priori*. Ends and means interact. Are the criteria good? What are the costs? What is the state-of-the-art? Are the objectives attained? Judgment must tell us whether we need to modify these things and run through another cycle or not.

Suppose the study has been done properly. Say the assumptions are reasonable, the chain of reasoning logical, the judgments as to the various inputs sound. This does not mean that the analysis is ended. As we have seen, the outcomes obtained from a model must be interpreted in the light of considerations which may not have been adequately treated by the model. Thus, in our example, the decisionmaker (or, for that matter, the systems analyst) may have established the requirement that a follow-on Minuteman worth considering have the capability to assure the destruction of, say, 95 per cent of a certain target list under a particular range of contingencies. But many questions occur. Perhaps the minimum cost of achieving this capability for all alternatives is too high; maybe the tasks of deterrence and limiting damage to the United States which we are trying to assure with our damage capability could be better done by spending less on strategic forces and more on air defense. The 95 per cent measure of effectiveness may be too high, or too low. Someone must translate the percentage of target destruction into its implications in terms of more meaningful criteria, such as the balance of military forces, the will to continue fighting, and the effect on our diplomacy. We can never know these things fully. For indicating the attainment of such vaguely defined objectives as deterrence or victory, it is even hard to find measures that point in the right direction. Consider deterrence, for instance. It exists only in the mind – and in the enemy's mind at that. We cannot use some "scale of deterrence" to measure directly the effectiveness of alternatives we hope will lead to deterrence, for there is no such scale. Instead, we must use such approximations as the potential mortalities that we might inflict, or the industrial capacity we might destroy. Consequently, it is clear that, even if a comparison of two systems indicates that one could inflict 50 per cent more casualties on the enemy than the other, we cannot conclude that this means the system supplies 50 per cent more deterrence. In fact, since in some circumstances it may be important not to look too threatening, we can argue that the system capable of inflicting the greatest number of casualties may provide the least deterrence!

The solution to a problem that has been simplified and possibly made amenable to mathematical calculation by drastic idealization and aggregation is not necessarily a good solution of the original problem. But even if the model and its inputs are excellent, the results may be unacceptable. The reason is obvious: Major decisions, in the field of military policy, are part of a political as well as an intellectual process. To achieve efficiency, considerations other than those of cost-effectiveness are important – discipline, morale, *esprit de corps*, tradition, and organizational behavior. The size, composition, location, and state of readiness of forces influence our foreign policy and the freedom of action we have. They also have a major impact on our domestic economy and public morale. The men who must somehow integrate these factors with the results of the study must necessarily deal with much that is nonquantitative, and their results may differ.

It is important for the user of analysis to distinguish between what the study actually shows and the recommendations for action the analyst makes on the basis of what he, the analyst, thinks the study implies. Some experienced and successful users of analysis hold even stronger views:

> Simply said, the purpose of an analysis is to provide illumination and visibility – to expose some problem in terms that are as simple as possible. This exposé is used as one of a number of inputs by some "decision-maker." Contrary to popular practice, the primary output of an analysis is not conclusions and recommendations. Most studies by analysis do have conclusions and recommendations even though they should not, since invariably whether or not some particular course of action should be followed depends on factors quite beyond those that have been quantified by the analyst. A "summary" is fine and allowable, but "conclusions" and "recommendations" by the analyst are, for the most part, neither appropriate nor useful. Drawing conclusions and making recommendations (regarding these types of decisions) are the responsibility of the decision-maker and should not be preempted by the analyst.[17]

When new minds – the decisionmaker's, for example – review the problem, they bring new information and insight. Even though the results obtained from the model are not changed, recommendations for action based on them may be. A model is only an indicator, not a final judge. While the analysis may compare the alternatives under a great many different assumptions, using various models, no one would expect the decision to be made solely on the basis of these comparisons alone – and the same would hold even if an immensely more complicated version of the study were to be carried out.

When should an inquiry stop? It is important to remember that, in problems of national security, inquiry is rarely exhaustive. Because it is almost always out of the question to collect – much less process – all the information that is required for exhaustive analysis, inquiries are partial, and the decisionmaker must get along without the full advantage of all the potentiality of systems analysis, operations research, and the scientific approach. Inquiries cost money and time; as we suggested earlier, they can cost in other values as well. They can cost lives; they can cost national security. This is not to say that some costs cannot sometimes be ignored; the point is rather that paradoxes arise if we allow ourselves to forget that almost all inquiries must stop far, far short of completion either for lack of funds or time, or a justification for spending further funds or time on them.

For these reasons, an analysis is usually far from finished when it is briefed to the decisionmaker or even when it is published. There are always unanswered questions that could be investigated further, even though the need for reporting requires a cutoff. And the decisionmaker's questions and reactions will usually involve an extension of the study.

---

[17] Maj. Gen. Glenn A. Kent, "On Analysis," *Air University Review*, Vol. XVIII, No. 4 (May–June 1967), p. 50.

Since we must often give our advice before we are fully ready, we may be wrong on occasion. But one cannot do useful work in the field of defense analysis unless he is willing to accept uncertainty. If, in the judgment of the analyst and those who use his analysis, the alternative ranked highest by the criterion is good enough, the process is over; if not, more and better alternatives must be designed or the objectives must be lowered. Analysis is helpful in reaching a policy conclusion only when the objectives are agreed upon by the policy-makers. In defense policy in particular, and in many other cases as well, objectives are not, in fact, agreed upon. The choice, while ostensibly between alternatives, is really between objectives, and nonanalytical methods must be used for a final reconciliation of views. Although the consequences computed from the model may provide guidance in deciding which objectives to compromise, such decisions are not easily made, and judgment must in the end be applied.

Chapter 4

CRITERIA AND THE MEASUREMENT OF EFFECTIVENESS

L. D. ATTAWAY

The central problems in the design of analyses to aid military decision-makers lie in selecting operationally useful objectives, measures of their attainment, and criteria. This chapter explores these problems. It attempts to show the relationship between costs, criteria, and objectives, and to point out common errors in their choice or use. It also shows the difficulties of definition and measurement which are introduced in going from simple decision problems concerning narrowly defined systems and operations to complex decision problems concerning broad defense systems, and provides some guidelines for proceeding in the broader context.

INTRODUCTION

This chapter has two aims: to show the relationship between costs, criteria, and objectives, and to point out some of the more common errors in their choice or use. It proceeds by first reviewing the various elements of a general decision problem and then discussing how they interact in a sequence of three examples, beginning with a rather narrow, well-defined problem of applied research, and continuing through a very broad, incompletely defined strategic problem. Examining this sequence of problems should provide insight into the techniques of measuring effectiveness, and some idea of the more difficult aspects of measurement. The discussion concludes by considering the character of the decision problem which remains after completion of such analyses.

ELEMENTS OF ANALYSIS

E. S. Quade has already outlined, in Chapters 1 and 3, the major elements of the typical decision problem in systems analysis. For our purposes here, however, we will find it convenient to express one or two of them somewhat differently, in order to avoid a common ambiguity of terms. Thus, in the following list of the elements of decision problems, we depart from Quade's usage by isolating something that we call an *effectiveness scale*, which, in turn, we use in defining *effectiveness*:

Objective:	What we desire to achieve
Alternatives:	Competitive means for achieving the goal
Costs:	Expenditures to acquire each alternative
Effectiveness Scale:	Scale indicating degree of achievement of goal
Effectiveness:	Position on effectiveness scale assigned to each alternative (by measurement)
Criterion:	Statement about cost and effectiveness which determines choice

The rationale of this change is straightforward.[1] Clearly, without a scale of effectiveness on which the position of an alternative will indicate its ability to achieve the goal, evaluation of alternatives would be impossible. The scale is a yardstick, along which we place our alternatives by means of some analytic or subjective technique of measurement; this position indicates the alternative's effectiveness. Now, people sometimes want to substitute the term "criterion" for "effectiveness scale," or replace "scale of effectiveness" with "measure of effectiveness." But to keep the following remarks unambiguous, and preclude some of the semantic difficulties often met in similar discussions, we will use the terminology and definitions just given.

Our breakdown of the decision problem is general enough to apply to problems as different as selecting a new aircraft engine; choosing the best operational mode for an interceptor force (for example, close versus broadcast control); designing a new interceptor aircraft force; allocating a budget between civil defense and active defense; and, finally, determining the size of the strategic budget and how it might best be distributed between offense and defense.

An example: selection of a new aircraft engine

As an example of how these elements of analysis figure in a relatively narrow decision problem, let us consider the selection of a new aircraft engine, and assume that the *objective* is simply to increase engine performance. Then the *alternatives* are obviously the various possible engine types that achieve this objective by such means as exotic fuels or novel design. The *costs* would be of two general kinds: the total capital resources (such as manpower and research facilities) that must be allocated to the research, and the time required to achieve a successful prototype. In this simple case, the *effectiveness scale* relates directly to the objective, and might be taken as the difference between the specific fuel consumption typical today and that achieved by further research, for fixed engine weight. The *effectiveness*

[1] Moreover, it implies no contradiction with anything Quade has said.

of a particular alternative engine type would then be its estimated reading on this scale. The greater the difference, the better the engine, since we desire to decrease the specific fuel consumption by research. In general, the amount of improvement will depend upon the amount of effort expended upon research, so that estimated costs and effectiveness might be related as shown in Fig. 4.1.

Such different levels of performance might result from a situation that H. Rosenzweig will discuss more fully later,[2] in which alternative 1 corresponds to a very conservative improvement over operational engines, and alternative 2, to a larger state-of-the-art advance.

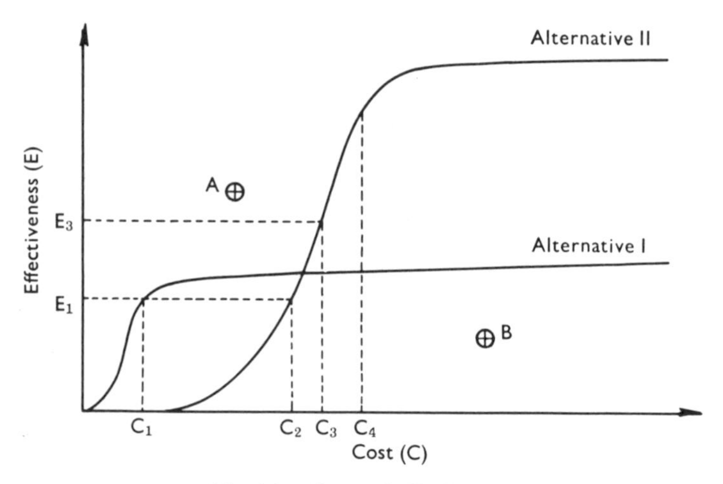

Fig. 4.1 – Cost and effectiveness

Note, however, that even if we assume that both these alternative research programs can be completed on time and are subject to essentially the same amount of uncertainty, we still could not decide between them. What is missing is some knowledge of why the improved performance is needed. Thus, although alternative 1 achieves only a modest level of effectiveness (E_1), it does so at one-third the cost of alternative 2. If the level E_1 is adequate, why not select alternative 1 and thereby minimize cost? Indeed, quite often cost will be limited by decree to some level such as C_2, in which case alternative 1 is the obvious choice. On the other hand, the goal of the research may be to achieve some minimal new level of effectiveness, such as E_3, no matter what the cost. Then alternative 2 is obviously the choice.

The point to be made is that, in general, it is not possible to choose

[2] In Chapter 6.

between two alternatives just on the basis of the cost and effectiveness data shown in Fig. 4.1. Usually, either a required effectiveness must be specified and then the cost minimized for that effectiveness, or a required cost must be specified and the effectiveness maximized. Clearly, the results of the analysis of effectiveness should influence the selection of the final criterion. For example, if C_3 is truly a reasonable cost to pay, then the case for C_4 is much stronger, in view of the great gains to be made for a relatively small additional investment. As a matter of fact, this approach of setting *maximum* cost so that it corresponds to the knee of the cost-effectiveness curve is a very useful and prevalent one, since very little additional effectiveness is gained by further investment.

Overspecification of Criteria

On the other hand, both required cost *and* effectiveness should not be specified; this overspecifies the criterion, and can result in asking for alternatives that are either unobtainable (point A in Fig. 4.1) or underdesigned (point B in the same Figure). An extreme case of criterion overspecification is to require maximum effectiveness for minimum cost. These two requirements cannot be met simultaneously, as is clear from Fig. 4.1, where minimum cost corresponds to zero effectiveness, and maximum effectiveness corresponds to a very large cost.

Maximizing Effectiveness/Cost

Somewhere in the middle are criteria that apparently specify neither required cost nor effectiveness. One which is widely used calls for maximizing the ratio of effectiveness to cost. This seems to be a workable criterion, since, in general, we want to increase effectiveness and decrease cost. Nevertheless, as we can see by examining Fig. 4.2, it has a serious defect. Since the effectiveness-cost ratio for either alternative is simply the slope of a line drawn from the origin to a given point on the curve for that alternative, and since, in this example, the ratio obviously takes on a maximum at the knee of the curve, our choice between the two alternatives seems to be settled at once. Thus, alternative 1 is clearly preferred with this criterion. However, if E_3 is the minimum level of effectiveness acceptable from a research program, then alternative 2 is the obvious choice. The point to be made here is that unless the decisionmaker is completely unconcerned about *absolute* levels of effectiveness and cost, then a criterion such as this, which suppresses them, must be avoided.

Theoretically, it is possible to escape this need for specifying either the required cost or effectiveness by expressing cost and effectiveness in the same units, such as dollars or equivalent lives saved. For if this can be done, then it is possible to subtract cost from effectiveness, and take as the crite-

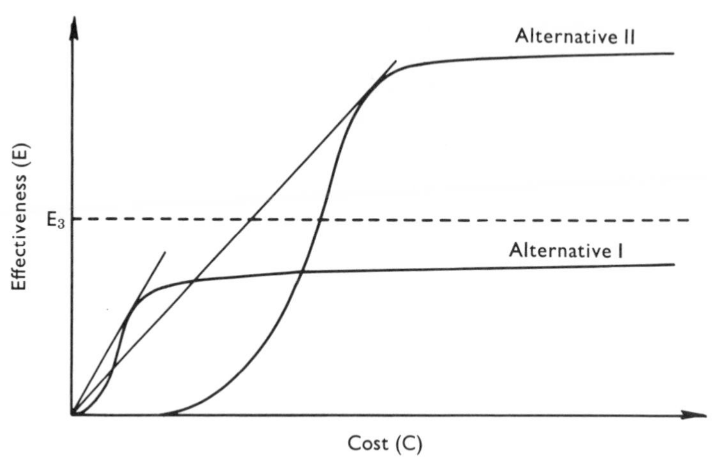

Fig. 4.2 – Effectiveness/cost ratio

rion the maximization of this difference. But seldom, if ever, can cost and effectiveness be expressed in similar units, and we may assume that the earlier description of a criterion applies.

Dominance

Infrequently it happens that selection between alternatives is easy. An extreme case of this is shown in Fig. 4.3, and occurs when an alternative

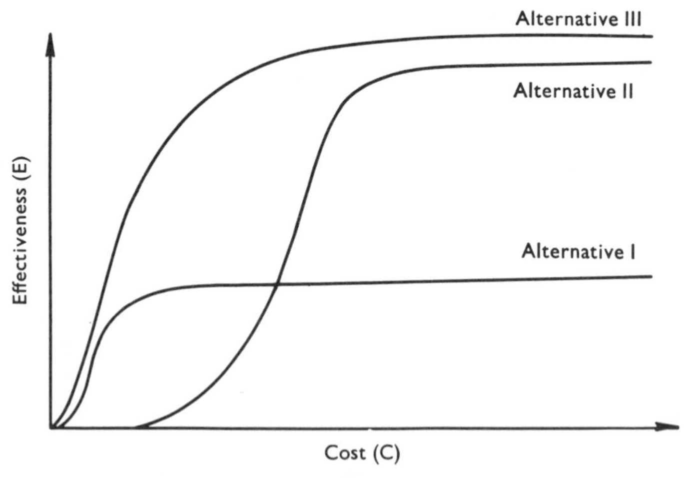

Fig. 4.3 – Dominance

such as 3 is more effective than any other at every cost. In such a case it is clearly advantageous to select alternative 3, which is said to *dominate* alternatives 1 and 2 at all levels of investment and effectiveness. Note that it is still not permissible to overspecify the criterion and require maximum effectiveness for minimum cost. For the situation of dominance only permits us to select alternative 3; minimum cost still corresponds to zero effectiveness for alternative 3, and so forth. Even though dominance designates alternative 3 as preferred, the required level of effectiveness must be specified before the preferred level of investment can be selected.

In this example of propulsion research, as in many others in advanced research or specific component design, the goal has been simple and obvious. Further, in such cases an appropriate scale of effectiveness is usually obvious and related directly to the goal. Finally, the measurement of effectiveness (that is, the location of an alternative upon that effectiveness scale) is straightforward in such cases. Since the example at hand will be discussed by H. Rosenzweig in some detail,[3] we can conclude our discussion of it here, and pass on to the more difficult, but perhaps more interesting, questions the analyst must face in identifying goals, selecting scales of effectiveness, and performing the measurement of effectiveness for some of the other problems which were mentioned earlier.

A SECOND EXAMPLE: CHOICE OF OPERATIONAL MODE FOR INTERCEPTORS

The second of these problems deals with selecting the best mode of operation for an interceptor force: close control versus broadcast control. Close control is defined as that mode of operation in which individual interceptors rely heavily upon vectoring commands generated by an air surveillance system external to the aircraft. Broadcast control is that mode of operation in which vectoring commands are generated within the individual interceptors, based upon air surveillance data provided by an area surveillance system external to the interceptors.

Selection of a Scale of Effectiveness

The ultimate goal of an interceptor force is to prevent damage to the United States, and a useful scale of effectiveness must relate to this ultimate goal in a meaningful way. In this case, there is a hierarchy of potential scales of effectiveness, which includes, for example, the probability of a bomber kill per interceptor attempt, average number of bomber kills per interceptor sortie, total number of bomber kills in a campaign, and number of U.S. survivors in a complete campaign. Which is most useful will depend upon many factors. A general guideline is to choose the narrowest goal possible in order to minimize analytic effort.

The usefulness of these scales depends upon what part of the interceptor force is fixed. If the choice between close and broadcast control is to be

[3] In Chapter 6.

made for an existing interceptor force for which aircraft types and numbers, aircraft and ground systems, and deployment are all fixed, then a simpler scale may suffice. Thus, we might expect that maximizing the average number of kills per interceptor sortie will also minimize the damage to the United States in a campaign. But the probability of a bomber kill per interceptor attempt is too narrow an effectiveness scale to use, for modern interceptors are capable of several attempts per sortie, and using this scale might lead to selecting an alternative which maximizes first-attempt performance at the expense of wasting most of the aircraft's sortie endurance and armament.

As long as the choice is between types of control, the average number of kills per interceptor sortie should be an adequate scale. On the other hand, including just one other aspect of the interceptor force in the choice can require the use of an even broader scale of effectiveness. For example, if the mix of interceptor types is allowed to vary, and the decision problem is to choose the preferred mix *and* control mode, then a scale such as the total number of bombers killed per campaign must be used. For in this case, the use of the narrower scale of average kills per interceptor sortie could lead, for example, to selecting a force of interceptors and mode of control which maximizes kills per sortie in a way that simultaneously reduces the total number of sorties, thereby reducing the total kills per campaign. The correct selection entails the use of an effectiveness scale that correctly relates to the ultimate goals of air defense, such as the total number of bomber kills per campaign.

If the choice is broadened further to include the number of aircraft and their mix and deployment, then an even broader effectiveness scale must be employed, such as U.S. survivors of a campaign. For the ability of an interceptor force to prevent damage to the United States depends upon the geography of the target system and the force deployment relative to the target system. In a choice involving deployment, the effectiveness scale must reflect that factor correctly. The number of bomber kills per campaign does not, since its use might lead to accepting a deployment which effects more bomber kills than a second deployment, but which permits greater damage to the United States than does that second deployment.

Selection of a scale of effectiveness is often a difficult job, requiring understanding of the problem structure and invariably calling for compromise among the several factors of the real world and the analysis. Unfortunately, about the only general guidance one can give is to select scales of effectiveness which reflect the essence of the problem at hand and simultaneously make the measurement of effectiveness as simple as possible. We have just considered the need for reflecting the "essence" of the problem; let us now turn to the measurement aspect.

Measurement of Effectiveness

In the measurement of effectiveness, as in the selection of the effectiveness scale, the means of measurement should reflect the essence of the problem and make measurement both feasible and as easy as possible. The aim is to obtain a quantitative relationship between cost and effectiveness, similar to that which we found for the engine research example (see Fig. 4.1). The more factors which are allowed to vary, or which are to be optimized, the more difficult the measurement, since the technique must then adequately represent the appropriate relationships between the several varying factors. And the dependence of effectiveness upon each such variable must be spelled out, which increases exponentially the number of cases which must be "measured." In short, fixing most factors permits the use of a narrower effectiveness scale, such as total bomber kills per campaign rather than damage to the United States, which is much simpler to calculate and therefore requires vastly less labor. Therefore, in our interceptor control problem, it will be easier to measure the effectiveness of close and broadcast control for the situation in which all the characteristics of the interceptor force are fixed.

The task of measurement can be simplified still more, however, since even for a force completely specified as to aircraft types and numbers, aircraft and ground systems, and deployment, there are many ways of achieving close and broadcast control. Clearly, it would be desirable to discover and then compare only what we think are the best ways of achieving each. This means, of course, that the final measurement of effectiveness of our two alternatives would then rest upon a previous analysis which selects the best method in each case. Such an analysis would investigate the many aspects of the defense system which, in the ultimate analysis, we have just taken as fixed. It would also have to consider the nature of the enemy's operational force and what he chooses to do with it. And if the enemy is permitted to attack defenses, then the resulting variation in the characteristics of the defense forces would also have to be included in the analysis; in effect, the mix and deployment of aircraft types and the amount and ability of the ground systems might, at some point in the analysis, be variables because of possible enemy actions against them.

A prior analysis such as this for finding the best way of achieving close control is usually designed to provide a parameterized estimate of the best performance of an interceptor as a function of what the enemy does in fact do. For example, the product might be in the form of the probability of a bomber kill per interceptor attempt as a function of bomber stream density, altitude and speed, bomber radar and IR cross section, and residual ground environment performance. The aim is to provide a generalized estimate of the performance of the best operation as a function of the specific character of the actual conflict. This estimate, which would involve operational and experimental data, technical extrapolation of equipment performance, and mathematical modeling of the dynamic interceptor-bomber encounter, could then be used to evaluate many specific possibilities.

In a larger sense, the objective of such analysis is to place the final estimate of performance upon as fundamental a basis as possible, in terms of components and operations which have physical and operational meaning, and to provide an understanding of the accuracy of such estimates. To go into greater detail about how this is done would be inappropriate here, and, in any case, the point that should be emphasized is that such analysis is an essential, irreplaceable part of all systems analysis if it is to be truly meaningful.

Assuming, therefore, that earlier estimates of performance for the best form of close and broadcast control are in hand, it is possible to measure the effectiveness of the two alternatives. In some cases, these estimates themselves might comprise such an evaluation. Usually they do not – at least not in terms of an effectiveness scale such as the average number of bomber kills per sortie – because they do not integrate the operational factors appropriately. For the average number of bomber kills per interceptor sortie depends upon such things as the actual density, altitude, and speed of a bomber stream and the number of interceptors that actually penetrate the stream – which, in turn, depends upon where the interceptors are located, which targets are attacked by how many bombers, how much warning is available, and many other *operational* factors. Thus, some technique is needed for reflecting these operational factors in the estimation of the average number of kills per sortie. This might take the form of a map exercise in which interceptors are actually deployed and then committed against attacking bomber streams, duels are fought on paper or in a computer, and a final over-all campaign estimate is made of the average number of kills per sortie.

Selection of an Alternative

This analysis must be performed for each significantly different but important case of enemy force, tactic, and competence. The final output might be of the form we saw earlier in Fig. 4.1, although each important case would, of course, have its own curves. Given such results, a criterion for use *in a particular case* would function as we noted before; that is, if one alternative were dominant at all levels of effectiveness and cost, it would be the obvious choice; if not, then either required cost or effectiveness would have to be specified before a decision could be reached. A criterion for use *across the cases* is a more difficult matter, which we will postpone until later, except to note that if either alternative is dominant across all cases then the decision is again straightforward.

This example has been discussed in detail in order to emphasize the need for reflecting in an effectiveness measurement those factors that are important *to the decision under consideration.* The point is that the particular questions we have considered are questions that must be considered in order to understand close versus broadcast control. Such detailed handling of these particular components is possible only because the problem is restricted to the narrow question of type of interceptor control. If a broader question is addressed, such as the choice between manned interceptors and local defense, or between civil defense and active defense, then the same detail may be possible, but *only relative to a higher level of system component.* That is, to analyze such broader problems, it is necessary to *suboptimize* and *aggregate.*

Suboptimization

When, as in the preceding example, we simplify the problem of selecting an alternative by completely fixing certain characteristics that might, in fact, vary, the selection that results is called "suboptimum." It is suboptimum because we could usually do better if we allowed some or all of these characteristics to vary simultaneously, and made our selection from the resulting large set of possible mixes. Clearly, however, suboptimizations must be performed often and widely, since it is both necessary and permissible to make many decisions independent of each other. For example, it should be possible to design ballistic missiles independently of COIN aircraft, and ballistic missile defense techniques independently of interceptor ordnance. However, care must be taken not to overdo suboptimization. For example, airborne ordnance and fire control systems must be designed with due attention to the ground environment, or else vectoring accuracy and airborne radar performance can end up badly incompatible.

Aggregation

Suboptimization permits the design of various components, such as wheels, engines, and ordnance, to be fixed. They can then be represented by a single over-all component, such as an aircraft; that is, we can then "aggregate" components into larger systems. Without this ability to aggregate, we could not study problems embracing many components, and the level of aggregation is an important aspect of any analysis.

For example, in the discussion of interceptor control, the level of aggregation did not go beyond such operational abstractions as these:

Deployment
Availability rate

Payload
Flight characteristics (speed, range, altitude, loiter)
Air-to-air detection probability
Probability of converting detection to bomber kill

But if the object of study were command-control of a controlled central war, then the relevant systems might be abstracted at a much greater level of aggregation, as in this list:

The national command
CINCSAC
CINCNORAD
CINCEUR
SHAPE

On this second level, manned interceptors would still be of interest, but they would be only one subsystem among many in a highly aggregated operational abstraction labeled "CINCNORAD." So manned interceptors might be represented by total kill potential and cost, a highly aggregated representation. However, if the broader analysis is to be truly meaningful, such aggregation must rest upon valid analysis (including appropriate suboptimization) of subsystems of the kind we have considered here in talking about interceptor control. This broader analysis, such as evaluating command-control of central war, will often have just as much detail as did the interceptor control example, but it will handle an equal number of major components at a higher level of aggregation.

A THIRD EXAMPLE: ALLOCATION OF A STRATEGIC BUDGET

Thus, as we go from narrower decision problems to broader, more inclusive ones, we are actually going from many specific component studies to their synthesis. This synthesis, and the difficulties of dealing with many plausible future conditions of the world, are perhaps the most challenging problems in systems analysis today. We shall address them next, by discussing the last problem on our list: that of selecting the size of the strategic budget and allocating it between offense and defense. An obvious difficulty with such a broad problem is the magnitude of the analytic effort needed just to uncover all its facets, much less treat them thoroughly. We are not without tools for this task, however, and the first of them is, as always, definition. What are our goals?

The Definition of Goals

In discussions of our strategic forces, it is generally recognized that the deterrence they provide does not apply to the entire spectrum of possible conflict, but only to those enemy actions provocative enough to warrant our involvement in a nuclear exchange and all that it entails. For example, our strategic forces should provide direct deterrence against attacks upon the United States or Europe, but only indirectly affect situations such as Vietnam. If, then, we restrict our attention to that part of the conflict spectrum which involves nuclear exchange between the United States and an enemy, we can say that the goal of the strategic forces is threefold:

> First, to deter direct attack upon the United States by guaranteeing that sufficient strategic forces will survive to inflict upon the attacker an unacceptable level of damage;
> Second, to limit damage to the United States should deterrence fail;
> Third, to prosecute the conflict to a conclusion favorable to the United States.

Another possible sub-goal, not obviously attainable, might be:

> Fourth, in certain situations to strike enemy military forces first with sufficient offensive force that the U.S. air and missile defenses can then limit to an acceptable level U.S. damage from the enemy's responding strike.

We thus see that an over-all strategic goal is actually a set of *multiple goals*, having to do with damage to the United States, damage to the enemy, prosecution and termination of a conflict, and the destruction or preservation of military forces.

Further, these goals are only *proximate*, in that they represent in only a suggestive fashion the true goal. For example, we all recognize that deterrence, to be credible, must rest upon our ability to damage the enemy as well as limit damage to our own country; but we are forced to handle these two goals almost independently of each other.

Also, these goals are *dissimilar*, since their achievement cannot be compared in equivalent units. For example, we simply cannot equate industrial damage and mortalities.

Finally, these goals can be *conflicting*, in that trying to achieve one may reduce our ability to achieve another. For example, expenditures devoted to damaging enemy industry conflict with expenditures to defend our cities, in that both compete for precious resources. Then, because their achievement must be measured in dissimilar units, these goals cannot be balanced against one another directly, but only subjectively.

In short, any idealized strategic goal must be replaced by multiple, proximate, dissimilar, and often conflicting goals. Therefore, the definition of effectiveness, and its measurement, can be made only in relation to these kinds of limited goals. The resulting decision problems can be very difficult. For example, consider two strategic systems – that is, alternatives – which have the performance indicated in Table 4.1. Is it better to buy alternative

1, which successfully limits damage to the United States, but with a rather low level of deterrence, or alternative 2, which is less effective in limiting damage to the United States, but presents a greater deterrent?

TABLE 4.1

The performance of two possible strategic systems

Alternative	Population Surviving (%)	
	United States	The Enemy
1	95	85
2	70	30

It is clear that such decisions are in the province of the decisionmaker proper, not the analyst, and that they must be subjective. Systems analysis should remove as much subjectivity from the decision as is legitimate – no more, no less. By no means should the basic nature of the decision portrayed in Table 4.1 be hidden or analytically camouflaged; that is, the decisionmaker should be left the job of balancing damage limitation against deterrence. Adequate professional guidance should be sought by the ultimate decisionmaker in balancing these conflicting goals, since no analysis can substitute for expert military, political, or scientific insight.

Limitations of Effectiveness Scales

A large part of the difficulty of having to base a decision on results like those given in Table 4.1 arises from the effectiveness scale used – that is, population surviving. But the need to use such crude scales is unavoidable, in part because at this time we simply do not know enough about the internal processes of the principal elements of the United States to evaluate how they would be affected by varying degrees and kinds of damage. In fact, to date we can handle in our analyses essentially only the physically measurable external attributes of the United States. For example, we can reflect population in its many physical aspects – number, location, occupation, age, sex, dwelling, and so on – and we can also estimate the effects that a direct attack might have on these attributes. Similarly, we can reflect certain external aspects of industry, agriculture, the military, utilities, and so forth. But when it comes to estimating the impact of damage upon the ability of any such element to pursue its fundamental goals by means of its internal processes, we are generally without adequate tools of analysis.

The results are several. First, in measuring the effectiveness of an alternative in preventing damage to the United States, it becomes necessary to specify effectiveness scales for each of the principal elements – social, economic, military – of the U.S. complex. Second, these scales, no matter how carefully defined, must always turn out to be proximate, in that they will represent the effectiveness of an alternative to reduce damage in only a suggestive fashion. For example, when we consider limitation of damage to U.S. society, we are forced to use effectiveness scales such as population surviving or radiation levels following the conflict. We are hard put to handle such a sophisticated effectiveness scale as the ability to regain our 1960 subsistence level. Third, these scales are likely to be dissimilar, in that they will probably be expressed in units that are not equivalent. For example, population surviving and industrial floor space remaining intact might be used as effectiveness scales for the social and industrial elements, respectively. However, they cannot both be expressed in the same units, since we cannot equate a death to some amount of floor space. Fourth and finally, such scales will be conflicting, in that trying to do well on one tends to decrease the ability of an alternative to do well on another. For example, population surviving and industrial floor space surviving are conflicting when a defense mixture of active defense and fallout shelters for fixed cost is considered. For money spent on fallout shelters to save lives is spent at the expense of buying active defense, which saves both lives and floor space.

In brief, the situation for effectiveness scales is somewhat like that for goals, and leads to similar difficulties for the decisionmaker. For instance, consider two alternatives which have the performance shown in Table 4.2. Is it better to buy alternative 1, which preserves almost all the population but apparently with little provision for future subsistence, or alternative 2, which preserves a more modest fraction of the populace but with greater provisions for the future?

TABLE 4.2

Hypothetical performance of two other strategic systems

Alternative	Population Surviving (%)	Basic Industrial Floor Space Surviving (%)	Agricultural Acreage Surviving (%)
1	95	40	20
2	75	75	75

This trend towards expressing over-all effectiveness in terms of a number of very simple, but highly specific scales is contrary to the need to select scales broad enough to integrate the effects of dissimilar subsystems into an over-all effect. For example, in the strategic problem we are considering, it is not possible to use as scales the expected number of bomber kills per interceptor sortie and the expected number of re-entry bodies destroyed per ABM engagement. In order to use the output of the analysis as an aid to decision, a higher level scale of effectiveness is needed, such as population surviving, so that the contribution of the two subsystems can be combined. This, in turn, raises the question, Is it really possible to combine performance estimates for dissimilar systems, such as ABM and manned interceptors, inasmuch as the estimates usually differ markedly in accuracy and reliability? We will return to this problem a little later.[4]

Alternatives in an Uncertain Future

If we assume that in our decision problem – that is, selecting a preferred strategic budget size and dividing it between offense and defense – we are looking to the future when a diverse menu of strategic systems will be available, we might then be considering the various systems shown in Fig. 4.4.

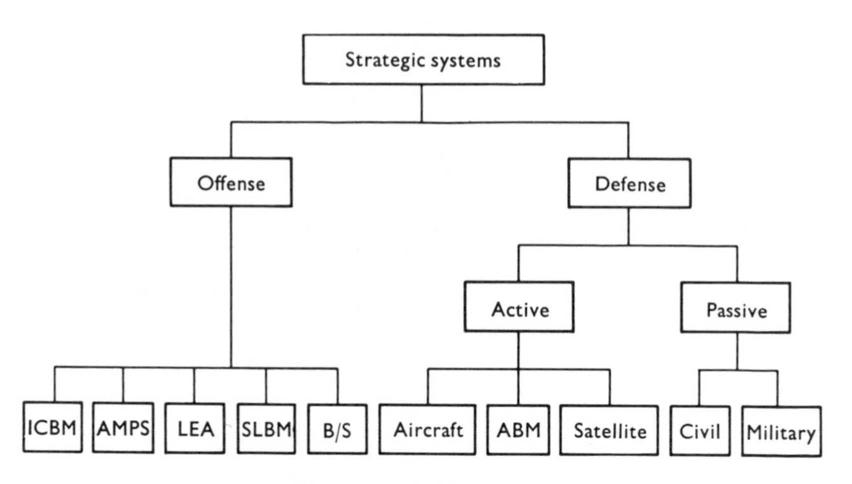

Fig. 4.4 – Strategic choices

The choice of an alternative in this problem is considerably more complex than in those we considered earlier. In this case, an alternative is identified by the amount spent in each lower box of Fig. 4.4 and a detailed specification of just how each amount is spent. The task for measurement is then to assign, to each such set of specifications, a value on each effectiveness scale being used. But each total budget can be spent in literally an infinite number of ways on the various systems. And since, as in the earlier examples, we want to spend each budget on these individual systems in the best way possible, each system considered must be subjected to analy-

[4] It is also discussed by H. Rosenzweig in Chapter 6.

ses which, like those discussed earlier, are designed to select its best form. If this could actually be done, then even in this complex problem we might be able to plot cost and effectiveness as we did earlier. But now each cost would refer to a different optimal alternative – that is, to a different specification of how the budget is best spent on the different systems (Fig. 4.5). Further, the best allocation for a particular system will usually depend upon the amount and manner of spending on other systems, as well as upon various factors, such as the threat, which are not under the control of the decisionmaker. We might explore this aspect of analysis a little.

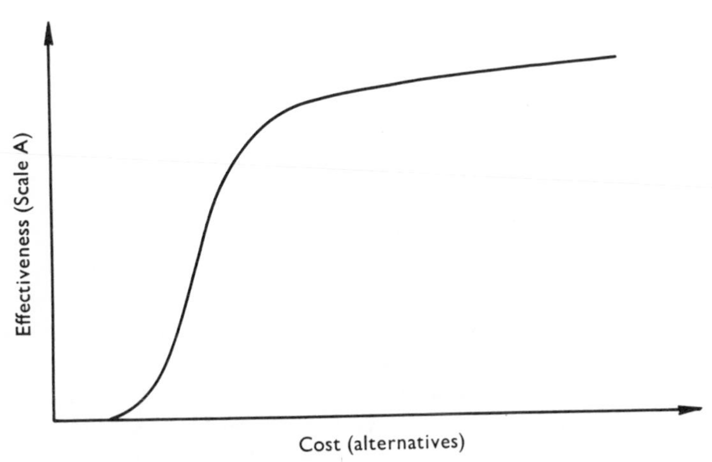

Fig. 4.5 – Effectiveness/cost for a series of alternatives

To do so, consider the ABM component of a possible defense system. In designing an ABM system, we would first like to resolve many uncertainties about the world in which it is to operate, such as these:

Uncertainties about the Enemy

Tactics

Technology (choice, level of achievement, quality)

Force size

Strategy (e.g., kind of war)

Uncertainties about the United States

Technology (level of achievement, quality)

System performance

Clearly, however, such matters are always largely unknowable, and only the last two are partly under our control. Let us look at each of them.

By "enemy tactics" is meant the detailed manner in which the enemy employs his forces. Will he apply ballistic missiles so as to saturate our

ABM? Will he use precursor attacks against defenses? Will he be smart, aggressive, operationally mature, or stupid, backward, inept?

As we all know, almost any level of technological sophistication and quality can be assigned to a *future* force. To what levels should the enemy be advanced in our design for ABM? Will he possess high-quality re-entry decoys? Electronic countermeasures? Advanced payloads?

The difficulty of defense is determined to a great extent by the size of the enemy's force. What he lacks in quality he may recover in quantity. Offensive force size determines needed attributes of ABM, such as rate-of-fire, total kill potential required, and extent of deployment.

The strategy pursued by the enemy can dominate all the preceding factors. For example, if he chooses to fight a controlled war, then force size allocated per hour will be significantly smaller than otherwise, which, in turn, should reflect upon ABM design.

"Level and quality of technological achievement" will mean for the United States roughly the same kinds of things as we noted for the enemy.

The last item in the list, "system performance," can be defined in the following way. If we have a set of assumptions about the preceding enemy attributes, and an assumed level of achievement and quality of United States technology, then a "best" United States ABM system, as "best" was used before, can be designed to match those assumptions. Then to perform the job of measurement it would be necessary to find some method of expressing that basic system in terms of operationally recognizable units such as a battery – units which could be used in operational models especially designed to integrate the various weapon system effects and to reflect the operational aspects of the problem. The basic operating performance of these units would be described by such factors as these:

Battery interception rate
Decoy discrimination rate (per battery)
Probability of kill
Number of interceptor missiles per battery
Battery hardness

The resulting set of values for these various factors will then be the "system performance." These performance figures, however, are never really known; they can only be estimated. Further, they cannot be estimated with equal accuracy; some will be pinpointed, and others will possess wide ranges of uncertainty. Moreover, they can only be estimated after such uncertain aspects of the world as those listed earlier have been specified.

Now, these six factors – enemy tactics, technology, force size, and strategy, and U.S. technology and system performance – determine the state of the world within which the effectiveness of each alternative is to be measured. Since we can visualize many possible future worlds, each significantly different in some of these six factors, we must design an ABM system for each important possible world, in order to understand fully the ABM problem. We can call each combination of these six factors a *contingency*.[5] Fortunately – given a reasonably defined contingency – we are usually able to design what might be called a "rubber suboptimization": a system fixed except for certain attributes, such as rate-of-fire or discrimination rate, which can then be specified as a function of cost. Rubber designs are needed to handle such difficulties as variations in defense deployment and enemy tactics within a contingency, as well as problems met in cross-contingency analysis. The product of such suboptimized design will then be a set of numbers, or a set of ranges of numbers, for basic performance factors of the sort listed a moment ago.

In addition to these characteristics, we would also have the ICBM described as to such items as CEP, payload, availability, and time to target; the manned interceptors described as to analogous factors; and so forth. That is, we need a set of *component studies*, each addressed to suboptimized system design within each of a set of contingencies. Such a component study, spanning all the important contingencies, is needed for each principal weapon or support system we might consider including in an alternative. Individually, these studies will differ markedly, as we noted before, in the accuracy and reliability of their estimates of best system performance. Together, however, they should provide a reasonably close appraisal of that accuracy and reliability, so that these shortcomings can be appropriately reflected in the measurement of effectiveness, as well as in the final decision.

No alternative is completely specified until each system is spelled out as to number of basic units, deployment, and so forth. However, the best such specification for manned interceptors, for example, can depend upon a similar one for local air defense, ABM, or both. For if we defend some cities only against ballistic missiles, and others only against manned bombers, we leave ourselves vulnerable to an obvious enemy tactic; if we defend against manned bombers with local defense only and provide no fallout shelters or area defenses, we leave ourselves vulnerable to fallout attack from weapons delivered just outside the local defenses. Such interdependencies also exist between allocations of offense and defense, and are a difficult part of the analysis. Their main effect is twofold. First, they limit the amount of suboptimization that can take place within a subsystem, independent of the other subsystems. Second, they give rise to the need

[5] The contingency problem is so central to the difficulties of systems analysis today that it needs careful development. As a matter of fact, most of the tried and true techniques of systems analysis are applicable only *within* a given contingency, and it is the need to handle analysis *across* many contingencies that gives rise to many current difficulties.

for a *synthesis* of the component studies, a synthesis that will result in an over-all strategic system – that is, alternative – which combines the individual subsystems in some best fashion. This need, in turn, prescribes that the output of the component studies will be most useful if it is parameterized.

This synthesis requires trading off the effectiveness of one subsystem for that of another, in order to achieve a best mix. For example, civil defense appears a better buy for initial investment than additional active defense, but a decision mechanism for determining the level of each is required. In such a process, differences in the accuracy and reliability of the results of the component studies must be recognized and taken into account. At least two means are available. The first is simply to treat an uncertain parameter as a contingency variable, and assign it an appropriate value for each contingency. The second is to perform *sensitivity analysis*. In this technique, the uncertain parameters are varied over their likely ranges to ascertain the sensitivity of the results to their actual value. If the sensitivity is slight, then a "typical" value of the uncertain parameter can be used throughout with little error. If sensitivity is great, then this analysis can be used to select the preferred values for use in several contingencies.[6]

Much can be done by means of permissible suboptimization in such a synthesis to arrive at a best alternative for a given cost. However, it eventually becomes necessary to compare competing alternatives in terms of the ultimate scales of effectiveness. This, then, comprises the final measurement of effectiveness. But because of the great expense in labor the final measurement involves, it should be performed only after as much suboptimization as possible has been performed, and upon as small a set of competing alternatives as possible.

Measurement of Effectiveness

Let us assume that component studies of the preceding kind have been performed for each system of Fig. 4.4, and that the competing alternatives have been synthesized for each contingency. What form, then, does the measurement of effectiveness take? For one thing, each measurement must be made inside a single contingency. It is necessary later to compare measurements across contingencies, but it is not consistent with the real world to vary contingencies (for example, from subsystem to subsystem) during a given measurement.

The measurement model will depend strongly upon the contingency under consideration. This can be illustrated by an example. Consider a contingency in which the enemy strategy is all-out, uncontrolled aerospace attack upon the United States.

---

[6] Two specific applications of sensitivity analysis are illustrated in detail in Chapters 8 and 9.

In such a case, much can be learned by assuming that both sides follow essentially predetermined strategies that call for the maximum rate of weapons use, in order to minimize the damage the enemy will inflict. The model then becomes basically a mechanism for tracing out in time the delivery of the various weapons from each side, and then converting this history into final estimates of U.S. and enemy population casualties, and surviving manufacturing capacity and agricultural acreage. The time history will depend upon the flight characteristics and deployment of the weapons of both sides; the final damage to both sides will depend upon whether or not a given weapon survives to launch, whether or not it penetrates to target, whether or not its target is still there upon its arrival, and whether or not it is capable of destroying its target.

While the general nature of this model seems rather clear, many decisions are still unspecified by our assumptions and call for analysis of various sub-cases. For example, how will the enemy divide his weapons between U.S. military and non-military targets, between bomber targets and missile targets, and so forth?

In this extreme example of a central war contingency, the only random variation permitted within the measurement of effectiveness is the operational kind. Since the performance of weapons is governed by probabilities, any integrated use of weapons, as in a conflict, must also be governed by probabilities. But since any conflict can occur only once, what possible meaning can probabilities have? Any future conflict, although starting in a specified contingency, can actually unfold in many ways and have many final results. In each such unfolding – which might be called a "play" – the component probabilities influence the results. We thus are led to the notion of a *probabilistic model*, the output of which is a probability distribution of effectiveness. Figure 4.6 shows illustrative distributions of effectiveness for two different strategic alternatives (systems) within the same contingency; the horizontal axis gives effectiveness, and the vertical axis gives the percentage of plays in which the effectiveness equals a specific value. For example, alternative 1 should achieve an effectiveness of about one-half in 40 per cent of a large number of plays.

How does one decide within a fixed contingency between two alternatives when their effectiveness is given by probability distributions?[7] Let us assume that alternatives 1 and 2 are of equal cost. As we saw earlier, the criterion might then be to select the alternative with maximum effectiveness. But which of the present two alternatives should we choose? Alternative 1 certainly achieves a high effectiveness, but only some of the time. In order to make a decision, we must specify some preference for different kinds of probability distributions.

---

[7] This question will be taken up at greater length by Albert Madansky in Chapter 5.

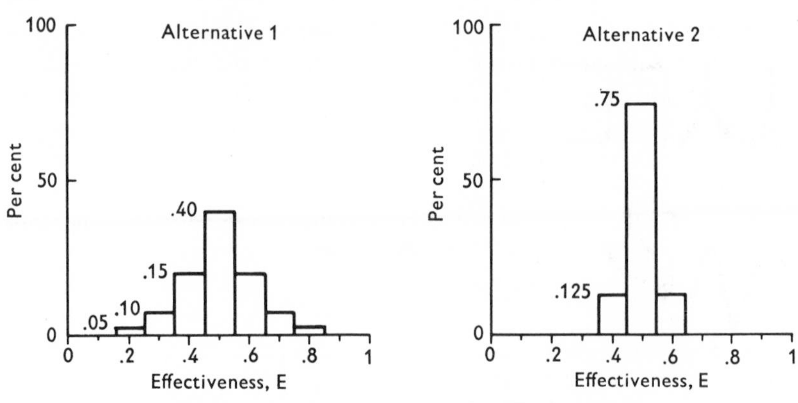

Fig. 4.6 – Percentage of plays with effectiveness, E

Often for reasons of convenience rather than accuracy, a common choice is to prefer the distribution that has the largest average value, because the model that gives such outputs is usually easier to build than one whose output is probabilistic. Clearly, even the average value tells us a good deal about the performance of alternative 2. Were we, therefore, to use average values, our criterion would then be to select the alternative with the largest average value of effectiveness. In the case shown here, such an average-value criterion does not determine a choice, since both alternatives have the same average effectiveness: one-half. However, even though an average value criterion cannot distinguish between alternatives 1 and 2, many a decisionmaker can. For example, if other inputs suggest that an effectiveness of seven-tenths is the minimum useful outcome, then alternative 1 could be the preferred choice.

With a large number of cases to be considered, it is often impossible to carry along distribution functions such as these, let alone expect the decisionmaker to digest them all. A common technique for reducing the distribution to a single number is to employ a criterion such as this: for fixed cost, select that alternative which maximizes the probability that a given effectiveness, such as seven-tenths, is exceeded. In the case at hand, we would choose alternative 1, which has a probability of about 10 per cent of exceeding a seven-tenths effectiveness.

The results of measuring the effectiveness of alternatives for a single contingency and fixed cost might then take the form shown in Fig. 4.7. Numbers to the left of the virgule express the average effectiveness; those to the right, the probability of exceeding a given effectiveness, such as seven-tenths. Both numbers are given for each alternative on each of the various scales, A, B, C, D, E, which might be the fraction of enemy population surviving, United States population surviving, and so forth.

Contingency k

Alternatives	EFFECTIVENESS SCALES				
	A	B	C	D	E
I	.4/.1	.6/.4	~	~	~
II	.5/.05	~	~	~	~
III	.3/0	~	~	~	~
IV	.7/.5	~	~	~	~

Entry code: Average effectiveness → .4/.1 ← Probability of achieving at least .7 effectiveness

(Fixed cost C)

Fig. 4.7 – Fixed contingency effectiveness results

When the table is completely filled in, we thus have an array of effectiveness numbers for each important contingency.

Although this kind of summary information can be useful at various decision levels, it is worth noting that it measures the effectiveness of each alternative at just one point in time – say, that corresponding to the end of a war. A decisionmaker will usually be interested in having additional information: How might the conflict progress? What decisions must be made during its course? How can it be terminated (and at what time and cost)? Analysis usually generates data relevant to these questions, although they are suppressed in Fig. 4.7.[8]

Criteria

Criteria are needed to select preferred alternatives within fixed contingencies. The principal problem arises from the need for multiple effectiveness scales. For example, for the specific contingency described earlier, in which the enemy strikes U.S. cities and military targets first and we respond in

[8] For example, the time history of delivery of weapons can be of use in both policy and design questions, such as what decision rate must be supported in the national command center, what data inputs are important there, and so forth. Similarly, a history of the variation of an effectiveness measure during a conflict can also provide insight into the problems of war prosecution and termination, even for an all-out war of the type we are discussing, where the strategy is predetermined.

an all-out attempt to limit U.S. damage, we might have the results shown in Fig. 4.8. How may a selection be based upon such results? We could decide immediately if there existed a situation of essential dominance; such dominance, however, occurs rarely. We could try to combine the three U.S. scales into a single scale, such as the expected level of subsistence of the United States ten years after the war, and then do the same for the enemy scales. This would require considerable but worthwhile research, similar to that we noted before as being necessary in making use of suboptimizations. But while some means are available in theory for making such a combination, today it is not possible.

Alternative	Enemy Surviving (%)			U.S. Surviving (%)		
	Population	MVA	Agriculture	Population	MVA	Agriculture
I	60	50	75	70	50	50
II	50	~	~	~	~	~
III	~	~	~	~	~	~
IV	~	~	~	~	~	~

Fig. 4.8 – Typical fixed contingency results (fixed cost)

The U.S. and enemy scales cannot be combined into a single scale as meaningful as post-war subsistence level. In general, therefore, judgment must be used to balance enemy damage and U.S. damage. If a decision cannot be made within a contingency, then the analyst simply must carry along several alternatives to the cross-contingency level of decision.

Criteria for use across contingencies will, in general, be more complicated, except when one alternative is dominant within and across all contingencies. Since such dominance is most uncommon, many contingencies must be considered. For example, an analysis to determine the proper allocation of a budget between offense and defense would certainly have to take account of most of the following major contingencies: all-out uncontrolled nuclear exchange initiated without warning by either the United States or the major enemy; such exchanges initiated out of a crisis; controlled exchanges of the preceding kind; accidental initiation; and, finally, various levels of conflict involving n^{th} countries and the United States. Needless to say, the large number of contingencies possible provides a strong motivation to develop techniques for eliminating some of them. Three in particular – "best estimate" analysis, "worst case" analysis, and "*a fortiori*" analysis – are widely used.

A "best estimate" analysis is one in which the uncertain factors describing a contingency are assumed to coincide with the analyst's best estimates thereof. This will sometimes be a valid technique; in most cases it will not, unless accompanied by appropriate sensitivity analysis.

A "worst case" analysis is one in which the factors used to describe the enemy are selected to make him exceptionally effective, intelligent, and aggressive. The philosophy is that an alternative which is effective in this worst contingency will probably be effective in all reasonable ones. Its shortcoming is that alternatives designed to be effective in worst cases tend to be either inadequate or prohibitively expensive, and thus are seldom procured. All alternatives then tend to get tested against this worst case attack, with the result that no alternatives get procured, even though other reasonable contingencies can be met by them. For example, our inability to handle an all-out attack by a highly aggressive and sophisticated enemy has delayed development of ABM systems able to handle less difficult contingencies, such as accidental or n^{th}-country attack. On the other hand, it is appropriate to use a worst case as the basis for designing a deterrent, for obvious reasons.

But even if a "best estimate" or "worst case" analysis points clearly to a preferred alternative, other contingencies should be examined, in order to understand the alternative's usefulness under what are possibly more likely contingencies. For example, the possibility that a given ABM alternative will effectively defend against most n^{th}-country attacks, as well as many accidental attacks by our major enemy, is important knowledge which should be reflected in deciding whether or not to buy ABM, and in evaluating its selection and design.

It is sometimes possible to exclude an alternative by using a device just the opposite of the "worst case" approach. The alternative under consideration is designed as optimistically as possible, and the contingency most favorable to that alternative is chosen. Then, if the alternative still performs badly, it can be discarded. Such a technique is called "*a fortiori* analysis."

To avoid using any of these three techniques to discard important contingencies inappropriately, the analyst should increase the level of aggregation of the final synthesis. This process, which reduces detail and broadens the spectrum of contingencies which can be handled, can legitimately be carried to the point of expanding the scope of the analysis to include national constraints which limit the number and quality of programs a nation can undertake. Thus, if the total defense budget of a major enemy has historically been allocated according to some traditional pattern, this pattern should be considered in estimating the enemy's budget levels for offense and defense, achieved levels of technology in many disparate fields, and so forth.

But it is no less true at the end of a study than it was at the beginning that systems analysis always involves human judgment. There is, of course, a natural desire to attach probabilities to contingencies, so that those of low probability can be ignored, or the results of analyses within several contingencies can be weighted and combined into a single measurement of effectiveness. It is also true that attempts to attach either absolute or relative probabilities to contingencies will in most cases fail, so that no such combination can take place. Therefore, the ultimate conclusions will have to be made by the subjective consideration of the important contingencies, taken individually and severally. The need for professional military, political, social, and scientific judgment in the ultimate decision process is thus clear.

In general, it is not desirable (or possible) to specify detailed criteria before the results of a study are in hand. Rather, the results should be used to determine what goals are in fact attainable, which contingencies it is feasible to meet, where large payoffs for small investment may occur, and so forth. Politics has been called "the art of the possible." So is military strategy and its support by military systems: We can readily identify *desirable* goals, but usually we can specify *attainable* goals only *after* research. Thus, criteria should not precede results, but should follow, and goals should not be static, but should change to conform to the realities of engineering, science, military operations, and politics.

Summary

This Chapter can be summarized very briefly. A principal aim of systems analysis is to find the relationship between cost and effectiveness, such as is illustrated in Fig. 4.9.

The uncertain nature of many aspects of the world forces the analyst to consider different contingencies, within each of which this relationship between cost and effectiveness must be estimated. Therefore, in many problems there are really *two* fundamental variables – cost and contingency – upon which the effectiveness of an alternative depends. This can be signified by a three-dimensional, rather than two-dimensional plot, as in Fig. 4.10.

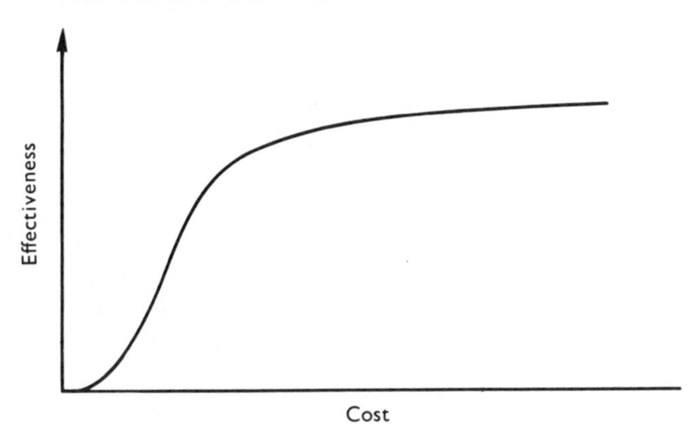

Fig. 4.9 – Cost and effectiveness

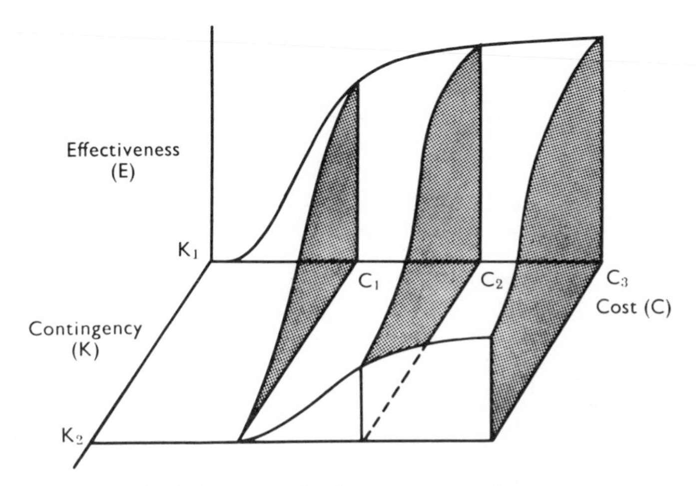

Fig. 4.10 – Cost, effectiveness, and contingency

Clearly, as the contingency and cost are varied, the effectiveness traces out a three-dimensional surface, giving effectiveness versus both cost and contingency. This surface can be sketched out by selecting several contingencies, such as K_1 and K_2, and estimating, within each, the effectiveness-cost relationship, as illustrated in Fig. 4.10. At the same time, the relationship between contingency and effectiveness for several fixed costs (C_1, C_2, and C_3) can be indicated.

The principal problems of systems analysis derive from the fact that the effectiveness scale is really many scales, and from the need to consider alternatives across contingencies. Because of the difficulties it introduces, analysts and decisionmakers tend to ignore this third dimension of the problem, often by restricting the analysis and decision to that contingency which fits their own preconceptions or biases. Since, essentially, the contingency analyzed determines the outcome of the analysis, this arbitrary restriction leads to decisions that support the bias of the analyst or decisionmaker, rather than objective results. This is a major source of error in the use of analysis today. Of course, to these difficulties must be added the need to synthesize already broad component studies into even broader studies, with the attendant need to combine performance estimates for dissimilar systems about which our knowledge is not uniform.

These problems are difficult, and it is clear that judgment still must play a large part in final decisions. But systems analysis can do a great deal toward placing that judgment on a firm, objective basis.

Chapter 5

UNCERTAINTY

ALBERT MADANSKY

This Chapter defines uncertainty, identifies its major sources in military analysis, discusses methods the analyst uses to handle it, and suggests some rules for minimizing its effects in decisionmaking. It points out that uncertainty is a central and crucial problem in systems analysis that must be treated explicitly; that it is not merely a difficulty in principle, but a considerable practical problem as well; that choice under uncertainty is different from choice among certain outcomes; and that, while there are no rules that are simple, complete, and universally agreed upon, reasonable choice is possible even where uncertainty exists.

The theory of decisionmaking under uncertainty is admittedly only an idealization of real life decisionmaking and is more normative than descriptive. Nevertheless, it does provide a conceptual framework into which the key aspects of a systems analysis, the criterion problem and the treatment of uncertainty, can be fitted. Though this chapter attempts to support this contention with some examples, it deals primarily with a mathematically simplified description of the theory, illustrated by a somewhat frivolous example containing ingredients of a real systems analysis.

PROBABILITIES – OBJECTIVE AND SUBJECTIVE

Before we even begin to contemplate the problem of how to make decisions in the face of uncertainty, we must have in mind a clear and complete description of all the events or contingencies which can possibly occur. This may be a tall order in a real systems analysis, but it is the only toehold we have on the uncertainties – a list of the possibilities.

For some of these contingencies there may be available either enough data or sufficient theory so that we *know* the probability of occurrence of each. For example, the Air Force has conducted sufficient exercises to determine the probability distribution of the actual impact point of a bomb aimed at a specified target from a given altitude at a given speed. Here is a case of the probability distribution of the actual impact point being objectively determined from enough data. For another example, suppose we had 20 missiles of a type whose reliability was well determined on the basis of a test program. Then we need not fire 1000 (or 1,000,000) volleys of 20 missiles of this type to determine the probability distribution of the number of successful firings. Mathematical theory tells us, without firing a single volley of 20, that the probability distribution of the number of successful firings in such a round is the binomial distribution.

These are two examples of *objective probability distributions*, probability distributions which are empirically or theoretically derivable and which are incontrovertible. There are, however, contingencies which we admit are governed by a probability distribution, but where we have little or no data or theory to enable us to determine the distribution incontrovertibly. The successful operation of a newly designed missile, not yet even fabricated, is obviously governed by a probability, the reliability of the missile. Yet with no data and little theory about its reliability, two people with equally good judgment might assess its reliability differently. Each, in a systems analysis involving this missile, would use his subjective assessment of the reliability, pointing out, of course, that the assessment was subjective but giving his grounds for that assessment. In short, this example illustrates the existence and use of a *subjective probability distribution*, a probability distribution which is not empirically or theoretically derivable but instead is imputed by the analyst to govern the contingencies.

Let me introduce here two technical terms, *situations of risk* and *situations of uncertainty*.[1] What characterizes both these situations is that in each there is an uncontrollable random event inherent in the situation. The distinction between a risky situation and an uncertain situation is that in the former the uncontrollable random event comes from a known probability distribution, whereas in the latter situation even the probability distribution is unknown. That is, a risky situation has associated with it an objective probability distribution; an uncertain situation has imputed to it a subjective probability distribution.

One operational distinction between objective and subjective probability distributions – that they are used in distinct ways – can be illustrated by our reliability example. Suppose we determined that ρ, the reliability of the missile, was 0.9. Then, using the binomial distribution, we could make the objective statement that "the probability of 3 or more successes in 5 shots is 0.99144":

$$\binom{5}{3}(.9)^3(.1)^2 + \binom{5}{4}(.9)^4(.1) + \binom{5}{5}(.9)^5 = .99144.$$

If we did not know or could not determine ρ, we might be willing to

[1] Note that this use of "uncertainty" is not to be confused with that in the title of this Chapter. Due to an unfortunate use of terminology, the word "uncertainty" has both a precise technical meaning, given here, and a loose, undefined, but nevertheless understandable meaning in systems analysis discourse.

make a subjective statement like, "I'd bet 3 to 1 that this missile's reliability is at least 0.5." Given such a statement, one might locate a representative point P on Fig. 5.1. Given enough such statements, one might obtain a curve like that which appears in the Figure.

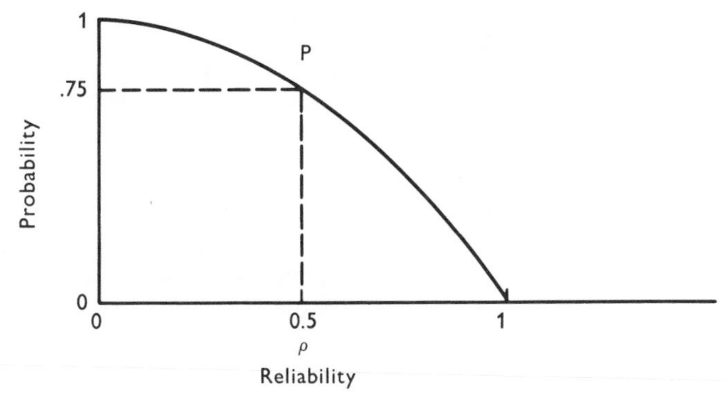

Fig. 5.1 – Subjective probability that reliability is at least ρ

Suppose the equation of this curve could be expressed as $F(\rho) = 1-\rho^2$. We could then make the subjective statement that "the probability of 3 or more successes in 5 shots, in my opinion, is 0.71429":

$$\int_0^1 \left[\binom{5}{3}\rho^3(1-\rho)^2 + \binom{5}{4}\rho^4(1-\rho) + \binom{5}{5}\rho^5\right] f(\rho)\, d\rho = .71429,$$

where

$$f(\rho) = d[1 - F(\rho)]/d\rho = 2\rho.$$

A more important operational distinction between objective and subjective probability distributions is that even when, upon calculation, they yield the same probabilities, few would want to use them interchangeably. This can be illustrated by the following set of examples.

Let me introduce you to Box I. It contains 100 balls, 50 black and 50 red. I would like you to engage with me now in a giveaway game. You are to choose a color. I am now going to draw a ball at random from this box. If we match, I will give you $1; if we mismatch, I give you nothing. Which color do you choose, or are you indifferent?

Now let me introduce you to Box II. It, too, contains 100 balls, some black, the remainder red. You are to choose a color. I am now going to draw a ball at random from this box. If we match, I give you $1; if we mismatch, I give you nothing. Which color do you choose, or are you indifferent?

Now that you are acquainted with the two boxes, let me propose a third giveaway to you. You choose the box from which I shall draw a ball at random. You also choose a color. If we match, I give you $1; if we mismatch, I give you nothing. Which box do you choose, or are you indifferent?

The first of the three giveaways is an example of a risky situation. The random event, the color of the ball I draw, is governed by a known probability distribution, 50:50 for red. The second of the three giveaways is an example of an uncertain situation, as the probability distribution governing the color of the ball I draw would be unknown to you. In both of these situations, everyone is indifferent in his choice of color. That is, in the uncertain situation, everyone acts *as if* the situation were a risky situation with even odds on red and black.

We'll get back to the implications of the third giveaway in a moment. But first let us see what the implications of the first two situations are for you as a decisionmaker. In the second giveaway, you would have acted as if you were in a 50:50 risky situation, even though it isn't necessarily so. You would have imputed a 50:50 subjective probability distribution to the colors in Box II. (For if you had believed otherwise, you would not have been indifferent in choice of color.)

Your reaction to the third giveaway, though, might contradict the view held by some that two situations, one risky and the other uncertain, with the same probability distributions, one objective and the other subjective, are for all intents and purposes equivalent situations. About 80 per cent of the people confronted with the third giveaway are not indifferent to the choices between the boxes. Most choose Box I. The uneasiness about one's subjective probability distribution makes most people opt for a situation they know more about, the risky rather than the uncertain situation.

There are two points to be made about these giveaways. One is that if the only basis for action in an uncertain situation is a subjective probability distribution, then use it in the same way you would an objective, or known, probability distribution. The other is that if you are at all uneasy about your subjective probability distribution, there are ways of alleviating this uneasiness. One way is to use sensitivity analysis on the distribution. Suppose I were to change the distribution of balls in Box I from 50:50 to 60 black, 40 red, and the payoff to $1.20 for a match on red and $.80 for a match on black. The expected value to you of either color in Box I is now $.48[2] instead of $.50. Keep the payoff for Box II the same. Now in the third giveaway, which box would you choose? If you switch to Box II, then you know that $.48 represents a lower bound on your subjective probability of either color in Box II.

Another way to alleviate your uneasiness is by data collection. If I let you draw some balls at random from Box II one at a time, replacing them as you go, then (though you still won't know the distribution of color in

---

[2] For red, $1.20 × .40 + $0 × .60 = $.48. For black, $.80 × .60 + $0 × .40 = $.48.

the box) you will incorporate the findings of your sampling into your subjective probability distribution and have a less uneasy feeling about the resulting subjective distribution.

Utilities – the theory of criterion selection

A hoary criterion for decisionmaking in risky and uncertain situations is the expected value criterion. In a situation in which there exists a measure of effectiveness and in which you have to make a choice between alternative systems, this criterion has you compute for each system the expected value of the effectiveness measure and choose that system which has highest expected effectiveness, assuming equal costs. The expected value calculation arrays the debits and credits, the effectiveness or lack thereof, of each system under the various contingencies, weights them by the probabilities of the contingencies, and determines the weighted balance between debits and credits.

Situations in which there is no natural quantifiable measure of effectiveness are, however, the rule rather than the exception in systems analysis. We have certain basic nonquantifiable goals, such as "deterrence" and "victory if deterrence fails." Yet in order to assess the alternative systems with regard to how well they measure up in achieving our goals, we typically use proximate quantifiable variables, such as "number of targets destroyed," or "number of targets destroyed per $10 billion cost," as criterion variables – as measures of the level of deterrence. But is the expected value of this proximate criterion variable, this proximate quantifiable measure of effectiveness, the appropriate criterion in a risky or uncertain situation?

To convince you in a simple way that the answer to this question is "no," let us suppose that you are confronted with the choice between two games of chance. A reasonable criterion variable by which to assess the two "systems," the two games of chance, is the number of dollars won or lost in the game. Now here are the two games of chance:

Game 1	*Game* 2
Heads: I give you $1	Heads: I give you $100,000
Tails: You give me $1	Tails: You give me $100,000

You choose the game. Then I toss a fair coin. If heads comes up, I pay you the amount appropriate to the situation. If tails comes up, you pay me.

It is clear that both situations are fair bets; that is, the expected or actuarial number of dollars you will get is $0, in either situation. Yet I dare say that most of you would prefer game 1 to game 2. The 50:50 chance that you will lose $100,000 in game 2 looms too large for you to risk it. In fact, even if I were to change to $110,000 the payoff to you if heads were to come up in game 2, thereby making the expected value of the game \$5000,[3] you would probably still prefer the first game.

We see from this that the expected dollar outcome of a game of chance is only a proximate criterion variable, and one which can lead to undesirable decisions. A more reasonable measure of the worth to you of the games of chance is some function of dollar outcome which reflects that winning \$100,000 is not worth as much to you as the negative of the worth to you of losing \$100,000. Such a function would look like that in Fig. 5.2. Given such a function, you can compute the expected worths of the two situations, and see that their ordering corresponds to your preference ordering. The expected worth of game 2 is smaller than that of game 1.

How then do you go about constructing – at least in principle – an appropriate criterion for decisionmaking in risky and uncertain situations? As you have seen from the above example, it was not the expected value calculation which might lead one astray, but rather that the expected value of the wrong variable was being calculated. What was appropriate was expected worth, not expected dollars. Thus, in a systems analysis we might say that there is an appropriate measure of worth or effectiveness such that, in a risky or uncertain situation, the appropriate criterion is expected worth or average effectiveness. This implicit definition leaves us merely with the enormous problem of finding an appropriate measure of system worth or system effectiveness.

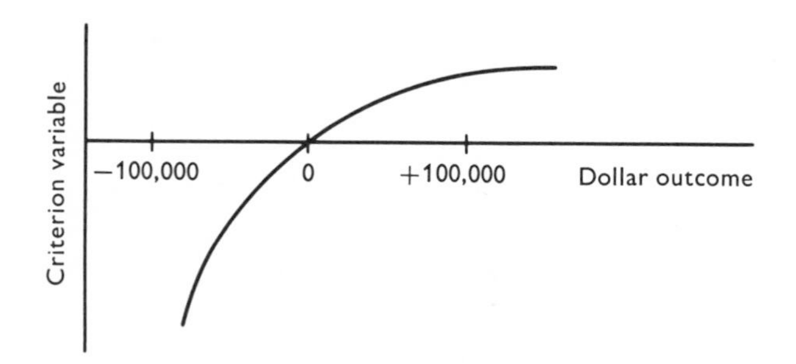

Fig. 5.2 – Worth as a function of risk

As an example of the method by which the mathematical theory of utility goes about constructing an appropriate measure of system worth, let us consider a situation in which the decisionmaker is confronted with a number of alternative systems, is sure about the strategic context in which the system to be selected will operate, and is in the enviable position of being able to rank them in order of preferredness. It does not matter

[3] $\frac{1}{2}$(\$110,000) + $\frac{1}{2}$(–\$100,000) = \$5000.

for the moment whether this ranking results from using a formal criterion of goodness as a yardstick or merely from applying the decisionmaker's visceral feelings about the systems.

To make things concrete, suppose I am about to buy a car. The alternative "systems" available to me are a Rolls Royce, a Cadillac, and a Chevrolet. My choice is dictated merely by the prestige value of the car, a non-quantifiable "effectiveness" variable, and its price. The "strategic context" in which this car is to operate is that of a family car. For the moment, suppose I am not interested in having a car so flexible that I can operate it in another "strategic context," by converting it into a hot rod, say, should I ever want to do that.

Now, how can I come up with a quantifiable variable which measures prestige? I might use a proximate measure like price, or horsepower, or the reciprocal of the number of owners of such a car in the United States. And any of these may be quite all right for my systems analysis. But let us see how utility theory produces a measure of prestige.

Suppose my ranking of cars in order of preferredness is, from most preferred to least preferred, Rolls Royce, Cadillac, Chevrolet. Suppose, though, that there are only two automobile manufacturers in the world, one who will build Cadillacs and only Cadillacs, and another, a sport who comes to you with the following proposition: "Pay me for a Cadillac. Now, let's toss a fair coin. If the coin comes up heads, I'll build you a Rolls. If the coin comes up tails, I'll build you a Chevy." These are the only alternatives contractually available, the Cadillac and a car I'll call the Rollsolet.

You must now decide between the two manufacturers *before* tossing the coin with our sport. If you choose Rollsolet, you are in a risky situation, not knowing what car you are actually going to get. We assume that you are willing to consider the sporting manufacturer's bid; that is, you are not adverse to gambling to determine the system you buy. After all, you have a 50:50 chance of getting a Rolls Royce, your preferred choice. The only thing that influences your choice between the Cadillac and Rollsolet is the odds offered. Now maybe 50:50 odds aren't good enough to make you choose a Rollsolet over a Cadillac, but suppose you are willing to negotiate with our sport on odds. Let's say that 60:40 odds are your break-even point; with better odds you'll choose Rollsolet and with worse odds you'll choose the Cadillac.

I am not interested in how you came up with 60:40 as the break-even odds. As a responsible and intelligent decisionmaker, you can be considered a heuristic computer who somehow digests the state of the world and the characteristics of the systems and comes up with these odds. But with these odds in hand we can arrive at a measure of prestige with the desired property that the prestige value of the car I get in a risky or uncertain situation is the average prestige value of the contingencies, the cars involved. The measure of prestige arbitrarily assigns 0 to the least preferred car, the Chevrolet; 1 to the most preferred car, the Rolls Royce; and 0.6, the break-even probability between Rollsolet and Cadillac, to Cadillac.

To see how and why this measure of prestige "works," consider the following situation. Suppose I am confronted with two automobile salesmen who propose the following.

Salesman A says, "Pay me for a Cadillac. After you do, I will draw a two-digit random number from a table of random numbers. If this number is less than 15, the car you get will be a Chevrolet; if between 15 and 69, it will be a Cadillac; and if between 70 and 99, it will be a Rolls Royce." Salesman B's proposal is similar, except that the cutoff numbers are different, being 22 and 54 instead of 15 and 69. Translated into probabilities, here are the proposals:

	Proposal A	*Proposal B*
Rolls Royce	.30	.45
Cadillac	.55	.33
Chevrolet	.15	.22

Let me first quickly calculate the average prestige value of each of the proposals. The average prestige value of A is $0.30 \times 1 + 0.55 \times 0.6 + 0.15 \times 0 = 0.63$. The average prestige value of B is $0.45 \times 1 + 0.33 \times 0.6 + 0.22 \times 0 = 0.648$. Thus B has higher average prestige value.

Now let us see why this average prestige value comparison reflects your feelings about proposals A and B. Considering proposal A first, suppose the Cadillac were not offered and the odds between the Rolls Royce and Chevrolet were the same, 30:15. Then proposal A is equivalent to a 55 per cent chance at a Cadillac and a 45 per cent chance at a 30:15 odds Rollsolet. Similarly, proposal B can be seen to be a 33 per cent chance at a Cadillac and a 67 per cent chance at a 45:22 odds Rollsolet. Since a Cadillac is the same to you as a 60:40 odds Rollsolet, proposal A boils down to a 55 per cent chance at a 60:40 odds Rollsolet and a 45 per cent chance at a 30:15 odds Rollsolet; proposal B boils down to a 33 per cent chance at a 60:40 odds Rollsolet and a 67 per cent chance at a 45:22 odds Rollsolet. After translating these into probabilities for the constituents of the Rollsolet, the Rolls Royce and the Chevrolet, we obtain the following results:

Proposal A

$$\text{Rolls: } (.55)\left(\frac{60}{60+40}\right) + (.45)\left(\frac{30}{30+15}\right) = .63$$

$$\text{Chevy: } (.55)\left(\frac{40}{60+40}\right) + (.45)\left(\frac{15}{30+15}\right) = .37$$

Proposal B

$$\text{Rolls: } (.33)\left(\frac{60}{60+40}\right)+(.67)\left(\frac{45}{45+22}\right)=.648$$

$$\text{Chevy: } (.33)\left(\frac{40}{60+40}\right)+(.67)\left(\frac{22}{45+22}\right)=.352$$

Thus, since proposal A is equivalent to a 63:37 Rollsolet and proposal B to a 648:352 Rollsolet, you should prefer B.

Suppose that the measures of worth of alternatives for which there exists a preference ordering are derived in the above manner, by introspecting and determining for each intermediate system a game-of-chance proposal between the most and least preferred systems which compares equitably with the intermediate system. It can then be shown mathematically that the result will be a scale of worths or utilities of the various systems with the property that the average worth of a system is also the appropriate measure of its worth in a risky or uncertain situation.

Notice that we do not look to average worth as our criterion variable because, on the average, over repeated trials, the average worth will be your actual outcome. This argument for using averages is phony, because no one ever said that you would be allowed to play any of the games of chance many times. Let me emphasize that we are led to average worth as a criterion only because it reflects your preferences in a risky or uncertain situation.

EXAMPLES OF THE USE OF UTILITY THEORY

The discussion by L. D. Attaway (Chapter 4) of the problem of deciding within a fixed contingency between two alternatives with their effectiveness subject to known random variation provides a good springboard for illustrating some of the concepts defined above. Briefly, Attaway is confronted with two alternatives and is given the probability distribution of a measure of effectiveness for each (which may be only a proximate measure of that elusive quantity, worth). These probability distributions are shown in Fig. 5.3.

Notice that both alternatives have the same average effectiveness, 0.5. Thus, if the measure of effectiveness truly measures the worth of the alternative systems, either alternative is satisfactory.

However, suppose, using Attaway's example, that the effectiveness scale represents fraction of population surviving, and that it is felt that unless we achieve seven-tenths effectiveness we are lost. What would then be an appropriate measure of worth? Two possibilities are given in Fig. 5.4. To a decisionmaker who accepts the measure shown on the left, effectiveness less than 0.7 is worthless, and any above 0.7 is equally good. The measure shown on the right assigns some worth to an effectiveness of 0.7, and

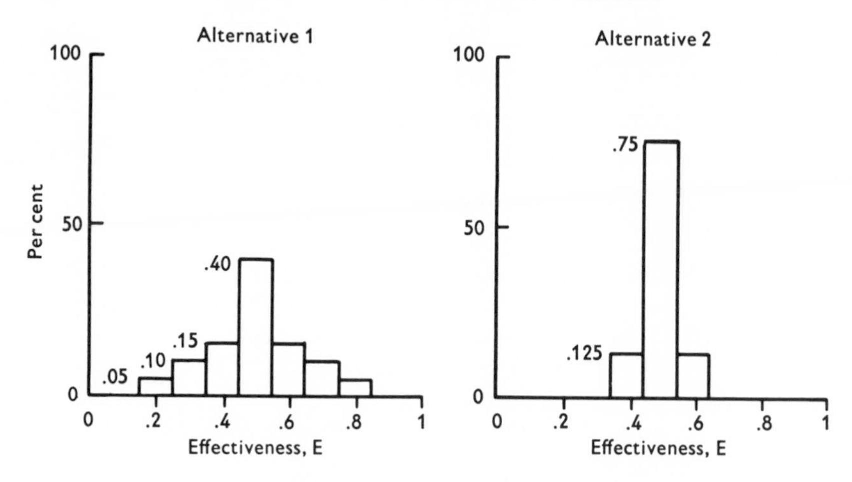

Fig. 5.3 – Percentage of plays with effectiveness, E

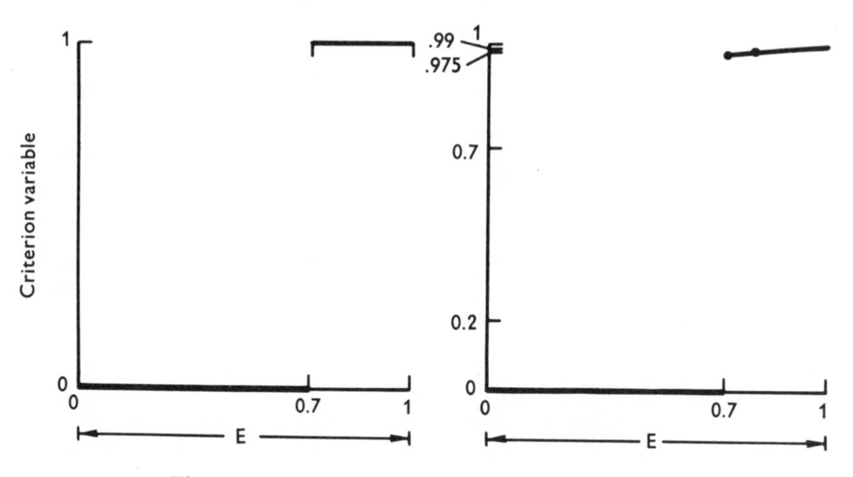

Fig. 5.4 – Worth as a function of effectiveness (case 1)

increasing worth (but at a decreasing rate) to higher levels of effectiveness.

Given either of these measures of worth, our choice would be alternative 1. It has higher expected worth.

Suppose, though, that Fig. 5.4 was altered so that 0.5 was the value of E for which worth was positive. Now (as we can see from the two curves in Fig. 5.5) the choice of measure of worth is critical. The following calculations make this clear. For alternative 1 in Fig. 5.3, the expected worth indicated by the curve on the left of Fig. 5.5 is:

$$0\times(.05+.10+.15)+1\times(.4+.15+.10+.05)=.7.$$

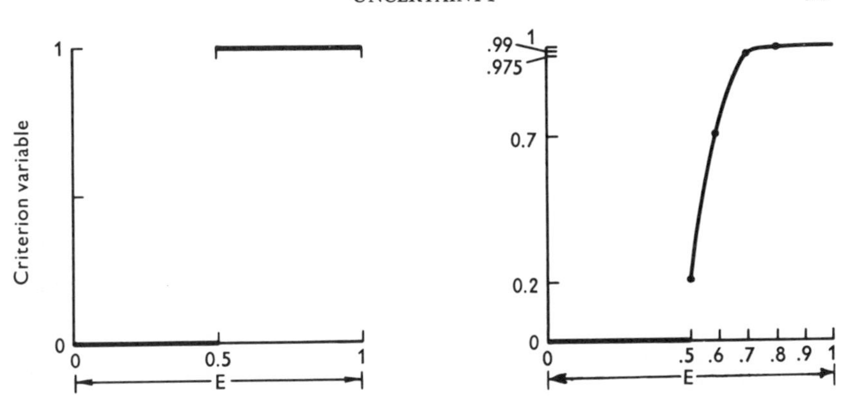

Fig. 5.5 – Worth as a function of effectiveness (case 2)

In contrast, the expected worth for alternative 1 as indicated by the curve on the right is:

$$.4\times.2+.15\times.7+.10\times.975+.05\times.99 = .332.$$

Similarly, for alternative 2, the expected worth indicated on the left is:

$$0\times.125+1\times(.75+.125) = .875,$$

and the expected worth indicated on the right is:

$$.75\times.2+.125\times.7 = .2375.$$

It is good to keep in mind that even though for a given measure of worth we can compare alternatives via expected worth, the expected worths of a given system based on two different measures of worth are incommensurate. Thus, the numbers 0.875 and 0.2375 are not comparable. This is obvious when you recall how we constructed a measure of prestige a few pages ago – by arbitrarily assigning the value 1 to the most preferred event. Two different people, with two different sets of preferences, need not feel equally strongly about their most preferred event.

You will see another example of the use of utility theory in R. D. Specht's description (in Chapter 10) of a model of the problems of hard point defense of missile sites. Here the measure of effectiveness is "number of our missiles that survive the attack," and the measure of worth is "number of missiles that survive the attack over and above those that would have survived if undefended." It is the expected value of this measure of worth that is used as the criterion variable.

Models of uncertainty in systems analysis

Two major types of uncertainty arise in a systems analysis, the technological and the strategic. "Technological uncertainty" is a fancy way of saying that the system, when it is finally produced, won't look like the system you have analyzed. The contractor may not have met specifications, the cost may have soared, the hardware characteristics of the system may have changed, and the operational concept of the system may have been altered in the meantime.

"Strategic uncertainty" is a broad term designed to encompass the fact that one is uncertain as to the state of the world or the strategic context in which the system is to be used, either because certain facts of life have been changed by unforeseen events (for example, a break-through in state-of-the-art which changes capabilities either on our part or on that of the enemy), or because the strategic context has been changed (for example, by the enemy reacting competitively to events, possibly even to the results of our systems analysis).

In our automobile example, a crude analogy to technological uncertainty would be your putting in an order for a Cadillac and not being sure whether the car you get will be a Chevrolet, Cadillac, or Rolls Royce. Our analysis of the two automobile salesmen's proposals is a simple model of technological uncertainty. It models technological uncertainty as a risky situation in which the final system characteristics are governed by a known probability distribution.

In a more sophisticated approach, though you would not want to model the "uncertainty" in the characteristics of a contractor's final product as a risky situation, you might be willing to model it as an uncertain situation and say that the final characteristics can be thought of as the result of a random set of events, but with unknown probability distribution.

Now let us model an example of strategic uncertainty, a situation in which there are two potential strategic contexts five years hence. Suppose you must choose between three systems, where in one strategic context you have provided them with worths of 1, 0.6, 0, and in the other, with corresponding relative worths of 0.4, 0, 1, arrived at by an introspective process similar to the one described above. But as I pointed out earlier in another context, these two sets of utilities are not comparable; you may have greater preference for system 1 if context A is the case than for system 3 if context B is the case. Thus, in comparing systems under different strategic contexts, you must come up with a composite ordering of all of the system-context combinations.

This kind of ordering, with a concomitant set of utilities, is not easy to arrive at. (In fact, asking you to believe that even a simple one-context ordering of utilities can be determined probably stretched your imagination.) So suppose you do what is done in practice and select a criterion variable to both order the system-context combinations and serve as proximate measures of system worth. For example, you might amalgamate accuracy, reliability, and cost into "number of targets destroyed per $10 billion cost," and use this variable as a criterion. You would then have to consider values such as the following:

	Context A	*Context B*
System 1	96	36
System 2	87	22
System 3	72	72

Systems 1 and 3 pose a problem which can be modeled as an uncertain situation. That is, there is some unknown probability that context A will prevail five years from now, and your decision will be based on an assessment of the expected utility of each of the systems, using as probabilities for the calculation your subjective probability of context A. If you are willing to bet 3 to 1 that context B will prevail, then the expected utility of system 1 is $(1/4)\times 96+(3/4)\times 36 = 51$, and that of system 3 is $(1/4)\times 72+(3/4)\times 72 = 72$. Thus you would choose system 3 in this model.

This model, though, omits a critical fact, namely, that the strategic context five years from now is not really a passive random variable obeying some unknown probability law. Though the model may represent an ingredient of the strategic context, by far the most important ingredient is the existence of an active competitor. Uncertainty about the strategic context arises primarily because it will be the joint result of your system choice and the decisions (and reactions) of the enemy.

A simple model of such a situation is that of the zero-sum two-person game, described later, in Chapter 11. In brief, this model assumes that the enemy's table of utilities of system-context combinations is the negative of your table. His interests are diametrically opposed to yours. The mixed strategy of the enemy is *his* choice of a probability distribution to govern the strategic context. This is in contrast with the earlier model as an un certain situation, in which the probability distribution, though unknown, was not a manipulative variable. The minimax theorem of zero-sum two-person game theory tells us what a "best" mixed strategy for the enemy should be. Your "best" mixed strategy is a concomitant product of the game theory model. Note that in the competitive situation your choice of system can no longer be characterized as that system which is best by some index, as it was in the risky and uncertain situations. Rather, it, too, is the result of a random draw from your mixed strategy probability distribution. And it does best against the worst the enemy can do to you (in the sense that it maximizes your minimum expected utility).

In practice, no one would want to make a decision on the basis of the toss of a coin. Moreover, it is difficult to model the multimove game that a five-year development period implies. One cannot simply choose a system and stick with it; the enemy's countermove once the system is chosen must be taken into account. But the mixed strategy derived from the simple static game described above can be useful as a guide to which group of systems to develop in parallel if one wishes to be flexible and hedge against the uncertainty of the strategic context. Also, despite the possible non-existence of a "dominant threat," to use J. R. Schlesinger's terminology (Chapter 20), or a "worst case," to use L. D. Attaway's terminology (Chapter 4), an examination of the enemy's optimal mixed strategy is quite useful in pinpointing just what are the important, though more modest, threats of the enemy.

It is instructive to reiterate here Attaway's observation about worst case analysis, that sometimes all alternatives are tested against the enemy's best strategy and found wanting, with the result that no alternative is procured (for example, no ABM system), even though other reasonable contingencies can be met by some of the alternatives. This is an outgrowth of, but not a consequence of, the zero-sum two-person game-theoretic approach to strategic uncertainty. All the minimax theorem yields is the best strategy for you against the worst the enemy can do to you. But the decision not to use the alternative dubbed "best" by the game theory model, because it is not good enough, may be a questionable practice, as in the case of the ABM.

Thus, the best one can say for the zero-sum two-person game theory model of the competitive nature of the system-context determination is that the game-theoretic solution will give some insight either into the reasonable subjective probability distributions to use in the model of an uncertain situation or into a reasonable system or set of systems to adopt. A more reasonable model, as indicated above, would be a multimove zero-sum two-person game. Even more reasonable would be a nonzero-sum game, which removes the assumption that the enemy's utilities are the negative of ours. However, this type of game is still in the research stage, and not yet usable as a model of strategic uncertainty.

Treatment of uncertainty in systems analysis

We have examined three models of uncertainty, the risky situation, the uncertain situation, and the competitive situation, noting how the first two can model decisionmaking under technological uncertainty and how the last two can model decisionmaking under an additional uncertainty, uncertainty of strategic context. We have seen that, though the appropriate measure of system worth is its "utility," this measure may not be readily assessable, and so proximate measures must be used instead.

To complete this survey, let me list, and in some cases recapitulate, certain special stratagems for treating uncertainty in systems analyses.

Buy Time

The models of technological uncertainty serve to illustrate one method commonly used to alleviate the effects of risk in decisionmaking. One can simply (or maybe not so simply) defer his decision until after the random number has been drawn. As between the contractor who makes Cadillacs only and our sport who offers the Chevrolet and the Rolls Royce at even odds, you would certainly wait, if you could, till the sport has tossed the coin, rather than take the Cadillac because the sport didn't offer your break-even 60:40 odds. The drawback, of course, is that because time is money, our sport would want to be paid to engage in this delayed contracting negotiation, since he knows that if tails comes up he will get no contract at all.

Buy Information

As I pointed out in the analysis of our three giveaways, a person can alleviate uneasiness about the subjective probability distributions necessary for an analysis of an uncertain situation by data collection.

Buy Flexibility as a Hedge

In our automobile purchase example, if I am unsure about whether I want a family car or a hot rod, I would choose the Cadillac rather than the Rolls Royce or the Chevrolet, in order to obtain flexibility as a hedge against "strategic" uncertainty.

In our two-strategic-context example, suppose you would prefer system 1, if context A prevailed, over system 3, if context B prevailed. You feel that the values of the effectiveness measure in the two contexts do not really reflect your utilities of the system-context combinations, in that you like system 1 in context A more than 1 1/3 times as much as system 3 in context B. Thus, you would hedge against the 3 to 1 shot, context A, prevailing five years hence by disregarding the expected worth comparison and choosing system 1. Or you might combine a hedging and buying-time stratagem and choose both for initial development, dropping one of the systems when you have better information on the odds on context A. Theoretically, hedging occurs only when your measure of worth is not the same as the utility of system-context combinations. Practically, this is always the case, as knowing the utilities in any situation is merely an idealization.

Use "A Fortiori" Analysis

In a comparison of alternative systems, one can eliminate systems by a "dominance" or *a fortiori* argument. We noted in our two-strategic-context example that in both strategic contexts system 1 has higher worth than does system 2. It is thus irrational to consider system 2 at all in the analysis.

If, in general, one were to find a system which had higher worth in all strategic contexts than that of any other system, our problem would obviously be solved. We would choose that system. In our example, though, as is usual in a systems analysis, system 1 is better in one, system 3 in the other of the two contexts. Thus, the principle of dominance cannot always lead to a preferred alternative. It can, however, eliminate inferior alternatives.

Use Sensitivity Analysis

You have seen earlier in this Chapter an example of the use of sensitivity analysis to determine your degree of uneasiness about a subjective probability distribution which you might consider as an input into a systems analysis. But there is a further point to be made here, namely, that this sensitivity analysis should be recorded as part of the analysis. It would be extremely misleading to present a subjective analysis based only on your "best guess," without varying the distribution to see where it would lead you.

This point is made in a different guise by H. Rosenzweig in Chapter 6, where he discusses sensitivity analysis as a means of dealing with technological uncertainty. He recommends "show[ing] the performance of each system as a band of different width instead of as a fine, single line." Our recommendation goes further, namely, to include subjective probabilities across the band, since the extreme of the band may not be as likely as is the "fine, single line" somewhere in the middle of the band.

Of course, if, as in Rosenzweig's Fig. 6.4 (p. 120), one system dominates over the other after a sensitivity analysis in which the spectrum of possibilities, but not their probabilities, is reviewed, one need not bother with an assessment of subjective probabilities for the spectrum of possibilities. It is only as a tool for coming to some reasonable decision in a situation like the one depicted in Rosenzweig's Fig. 6.5 (p. 122) that an assessment of subjective probability distributions is worthwhile.

POSTSCRIPT

The prevailing viewpoint in this Chapter is that the cornerstone of a systems analysis under uncertainty is a critical assessment of an appropriate measure of system worth and educated guesses about appropriate probabilities. However, though the theory of decisionmaking under uncertainty described above is neat and intellectually elegant, and though the theory has relevance to the real world and offers insight into ways of handling uncertainty, it is important to keep in mind that this theory is only a first approximation to reality. If this theory is used blindly, reality can overwhelm you with realities.

Chapter 6

TECHNOLOGICAL CONSIDERATIONS

H. ROSENZWEIG

This Chapter briefly reviews the typical sorts of decisions involving technology that are made during the life of a weapon system, particularly those that fall under the provisions of DOD Directive 3200.9, Initiation of Engineering and Operational Systems Development (*July 1, 1965*). *Three general problems seem to recur in all of these decisions: establishing technical feasibility, selecting the best technical approach, and comparing the cost-effectiveness of the proposed system with that of its competitors. The Chapter examines each of these questions in detail, as well as the role of technological considerations in resolving them. Among the topics considered in this discussion are the nature of parametric and trade-off analyses, and the influence of technological uncertainty in determining the existing state-of-the-art and the risk associated with a development program.*

INTRODUCTION

Earlier authors in this book have made it clear that, in its simplest terms, the purpose of systems analysis is twofold: to provide the decisionmaker with the information necessary to arrive at a well-informed decision, and to present this information in a well-organized, concise, and intelligible form.

On this view, systems analysis is essentially concerned with a decision process. If the decision involves hardware (that is, pieces of equipment which must be engineered), it "involves" technology. Thus, if we are to investigate the role that technical considerations play in systems analysis, one way to proceed would be by examining the types of decisions which are made regarding hardware items and the kinds of analyses which are made to assist the decisionmaker with these decisions.

Since there are, however, so many different types of hardware to be found in the defense industry, and since the issues on which decisions must be made are so varied, it is virtually impossible to effect an exhaustive survey in this manner. Yet many of the analyses which are conducted have common features, some of which can be abstracted from particular analyses by examining the types of decisions they are conducted to support during the life-span of a relatively large unit of hardware – a weapon system.

TYPES OF DECISIONS

During the life-span of a weapon system, many (although not necessarily all) of the following types of decisions will be encountered.

1. Should money be spent on "Concept Formulation" for this weapon system?
2. Should the "Contract Definition" phase of the development process be initiated?
3. What is the best technical approach?
4. Should full engineering development of the system be undertaken?
5. How many units should be produced and deployed?
6. Should modifications be made?
7. Should the weapon system be replaced or phased-out of the force?

Let us examine each of these decisions.

Concept Formulation

Weapon system concepts may be generated in many different ways. Sometimes the idea is generated as a result of examining ways to meet a desired operational capability, as the idea of a railroad-mobile ICBM may have resulted from studies of ways to reduce the vulnerability of the Minuteman force. Sometimes it develops as a way to exploit a new technological capability, as the Aerospaceplane concept arose with the idea of extracting oxygen from the atmosphere with cruising airplanes. In most cases, however, it is not possible to trace the roots of the idea with any precision. So many studies are continually being made that ideas somehow evolve in them, get accepted, and then are so rapidly followed down so many paths that even an astute historian would find it difficult to track his way back to their source. However, regardless of how they are generated, they are usually followed by a period in which their proponents must seek money to pursue them further through more detailed studies and, perhaps, experimental research. This, then, generates the first decision point in the formal Research, Development, Test, and Evaluation (RDT&E) process. Should any money be devoted to further studies and experimental research on the new concept and, if so, how much?

In order to reach a decision on the allocation of research funds, the decisionmaker must have information regarding, first, the technical aspects of the concept and, second, the potential payoff if the weapon system is developed. On the first account, if it is an advanced weapon system, one issue of prime importance will be the *technical* feasibility of the concept; that is, it must be determined what types of technological advances will have to be made to develop the weapon and what the prospects are for achieving them. Toward this end, the decisionmaker must be provided with an analysis of the important technological features of the concept, including a statement of the current *state-of-the-art* and the types of experimental programs needed to provide a high degree of confidence that the advances in technology can in fact be achieved.

But there are usually more ideas seeking funds than there are funds available. In this case we must be selective. This gives rise to the second requirement for this first sort of decision – the need to display the potential payoff of the proposed concept. To this end, the analyst must indicate how the new weapon system will be an improvement over currently available ones or alternatives which could be available in the same time. Will it perform a mission that cannot be done any other way? Will it do the same job as existing or other systems, but do it cheaper? In short, why is this a good idea?

To answer these questions, the analyst will have to present the estimated performance of the system. He will also have to indicate the *variation in the performance* as a function of the level of technology attained in critical areas. For instance, does the concept depend on making large advances in the load-bearing capability of the structure, or is performance relatively insensitive to this parameter? Such data indicate the technical risk associated with the concept; they indicate how great an advance in the state-of-the-art is required to achieve a desired level of performance.

Given these data by the analyst, the decisionmaker is then in a position to evaluate the request for funds for further study and experimental research. He must weigh the potential payoff against the cost of the program. At this stage, however, the data available for a decision will be fairly sketchy, since no significant amount of funds will have been expended in the study of the concept. The request for funds will also be fairly limited, since preliminary studies and the early phases of applied research are relatively cheap (as compared with those of a full weapon-system development).

This part of the RDT&E process can be described more precisely by reference to Department of Defense Directive 3200.9,[1] the provisions of which apply throughout DOD to the development of all new systems (or major modifications of existing systems) estimated to require either a total RDT&E cost in excess of $25 million or a total production investment in excess of $100 million. This Directive breaks the development of a weapon system into three phases: "Concept Formulation," "Contract Definition," and full development. What we have been discussing so far is the Concept Formulation phase, which consists of the "experimental tests, engineering,

[1] Office of the Secretary of Defense, *Initiation of Engineering and Operational Systems Development*, DOD Directive 3200.9, July 1, 1965.

and analytical studies that provide the technical, economic, and military bases for a decision to develop the equipment or system."[2] The primary objective is to establish firm and realistic design specifications and realistic cost and schedule estimates so that an adequate framework is obtained for management decisions to proceed with, cancel, or change the project.

What information must Concept Formulation provide? According to the Directive, it must demonstrate that the following prerequisites to entering Contract Definition have been accomplished:

1. Primarily engineering rather than experimental effort is required, and the technology needed is sufficiently in hand.
2. The mission and performance envelopes are defined.
3. The best technical approach has been selected.
4. A thorough trade-off analysis has been made.
5. The cost-effectiveness of the proposed item has been determined to be favorable in relationship to the cost-effectiveness of competing items on a DOD-wide basis.
6. Cost and schedule estimates are credible and acceptable.[3]

The first requirement relates to the technical feasibility of the concept. The experimental research programs conducted prior to the initiation of Contract Definition must have demonstrated that the technology needed is "sufficiently in hand." We do not want to start into Contract Definition still needing breakthroughs.

The second, third, and fourth requirements indicate another role of technical considerations in systems analyses. They are summed up essentially in the third requirement – the selection of the best technical approach.[4] Here, however, to obtain the "best" approach we may have to effect a compromise among performance, cost, technical risk, and development time.

In addition to selecting the best technical approach for the particular weapon system under consideration (say, the wing geometry or structural concept or fire control system of an airplane), the fifth requirement indicates that we must compare this selected approach with other alternative ways of achieving the same end. If the weapon system under investigation is a low-altitude penetration bomber, for instance, we may have to compare it with missiles and with stand-off missile-launching aircraft. The comparison is on the basis of over-all cost-effectiveness. Technical con-

---

[2] DOD Directive 3200.9, p. 2.
[3] DOD Directive 3200.9, p. 4.
[4] As we shall see later, it might be difficult to formulate precisely this notion of a best *technical* approach.

siderations will enter in estimating both the cost and the effectiveness of the selected and competing weapon systems.

Contract Definition

If the prerequisites are satisfied, Contract Definition – the second phase – may begin. Initiation of Contract Definition represents at least tacit approval on the part of the service and DOD that the weapon system chosen is the preferred approach to achieving the desired objective. The system that goes into Contract Definition has been selected over competing systems and thus already has a considerable amount of "push" behind it to support a decision to proceed into full-scale development. During Contract Definition, the effort is to verify the analyses completed in Concept Formulation and to write a definitive contract – one that has a firm fixed price or a fully structured set of incentives.

Two major decisions must be made at the conclusion of Contract Definition: the selection of the best technical approach and the decision as to whether or not to proceed with the last phase of development. In a sense, these are refinements of decisions made earlier. As we saw, to justify go-ahead for Contract Definition we have to be sure that the best technical approach has been chosen. Since this is done on the basis of studies conducted in the Concept Formulation phase, which can be quite long for a major weapon system, there will probably be a reasonably good framework of design studies on which to base this decision. In Contract Definition the design is narrowed still further and refined in an attempt to obtain realistic cost and performance data on which to base a final development decision. The problems involved in the selection process, however, are essentially the same as before.[5]

Full Development

The go-ahead to undertake full engineering development of the system is again similar to that made after the Concept Formulation phase. At this point, however, there is a chance to review the decision on the basis of the refined cost and performance estimates obtained during Contract Definition and in the light of the current military, political, and economic situation. This review is important since the decision to begin full development signals the expenditure of relatively large sums of money. The problems involved in the decision process – that is, in comparing the competing systems on a cost-effectiveness basis – are the same, however, as those encountered in the Contract Definition decision.

---

[5] For a comprehensive introduction to Concept Formulation and Contract Definition, see the discussion prepared by Peat, Marwick, Livingston & Company, *A Report on Contract Definition*, Boston, Massachusetts, January 2, 1967 (DDC No. AD 646240).

The Number of Units Deployed

Perhaps the most difficult problem to analyze is that of the number of weapons of each type which should be deployed. The difficulty arises because, while we can compute the incremental cost of, say, increasing the Minuteman force from 1000 to 1200 missiles, we find it very difficult to estimate the incremental capability provided by an extra 200 missiles or to measure it quantitatively. Yet the decisionmaker must decide how many Minutemen to deploy and, given a limited budget, he must decide whether he should spend his money for x many Minutemen and y many F-111's or $x+\Delta x$ Minutemen and $y-\Delta y$ F-111's. In these decisions he is aided by analyses, and these analyses are affected by technical considerations.

The type of analysis required here is similar, in a sense, to the cost-effectiveness comparison made in support of the initial development decision. In fact, the previous analysis provides the basis for initial scheduling of the number of units in the force. In justifying the development of the F-111, it was necessary to show that a force of x airplanes would provide such and such a capability relative to, say, the existing force of y F-105's and a proposed force of z new airplanes of a different type from the F-111. We cannot decide to go ahead with development without a reasonable idea of the number of expected production items.

After the system is in development and is being deployed, however, we review this decision periodically. It is here that we are forced to make further analyses and comparisons which may be even broader than the original study. Given a budget squeeze, the decisionmaker may have to decide whether to reduce one wing of F-111's, cut back on the number of Minutemen being deployed, or stop development of an advanced technical program such as DynaSoar, whereas in the original development decision we had probably compared the F-111 only with competing systems, that is, systems designed to perform essentially the same task.[6] Aside from the wider scope of the comparison, however, the problems encountered in the analysis and the effect of technological considerations on it are similar to those of the cost-effectiveness analysis made to support the development decision.

Modifications

As soon as the development go-ahead is given and the design of the weapon system is set (sometimes even before this), modifications of the system are being studied, modifications both small and large. If the development process takes from five to seven years, there is plenty of time for examining

[6] We shall see later that the question of defining a competing system will depend on how we define the task.

and proposing modifications even before the original design is operational. The fact that we reached the "F" model of the Atlas before it was phased out and that we have now reached the "G" model of the F-104 indicates that we can expect several major modifications during the operational life span of a weapon system. Indeed, it is not too unusual for the "B" model of an airplane to be in development while the "A" model is still in flight test. How does analysis assist in making decisions on these modification programs?

The problem of justifying a modification is essentially the same as that of justifying a new development. We must ascertain the technical feasibility of the proposed modification, we must select the best technical approach to achieve the desired capability, and we must show that the modification compares favorably with alternative systems on a cost-effectiveness basis. However, it is usually somewhat easier to "justify" a modification once the system is in development than to justify the development initially, primarily because the large investment we will have already made in the development of the system must now be considered a "sunk" cost. From this base, relatively modest improvements in performance can usually be made at relatively modest cost. For instance, we may be able to justify an engine modification for the B-52 (from the F to G model) to increase its range by 10 or 15 per cent, but not be able to justify the development of a new aircraft at $1 or $2 billion to provide a comparable increase in capability. For this reason, modifications to existing systems often are a more attractive way to attain a new capability than new weapon systems.

We can, of course, take this too far in several respects. On the one hand, we may try to push the life of the system too far by continual modifications. It may be better to start a new line of fighter airplanes than to continue to push the old design. This would be especially true if we had a high degree of confidence that the new system would itself enjoy a long life span. On the other hand, we can tend to fool ourselves into thinking that the modification will cost less than it really will because we are using the same basic airframe. If we need a new wing, a stretched fuselage, and new engines, the "modified" airplane may resemble the original only in name and the cost may be just as much as for a new design. Thus, there is no easy way to say just when a modification is to be preferred to a new design. Each proposed modification must be considered individually and as if it were for a new development. We have to consider the alternatives on a performance, cost, availability, and risk basis.

Replacement or Phase-out

The decision to replace a weapon system with a new one or to phase it out of the force entirely is the mirror-image of the initial development decision.

The same types of analyses are required, except that now our weapon system is one of the alternatives with which a new system must be compared. Thus, the effect of technological considerations on this decision will be essentially the same as before.

Summary

If we review the types of analyses that must be made to support decisions regarding a weapon system throughout its life, we can find three general problems which seem to recur in each. These can be categorized as the problems of:

1. Establishing technical feasibility;
2. Selecting the best technical approach;
3. Comparing the cost-effectiveness of the proposed system with that of its competitors.

In what follows, we shall examine these problems individually and in detail, specifying the types of analyses they require and the effect of technological considerations on these analyses.

TECHNICAL FEASIBILITY

The expression "technical feasibility" is both widely used and highly ambiguous. In the loosest interpretation, a system is technically feasible just so long as it lies within the current framework of physical laws. In this sense, an Aerospaceplane which calls for cruising flight in the atmosphere at 22,000 ft/sec is technically feasible, even though we do not know how to provide it with a propulsion system that can both produce a positive thrust coefficient and withstand the aerodynamic heating. On the other hand, in a more stringent interpretation of "technical feasibility," we might say that it is technically infeasible to develop an airplane to cruise 20,000 n mi with a payload of 50,000 lb, not so much because such an airplane would weigh over one million pounds and we just cannot construct a vehicle of that size, but because it would be impractical or unreasonable to do so. Between these two extremes fall many other meanings. Such ambiguity warrants a more careful investigation of the notion.

The question of technical feasibility arises when the proposed system requires an advanced technology or, to use a different terminology, an advance in the state-of-the-art. It is this concept of "state-of-the-art" which causes the confusion. Just when is a development within the state-of-the-art? How much of an advance in the state-of-the-art is required? In order to examine these issues, we can make use of a specific example. Let us consider a series of aircraft engines varying from what we would call the existing state-of-the-art to an extremely advanced state-of-the-art (perhaps even to what we might call technically infeasible). First, we would start with an engine that is currently in production, an engine that perhaps represents the existing state-of-the-art.

Second, we can consider a proposed engine which is somewhat larger than the one currently in production, but no higher in turbine temperature, no greater in compressor blade loading, no higher in combustion chamber heat release, and so on. As long as the increase in size itself is not very great,[7] we would certainly say that this engine is within the current state-of-the-art and have a high degree of confidence in our ability to design and fabricate it so that it would perform as predicted. While we could expect technical problems to arise in the development process, as they do with any complicated piece of new machinery, we could be confident that these problems would be readily resolved without undue expense or delay.

Third, we can consider a proposal for a more advanced engine, one which has a somewhat higher turbine temperature, combustion chamber heat release, or compressor loading than an engine currently in existence or in production. In this case, however, let us assume that test compressors, combustors, and turbine rigs have been run in the laboratory at these newer conditions for short periods of time, but that they have not been made in flight-weight versions or assembled into a complete engine. In this case, we might have somewhat less confidence in our ability to design and fabricate the proposed engine and yet still come very close to target performance within the predicted time and cost of development. We might say that such an engine required a slight improvement in the state-of-the-art.

Fourth, and one step further, we have a proposed engine calling for compressor loadings, combustion heat releases, or turbine temperatures which have not even been obtained in test rigs, although the design of the engine is reasonably sound analytically. In this case, the proposal might involve a scheme for varying the flow area of the turbine in a way that had never been done before. It might call for the operation of the turbine or compression over a broader region of its performance map than ever before, or for a turbine cooling scheme involving the machining of intricate passages. Here we might say that "we can do it on paper" but not be sure that we could do it in the shop. Although analytically it should work, we might have some doubts about our ability to produce, say, intricately machined turbine blades or to make the by-pass valves within the specified leakage tolerances. Here we might say that the concept needs technical verification. If we use the term "technical feasibility" in a relatively severe sense, we might say that the technical feasibility had to be demonstrated.

Finally, we can imagine a proposal for, say, a supersonic combustion

---

[7] Granted, this notion is somewhat hazy.

ramjet about which we cannot even be sure that our designing analysis is correct. Not only has there been no testing of similar components, but there has not even been enough basic research in hypersonic aero thermodynamics to assure us that the theoretical work on which the design is based is sound. In this case, we might say that the proposal is based on an advanced state-of-the-art in two senses: we have not demonstrated the technical feasibility of the concept (as we had before) and we have not verified the theory on which the design is based.

For convenience we can briefly identify various rungs in the state-of-the-art ladder which are brought out by the example. From the bottom up, these can be recognized in terms of:

1. Existing technology;
2. Scaled version based on existing technology;
3. Limited component tests available;
4. No laboratory or component work available;
5. No laboratory or component work and limited theoretical basis.

This series does not include all the possibilities. But it does serve to show the possible ambiguity that could result from uncritical use of the expression "technically feasible." Different people will draw the "line of feasibility" at different "rungs" in the ladder. Some might stop at the third and others would consider that all of the proposed engines were technically feasible. The important point is not so much to achieve agreement as to where to draw the line as it is to make clear the distinctions between the various rungs.

We are now in a better position to see how issues of technical feasibility enter into the decision process at various stages in the development of a weapon system and how they affect analyses aimed at supporting decisions. Some guidance is provided in the wording of Directive 3200.9, which, as we have seen, indicates that during the Contract Formulation phase, we must show that a proposed system primarily requires engineering rather than an experimental effort.

We might interpret this to mean that we must ascend the ladder to about rung 3 before the system can be considered for development. Thus, a major objective of a program in its early stages would be to attain this level of technology, and analysis offered in support of requests for funds for applied research should indicate what is necessary to put the needed technology "sufficiently in hand."

Different types of research and experimentation may be appropriate for taking different steps in the ladder. For instance, we may want to suggest shock tunnel runs on bodies of revolution in an attempt to verify the aerothermodynamic theory on which the performance analysis of the supersonic combustion ramjet is based (step 5 to 4). We may want to fabricate turbine blades and make up a turbine rig to demonstrate that the fabrication techniques can be developed and that the turbine can perform as predicted under the temperature and pressure conditions it will encounter in the operational weapon system (step 4 to 3). We may feel that it is necessary to put the components together to see whether or not flight-weight compressors, turbines, and combustors similar to boiler plate designs already tested in the laboratory can be satisfactorily integrated into an engine which meets both weight and performance specifications. This is often achieved by putting together what is called a "demonstrator engine," usually a small-scale version of the engine which would be used in the operational weapon system (step 3 to 2). These are typical of the means available for advancing the state-of-the-art in engine development.

In supporting his request for research funds, then, the analyst must first present a detailed analysis of the "current" state-of-the-art. He will have to discuss the work done to date and describe what types of systems could be designed and put into production without further experimentation. Second, he must indicate what types of advances are required in which component areas for the proposed weapon system and suggest programs to advance the state-of-the-art in these areas.

Up to now we have spoken of the technical feasibility of components. The over-all technical feasibility of the weapon system is a function, of course, of the technical feasibility of its components. We must distinguish, however, between weapon systems that require advances in just one or maybe two component areas and those which require several different advances at once. Since the notion of technical feasibility is very closely related to the confidence (or subjective probability) that we have in being able to meet a given performance specification at a given time with certain unds, the technical feasibility of the over-all system decreases rapidly as the number of advances that must be made in component areas increases. Consequently, the recent tendency has been to try to limit new development to major advances in one area at a time. For example, it has been suggested that the avionics requirements for the F-111 were intentionally set *within* the current state-of-the-art, since the major advance in that weapon system was felt to be the variable-sweep wing. By relaxing the requirements for the avionics, a possible source of delay was removed, the idea here being that, if we push the state-of-the-art in many different areas at once, *at least one* will come back to haunt us. It takes only one to delay an entire program. With this in mind, the analyst and designer must attempt to compromise between the sort of technical advance that will result in a system which is an improvement over existing systems (modified or not) and that which requires so many basic changes that the probability of meeting performance, development, time, and cost schedules is extremely low.

One final caution should be injected here. Invariably, technical specialists will differ both in their interpretation of what the current state-of-the-art is and what the probability of success is for advancing it through some specific program of experimentation. Thus, the analyst and decisionmaker should not expect to get a unanimous opinion from the technical community. It is important, however, to get at the sources of these differences, three of which are particularly worth noting. First, we must, of course, be wary of special interests. A person advancing a pet technical project may tend to play down some of its difficulties in order to get funds to try it out. On the other hand, other experts, perhaps from different branches of the military or from different companies, may tend to play down an idea for financial reasons or because of a built-in NIH (Not Invented Here) factor. In short, everybody seems to like his own ideas or his own company's proposals better than the other guy's. This can blind him to the problems the idea involves.

Second, some people (and even companies) are just more optimistic than others. While one fellow may fix on the problems and see dozens of places where an experiment can fail, someone else will be sure that a way will be found to go from here to there.

Finally, because of the uncritical use of expressions like "technical feasibility," analysts often encounter apparently serious contradictions which, when explored further, turn out to have no basis in fact. For instance, Expert A might say that something cannot be done and Expert B that it can, where what A really means is that it would be a long, expensive program with many possible pitfalls along the way and what B really means is that, given enough time and money, you can do just about anything. The argument is purely verbal.

There is no easy formula for mediating disputes which are basically over the expected rate of progress in a field. All the analyst can do is to solicit opinions from those he feels are competent, listen to the conflicting claims, attempt to weed out special interests, reduce what is purely verbal in the dispute to a minimum, and then present the substance of the situation to the decisionmaker in as clear, concise, and intelligible a form as possible.

Selection of best technical approach

Scope of the Problem

The problem of selecting the best technical approach can arise in several different contexts. Perhaps an example will help to explain this point. In considering an air defense system, we may be interested in selecting the best technical approach among alternative interceptors designed to fly at Mach 3.0 at 70,000 ft with a range to intercept of 500 n mi. Here we would be interested in selecting the type and size of engine, the wing planform and size, the structural concept, and so on.

We may, however, take a somewhat broader view of the air defense problem and attempt to select the best technical approach among several different types of interceptors. For instance, we may be interested in comparing a new Mach 3.0 airplane with an existing design, such as the F-106 or F-111. In this case, we would attempt to see how well the different airplanes perform the air defense mission rather than how well different airplane designs meet some prescribed performance requirements. In this type of comparison, we are concerned with what the performance requirements (in terms of speed, range, and so on) should be for an air defense interceptor.

Finally, on a still different level, we may be concerned with deciding whether we should employ high-speed interceptors for air defense or resort to a mobile air platform (a big, slow airplane) which relies on long-range missiles to effect its kills. Again there is a sense in which we would want to select the best technical approach. Here, however, it is the best technical approach to the air defense problem rather than the best technical approach to the design of an interceptor.

Very often, in speaking of selecting such an approach, people have in mind the first, or the narrowest, type of analysis. In fact, the very use of the phrase "best *technical* approach" implies that some very rigid restrictions are placed on the notion of "best," that only *technical* factors are involved in the choice. It implies that the best *technical* approach may not be the *best* approach, perhaps because it costs too much or it has undesirable political implications. While at first glance such a distinction seems perfectly valid, we must be wary of it. Quite clearly, important technical factors enter into the selection of the best approach at each of the levels of decision indicated above. At the third level, however, it becomes difficult to make sense of a notion which, for instance, does not include at least cost considerations. For this reason, we might shift the emphasis here and speak not of the selection of the best technical approach, but rather of the effect of technical considerations in the selection of the best approach. In doing this, we will still have to be concerned with all three of the levels mentioned above. Now, it might seem appropriate to start with the more general problems to see how technical considerations enter into the selection of design requirements before we consider the narrower problem of selecting the best configuration to satisfy some fairly specific design requirements. Nevertheless, perhaps the best way to bring out the effect of technological considerations in the selection of the best approach is through a specific example. To this end, let us continue with the air defense illustration and leave the general problems for the next section.

Formulation of the Problem

In order to select the best approach to air defense, we must first formulate the problem and select criteria for evaluating alternative systems. Since the primary purpose of a U.S. air defense force in general war is to keep enemy bombers from reaching their targets, and hence from killing U.S. citizens or destroying U.S. weapons, one criterion for evaluating alternative designs is immediately apparent: the number of U.S. fatalities (or weapons destroyed) from bomber attacks. If we compare alternative systems on the basis of the same budget allotments, then it would seem that for any given level of expenditure on air defense we would want to pick the system which minimized the U.S. fatalities (or weapons destroyed) from enemy bomber attacks.

But wars can start under many different circumstances and the performance of the air defense system will certainly vary accordingly. In addition, there will be some degree of uncertainty regarding the number and type of bombers the enemy will have and the tactics he will employ. Starting with current intelligence estimates, we can estimate at least a possible, if not probable, enemy bomber threat both quantitatively and qualitatively. But we will want to know how the various proposed defense systems will perform against a variety of threats.

Thus, we can start to see the formulation of the problem according to a pattern that other authors in this book have already defined. We shall have to consider the performance of each alternative air defense system against a variety of enemy threats in a variety of wartime situations. In each case, the systems will have to be compared on the basis of the number of U.S. fatalities (or weapons destroyed) through the bomber attacks. Various budget levels will also have to be considered. From the data obtained, a decision can then be made as to the "best" approach to the air defense problem. Since it is possible (even probable) that different systems will minimize U.S. fatalities under different sets of assumptions about the enemy threat or enemy tactics, or how the war will begin, or what the air defense budget levels will be, some sort of compromise will obviously have to be accepted in the selection of the "best" approach.

Design Parameters and Parametric Analysis

After formulating the problem and establishing the criteria for comparison, we are in a position to examine alternative systems for accomplishing the mission. For the purposes of our example, we will limit ourselves to one class of *similar* systems – manned interceptor aircraft. Our first task would probably be to delineate the basic aircraft design parameters affecting the over-all performance of the system. In the present case, these would be cruise speed, combat altitude, range, payload (both quantity and type), and type of subsystems.

In establishing the design requirements, we would want to know the effect of each of these parameters on the over-all system performance; we would want to learn the answers to such questions as these: How many intercepts can be made with a Mach 2.0 airplane as opposed to a Mach 3.0? How many intercepts are made if each aircraft carries 4 as opposed to 8 air-to-air missiles? One way to narrow in on these questions is by means of a *parametric analysis*. To conduct this type of analysis we would start by describing the performance and cost of the system with relatively simplified equations or a model. This model would be complete enough that specifying values for the basic parameters would allow us to obtain first the design characteristics of the airplane in terms of gross weight, wing size and planform, structural weight, and so on. From this information, we would then be able to determine the cost of the airplane and its associated equipment. Given a budget level, we could then determine how many aircraft of this type could be purchased and operated over a specified time. Finally, we could compute the number of intercepts this force would make against various enemy attacks in various sorts of wars. Thus, in the parametric analysis we would have a system of equations (a model) which related the effectiveness of the system, as measured in terms of the number of intercepts, to independent design parameters such as speed, altitude, and range. By varying the values of these independent parameters over a wide range, we could determine the effect of each on the over-all performance of the system and be in a position to select the design requirements for the airplane.

In conducting this analysis, we would very probably take advantage of some fundamental technical *scaling laws* to generate a family of "rubber" engines and airplanes. These engines and airplanes would be "rubber" in the sense that they would vary primarily in size while maintaining a similar geometry. For instance, we know that the design-specific fuel consumption of a turbojet engine is relatively unaffected by the design thrust of the engine as long as such characteristics as compressor pressure ratio, turbine temperature, and by-pass ratio remain the same. We also know that the weight of the engine varies roughly as the design thrust (over a reasonable range of thrust) and that the engine diameter varies roughly as the square root of thrust. These data can be used to compute the performance and weight of a family of similar engines (that is, engines which employ the same basic technology) as a function of design thrust. These "rubber" engines can then be used to compute the performance of a family of airplanes, similar in such characteristics as the ratio of fuselage length to diameter, wing loading, and wing planform, but different in gross weight.

Parametric analysis is invaluable in closing in on design requirements. Because the analysis is usually conducted on a computer, we can investigate a very large number of values for each of the independent design parameters. By using digital computers we may even make suboptimizations to select the best wing area, wing geometry, and engine size for a given range-speed combination. Thus, the "airplanes" which just represent points on a curve may each include several suboptimizations.

The output of a parametric analysis might be presented as in Fig. 6.1. For a given budget, a given attack situation, a given base structure, we might show the number of intercepts as a function of interceptor range for various airplane speeds. In the illustration, for instance, we show an optimum in both speed and range. There will be an optimum range since a continual increase in the range would increase the size and cost of the interceptors, reduce the number that could be purchased for a given budget, and, hence, reduce the number of kills. An optimum speed might be found at, say, Mach 2.2 because at this speed the number of intercepts is not substantially affected as long as the threat is a subsonic bomber, while the cost of the airplanes designed to fly at speeds above Mach 2.2 is high because of the need to use steel or titanium construction rather than aluminum.

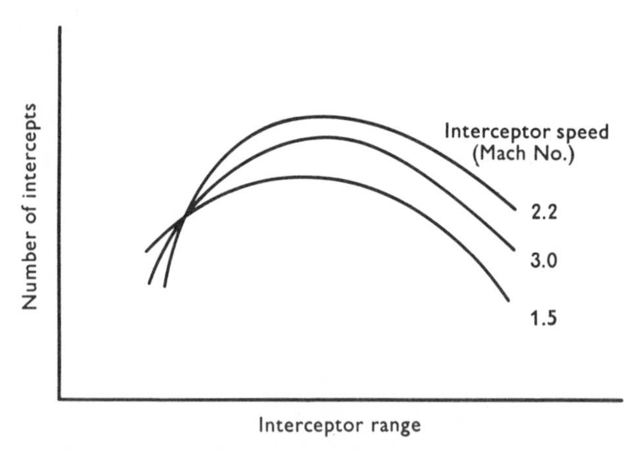

Fig. 6.1 – Parametric analysis output (constant budget)

By means of a parametric analysis we would generate the data needed to select the "best" approach – or at least get a first cut at it. In conducting an analysis of this type, however, there are several things we must watch out for. First, technical factors could introduce some discontinuities in the curves. In the examples we are considering, for instance, there would be a discontinuity at the speed of about Mach 2.2 since we must shift from aluminum to either titanium or steel at this point. If we consider very high speeds, perhaps up to Mach 5, we will get discontinuities in the type of engine that can be used. Up to about Mach 3.5 we can use a turbojet, but beyond that speed we need a turboramjet or a separate ramjet engine in addition to the turbojet. In view of these natural breakoff points caused by technical factors, we often take several different "point" designs for analysis rather than a continuous series or family of airplanes. In the present case, for instance, we might examine just Mach 2.2, Mach 3.0, and Mach 5.0 designs, if we considered each as representative of a certain class of airplanes.

A second factor we must watch out for is oversimplification. Often because of a strong desire to use the computer or to get closed-form analytic solutions, we will make simplifying assumptions in the equations used to compute performance. But to be sure that our parametric data have not deviated too far from reality, we should supplement the parametric study by a few more detailed "point" designs. Here we would perform more detailed layouts of various configurations aimed at meeting the requirements. We would try different structural concepts, different placing of subsystems, different types of landing gear designs, canards as opposed to conventional tail surfaces, and so on. In the process, we would turn the "rubber" airplane into a reasonable "paper" one. We would replace the approximate performance calculations with much more detailed analyses, and the results of the more detailed design studies would be fed back into the parametric analysis as changes in input values for the performance models. In this way, we could get more realistic outputs from the parametric analysis and do a better job of selecting the design requirements and the "best technical approach."

Trade-off Analyses

At this point, it might be appropriate to ask how parametric analyses are related to the trade-off studies which are called for as prerequisites to Contract Definition and in Contract Definition itself. To answer this question, we might find it useful to examine a distinction sometimes made in explaining Directive 3200.9, the distinction between inter-system trade-offs and intra-system trade-offs. These expressions seem to suggest that analyses of the type we have described would provide the inter-system trade-offs, since they show the effect of such characteristics as design speed, altitude, and payload on over-all performance. In a sense, different values of design speed represent different systems and the "trade-off" is between design speed and performance, between design payload and performance, and so on.

The interpretation of what is meant by "intra-system trade-offs" is somewhat more difficult. At least two types of analyses might fit in this category. The first concerns those trade-offs that can be made in the mode of operation of the system. For instance, for any given airplane we have the option of trading off fuel for payload (assuming that there is room in the airplane for the extra payload). Thus, we can elect to fly for shorter distances with more payload or for longer distances with less pay load. This trade-off can be represented as in Fig. 6.2.

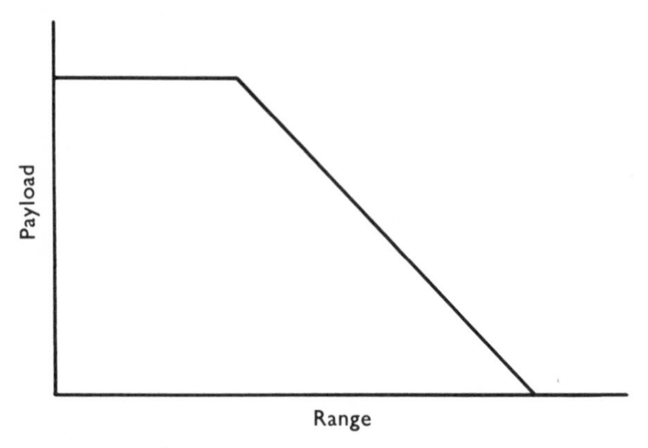

Fig. 6.2 – Range-payload trade-off

Another illustration of this type of intra-system trade-off concerns the flight path flown. We can, for instance, fly for a relatively long distance at high altitude and low speed or trade off range for either increases in speed or reduced altitude flight. We might show this trade-off as in Fig. 6.3, where the total range is plotted versus the portion of that range flown at high speed or low altitude (sea level).

These types of trade-offs indicate the basic flexibility that is available in operating the system *after it is designed.* Given any *particular* airplane, there is a wide latitude in the way it can be operated. As we have shown, speed, range, altitude, and payload can be traded off to obtain a wide variety of operational capabilities. These trade-offs should be indicated in the over-all systems analysis.

The second type of intra-system trade-offs we can speak of are those which are available *in the design stage.* For instance, we may want to accept a higher gross weight to improve an airplane's take-off capability or reduce the time it takes to climb to combat altitude. Here there is a close similarity to the inter-system trade-offs since, in a sense, the trade-offs are changes in the design specifications of the system. Perhaps the difference is one of degree rather than kind.

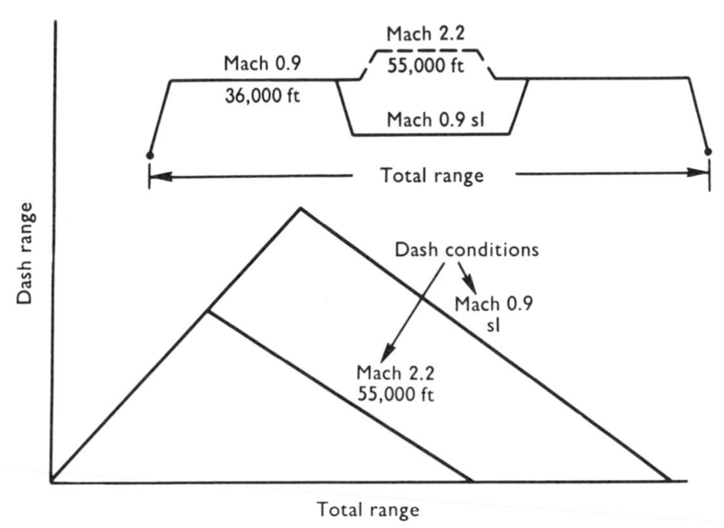

Fig. 6.3 – Cruise versus dash range trade-off

This similarity is brought out in another possible type of intra-system trade-off, which comes about because there are many design requirements and they are usually expressed in terms of minimum (or maximum) acceptable values. For instance, we might specify a minimum speed of Mach 2.0, a minimum combat altitude of 60,000 ft, and a maximum take-off distance of 6000 ft. We can then imagine three different designs with the following characteristics:

Characteristics	*A*	*B*	*C*
Speed ~ Mach No.	2.2	2.0	2.2
Combat Altitude ~ ft	60,000	70,000	75,000
Take-off Distance ~ ft	4,500	4,000	6,000

In evaluating these designs we will have to make trade-offs among the various parameters. Is the extra speed of A more important than the higher combat altitude of B? Is the shorter take-off distance of A and B more important than the higher combat altitude of C? These types of trade-offs will be important in selecting the final design.

Technological Uncertainty

Earlier we saw how the problem of technological uncertainty was involved in establishing the existing state-of-the-art and the risk associated with a development program. The question of uncertainty also arises in selection of the best approach. Perhaps the best way to illustrate this would be to continue with our air defense example. In this case, we can expect some difficulty in estimating the flight performance or weight of the aircraft and its subsystems. Again, this is just natural for any complex weapon system. Generally we are trying to make *some* advance from previous designs in speed, type of subsystems, aircraft design, and so on; otherwise we would have trouble saying why a new weapon system was needed. Although it may be agreed by almost all technical specialists that the task is primarily one of engineering rather than of experimentation, there is almost always some margin of uncertainty regarding the actual performance or the weight to get a given performance. Even after an extensive attempt to "define" the F-111, for instance, the airplane "grew" during development, so that the gross weight was somewhat higher than had been predicted. Because we can never really reduce this type of uncertainty to zero, it should be explicitly recognized in the early analyses. This can be done by including *sensitivity analyses* to show the effect of not meeting various performance requirements. For an airplane, we can do this in two ways. First, we can show how its performance (or weight) will vary with the design weight of the electronics equipment, the lift-drag ratio, the engine sfc (specific fuel consumption), and so on. In this sense, the sensitivity analysis will be similar to the trade-off studies discussed earlier.

A second type of sensitivity analysis, however, can be directed at showing the effects of not meeting (or exceeding) design requirements *after the aircraft is already in development*. Here, for instance, if we are comparing alternative airplane designs we might want to know what will happen to each if the electronics subcontractor cannot meet his weight or size requirements. Is there room in the design for a larger unit? Is there enough thrust to accept an increase of 20 per cent in the sensor weight and still meet design speed requirements? These types of considerations can be important in selecting a design. We may, for instance, select a design that does not have the best performance if everything goes as planned, but which declines in performance rather gracefully if subsystems grow or the engine does not perform as well as predicted.

Along these lines, this may be a good place to introduce another notion which, although not appropriate for discussion under the heading of "technical uncertainty," plays a similar role in the decision process. That is the notion of *growth potential*. Since almost every weapon system that is produced undergoes a series of modifications to improve its capability after it enters service, it is a good idea to see to what extent these subsequent improvements can be anticipated in the early phases of analysis. In our air defense case, for instance, we should point out that an aluminum airplane with a Mach 2.2 speed capability has very little growth potential in speed without major structural modification. This is due to the inherent temperature limit of the aluminum structure. On the other hand, a titanium or steel aircraft designed for the same speed or a higher speed (say, Mach

2.6) may be able to be "stretched" to Mach 3.0 or 3.5 with relatively minor modifications if the basic structure of the aircraft is designed for this growth potential. While the aluminum aircraft may be the "best" design on the basis of an existing estimate of enemy threat, the fact that the titanium or steel airplane has this inherent growth potential will make it more adaptable to meeting changes in the threat (in terms of bombers of higher speed). These factors should be pointed out by the analyst so that they can be included in the over-all evaluation.

Another type of technological uncertainty that we face in the interceptor problem is that of predicting the kill capability of the airplane against various types of bombers. Here, as opposed to the case of predicting aircraft weight, speed, and so on, this uncertainty can be very large. In predicting the aircraft speed and weight, we can have a high degree of confidence in our predictions, largely because, prior to development, we can reduce the known technological uncertainties of the major components through wind tunnel tests, structural tests, and so on, and the models used for predicting aircraft performance have been checked against reality many times (every time an airplane is built).

In the case of predicting how effective the airplane will be in shooting down bombers, however, the uncertainty might easily be measured in terms of factors of 2 or 3. Our ability to simulate air combat situations is not nearly so good as our ability to simulate, say, free flight in a wind tunnel. In addition, we get so little feedback from *actual* experience that we cannot check our models as we can in predicting aircraft speed and range. While we do conduct exercises, there is always some question about whether or not the exercise is a realistic model of an actual combat situation where we may be facing different types of electronic countermeasures or operating without information from ground control centers.

Faced with this type of technological uncertainty, the analyst can do several things. First, he can perform sensitivity studies to discover the effect of changes in assumptions regarding the various parameters used in computing the kill capability. Here it is very important to recognize the places in which relatively unsubstantiated assumptions are being made and to show how the answer depends on these assumptions.[8] Secondly, he can seek solutions that are relatively insensitive to variations in the unknown parameters. Finally, he can try to suggest ways for removing some of the uncertainties. While it is not always possible to remove them completely, there may be ways to bound the problem.

Clearly, we must live with uncertainty in the force planning business. While it is often difficult for the decisionmaker to decide what to do in the

[8] This point will be discussed in more detail in the next section.

face of uncertainty, it is imperative that he understand the sources and effects of the uncertainty and possible schemes for reducing it if he is to arrive at a well-informed decision.

Most of the examples we have used here have dealt with the problem of selecting design requirements. We have seen that this selection implies an iterative process which cannot be done properly without considering the types of designs that can satisfy these requirements. In parametric analysis we are likely to rely on relatively crude approximations to the design. As we get into the problem in more depth, however, we must refine our design data by taking into account the specific technical problems encountered in specific designs. Thus, by the time a "best" approach is selected in the Concept Formulation phase, it should have been based on a reasonably thorough understanding of the interrelations between the requirements and ways to meet them. The next phase in the development process is the refinement of the design in the Contract Definition phase, where now relatively narrow limits can be placed on the specifications.

Certainly the emphasis is placed differently in this phase of the analysis. We are, for instance, more concerned with discovering good, sound technical solutions to problems involving subsystems than with comparisons among the over-all designs on the basis of number of kills made. We are also interested in much more detailed analyses, such as comparing alternative manufacturing methods for a particular component. Yet the general guidelines are the same. We must indicate the trade-offs that can be made, the sensitivity of our answers to assumptions, and the uncertainty. We must encourage imaginative work in the selection of alternatives and not hinder this activity by imposing arbitrary restrictions. We must make the result of our analysis as "transparent" as possible, so that the decision-maker can see through from the start to the end of the problem and not just be presented with input and output numbers without knowing how we got from one to the other. In view of these similarities, we might skip from this problem of selecting a technical approach to meet given design specifications to that of comparing alternative systems.

Comparison with competing systems

One of the prerequisites to Contract Definition is a cost-effectiveness comparison between the proposed system and competing items on a DOD-wide basis. If we continue with the example of the preceding section, this means that after selecting the best interceptor for the continental air defense role we must compare this system with other ways to do the job. As a minimum, we would have to compare it with the existing air defense force and with other air defense concepts. However, it is possible to take an even broader interpretation of the "job" that is being done and insist on still other comparisons, one of which might be between area defense and point defense against bombers. In addition, if the job is defined as that of protecting U.S. citizens from Soviet weapons, we might insist on a comparison between the interceptor and a ballistic missile defense system. In a sense, as earlier chapters have made clear, these are systems "competing" for portions of the defense budget. But we can go even further. We can protect U.S. citizens from Soviet weapons by building blast or fallout shelters. Thus, we would have to compare our proposed interceptor against this mode of protection. Finally, since we can protect U.S. citizens from Soviet weapons by destroying these weapons before they are launched, it seems that we would have to compare the interceptor against U.S. ballistic missiles and bombers which can be used in counterforce attacks. But since Directive 3200.9 is not too specific about what is meant by "competing items" on a DOD basis, we are not given too much guidance as to how far up this hierarchy of comparisons we must go.

There are indications lately that requests for major weapon system developments will have to be supported by analyses that include rather broad comparisons. One Air Force study, for instance, was aimed at showing trade-offs among all Program Package I (strategic offense) and Package II (continental defense) weapons, plus civil defense and antisubmarine systems. Although in a sense the procedures used in these comparisons are similar to those used, for instance, in comparisons among interceptors, there are a number of different types of problems that arise. In this section, we shall consider some of them.

One problem that must be faced in making broader comparisons is that of comparing systems of varying state-of-the-art. In our example, for instance, we might compare the proposed interceptor to both the existing interceptors and to AICBM systems. Yet there is a difference in our ability to compute the performance or capability of these systems. We know what the existing system can do and how much it costs because we are, in fact, operating it. But because we can only predict the performance and cost of the more advanced systems, we just cannot have the same degree of confidence in predictions of the single-shot kill probability of an AICBM system and similar predictions for the effectiveness of existing interceptors. At least in the latter case there is some basis in test experience for our predictions. While even they are subject to uncertainty, they are based on a much more solid foundation than predictions of AICBM performance.

There is nothing that can be done to remove this disparity, except to make it explicit by conducting sensitivity analyses. We should not try to assume away the technological uncertainty by making a "best" guess for the value of every variable. Rather, we should show the performance of each system as a band of different width instead of as a fine, single line.

The decisionmaker should not be shielded from the technological uncertainty by best guesses.

Nonrelative Comparisons

In many of our analyses, such as our air defense analysis, we may not be too confident about the *absolute magnitude* of the answers we get, say, for the number of kills made in a given attack. Perhaps the uncertainty may be as much as a factor of two or three. Yet despite this fact, we may have a reasonably high confidence in the *relative* comparison we make between two interceptors, primarily because the two systems are very similar and the uncertainties which affect the performance of one will also affect that of the other. For instance, if there is some uncertainty regarding the ability to acquire the target in a situation involving electronic countermeasures (ECM), the problem will be the same for both interceptors, since they employ similar subsystems and perform the mission in the same way. Thus, if we make the same assumptions regarding the ECM problem for each, the relative performance of the two will be indicative of their differences in capability, even though the description of actual differences may be considerably off. Thus, as Fig. 6.4 indicates, although we may not know whether the effectiveness of the ECM will be 20 per cent or 80 per cent, and although the number of intercepts made by System A may vary by a factor of three, depending on the effectiveness of the ECM in this range, we can still say that System A is preferred to System B because it provides more intercepts *regardless of the effectiveness of the ECM.*

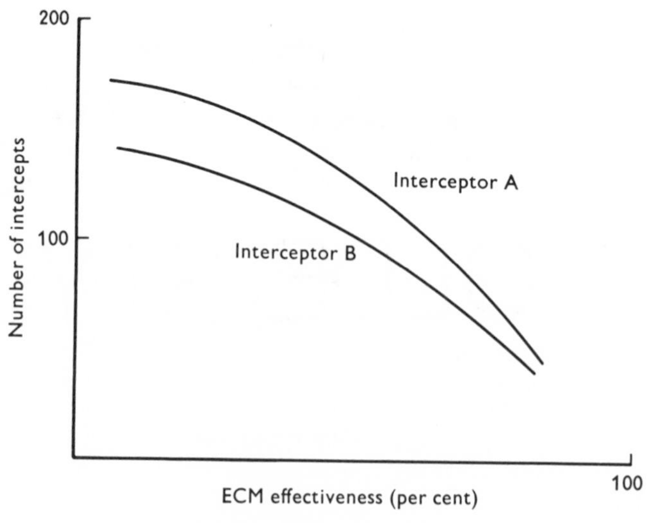

Fig. 6.4. – Effect of uncertainty

As we extend our comparison, however, we can no longer have this confidence in pure estimates of relative performance. Consider the problem of comparing the existing interceptor system to a proposed air defense system which uses different types of equipment and which thus handles ECM in an entirely different way. Clearly, our estimates of the actual capability to handle ECM will be much better for the existing interceptor system than for the proposed one.

This point is more clearly brought out when we compare the interceptor to AICBM or strategic offensive missiles. In this case, our criterion for comparison may be the number of U.S. lives saved in such and such a war for each dollar invested in the system. In the interceptor case, there may be uncertainty regarding ECM; in the AICBM, there is uncertainty regarding the effects of precursor blasts on target acquisition capability; in the strategic offensive missile, there is uncertainty regarding the CEP. In each case, the uncertainty could lead to differences in absolute performance of as much as a factor of 2 or 3. While we had this much uncertainty in Fig. 6.4, we were still able to say that System A was about x per cent better than System B because a similar uncertainty affected each system. Now, however, the situation may be represented schematically as in Fig. 6.5. If we are toward the left side of the curve on CEP, toward the center on ECM effectiveness, and toward the left on ICBM penetration, we would get one answer. If we are toward the right end of the uncertainty band on CEP and toward the right on ICBM penetration, we would get another answer. In other words, the answer is dependent on the *absolute magnitude* of predictions made on widely different technical issues. Thus, it would be very difficult to make meaningful comparisons if there were a large degree of uncertainty on each issue (as there would be in this case). Again, however, there is very little the analyst can do to reduce these uncertainties beyond trying to be scrupulously careful to point them out and to present the sensitivity analyses, along with technical information on the reason these uncertainties exist. In these circumstances, it would be extremely misleading for him to present only "best guess" data.

External Constraints

Another pitfall which we must watch for is the effect of arbitrarily established external constraints. An example will illustrate the point. Suppose we are comparing such vertical take-off space boosters as the Saturn or Atlas to airbreathing horizontal take-off recoverable boosters. Usually the criterion employed in such a comparison would be the cost of placing a given payload in orbit. Just as usually, however, there are some external constraints on the systems under consideration. For instance, we might require that the vertical take-off booster be launched from Cape Kennedy

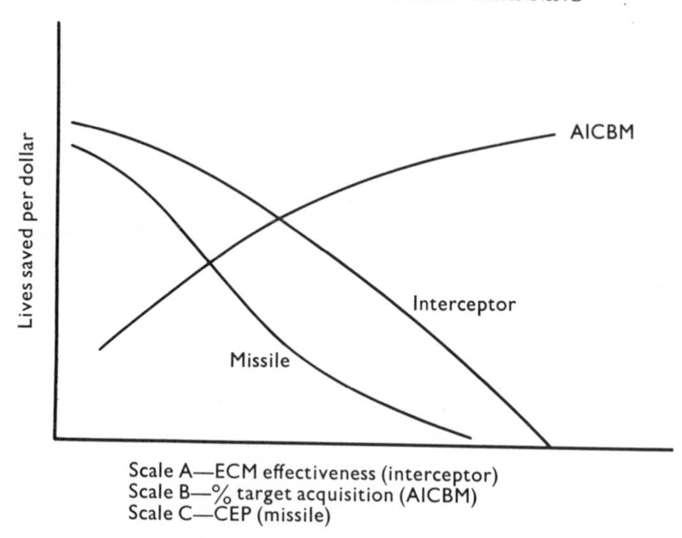

Fig. 6.5 – Nonrelative comparison

and that the horizontal take-off booster utilize existing SAC runways.

By placing a requirement for runway strength and length on the air-breathing recoverable booster, we are placing some definite design limitations on the systems that are considered. This may, for instance, limit the maximum gross weight of the booster and severely restrict the performance of the system. It may be that the performance as a function of gross weight is as shown by the solid line in Fig. 6.6. By placing an arbitrary weight requirement, we limit the capability of the system. If, on the other hand, we remove this restriction and add the cost of modifying a few runways to take a heavier aircraft, we may get the performance shown in the dashed curve. The heavier airplane would be the best choice, even though new runways had to be constructed, while the vertical take-off booster might have been preferred if we had stuck to the runway requirements.

The lesson to be learned from this example is that we should carefully examine the external constraints placed on an analysis. Sometimes they cannot be avoided. The sonic boom problem may impose severe constraints on supersonic transport design. There may be legitimate political constraints which preclude, for instance, the use of certain satellites, even though they may be attractive from a cost-effectiveness standpoint. There may be economic constraints that force us to use an American-built engine, although a foreign design would give better performance. These are constraints that we must live with. As analysts, however, we should not accept them passively. At *some* level of decisionmaking they are still

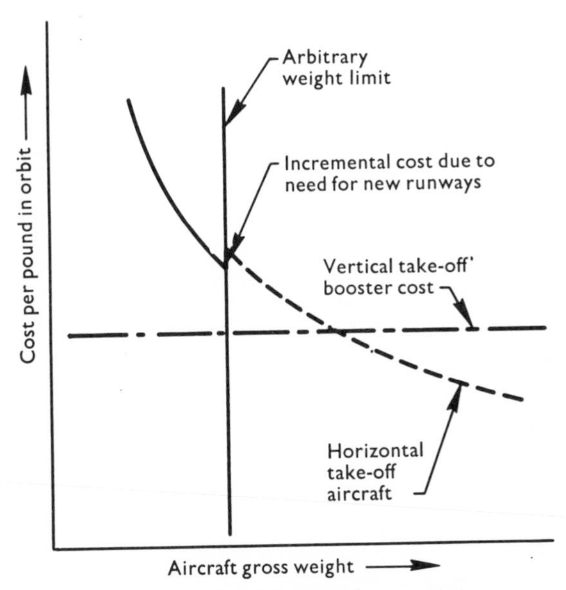

Fig. 6.6 – Effect of external constraints

subject to reversal. In effect, they are just one input to the decision process, which, for the time, appears to be of overriding importance. But since the decisionmaker should know how much he is paying in terms of reduced system effectiveness by insisting on these constraints, our analysis should indicate whether or not they have influenced the selection of the best approach, and, if so, to what extent.

Chapter 7

RESOURCE ANALYSIS

G. H. FISHER

This chapter indicates the nature and role of cost considerations in systems analyses, and discusses in particular the basic concepts and procedures of resource analysis as it has been developed over the years at RAND.

INTRODUCTION

Before discussing resource analysis in any detail, it may be useful to make a few preliminary comments. First, and most important, we should note that resource analysis is rarely (if ever) an end in itself, but serves rather as a part of, or an input to, the more general type of analysis we call systems analysis. To see clearly how resource analysis fits into systems analysis, let us consider briefly the main conceptual approaches that may be used in the over-all analytical process.

Earlier authors in this book have pointed out that the crux of the long-range military planning problem is the systematic examination of alternative system or force proposals, with a view to finding that alternative (or combination of alternatives) which seems preferable to others. While there are numerous and complex considerations that must be taken into account in a typical systems analysis, two key ones are effectiveness (utility) and resource impact (cost). Conceptually, the analytical process may take either of two basic forms:

1. For a *specified level of effectiveness* in the attainment of a certain national security objective, an attempt is made to determine that alternative or combination of alternatives which is likely to do the job with minimum resource impact.
2. For a *specified budget level* to be devoted to a certain area of national security, an attempt is made to determine that system or force proposal (or combination of proposals) attainable from the specified budget which is likely to achieve maximum effectiveness.

From these two conceptual approaches, we see how resource analysis fits into the total picture. In the first case, campaign or effectiveness analyses determine the configuration and quantity of each of the alternatives required to achieve the specified level of effectiveness. Then resource analysis estimates the resource impact (cost) of each of the alternatives. Here, resource impact is, in effect, the criterion of choice.

In the second case, resource analysis is required to determine the quantities (force sizes) for the alternatives that may be attained within the stipulated budget level. Given these quantities, effectiveness analyses then determine which alternative gives maximum effectiveness for the given budget level.

In either case, resource analysis is clearly essential to the total analytical process. Since our national resources are limited, rational choice among alternatives demands that resource impact be taken into account.

One further preliminary comment. The present Chapter attempts merely to outline the basic concepts and principles of resource analysis as it has been developed over the years at RAND. In the following Chapter, R. L. Petruschell and A. J. Tenzer will continue the discussion by illustrating the application of sensitivity analysis techniques to a problem in resource analysis. W. E. Mooz, in Chapter 9, will then show how various concepts and methods of resource analysis may be applied to a single large problem, that of costing a future manned aircraft system.

Some definitions and concepts

The fact that the term "resource analysis" can have several different meanings is worth emphasizing at the outset – and not simply because usage can vary with the user. The meaning can also shift considerably as a result of the context of the particular problem at hand, since the context will determine the concepts and techniques used in the analysis. Moreover, the meaning will often be very sensitive to the form in which the results of the analysis are presented.

Generally speaking, however, the expression "resource analysis" is taken by most analysts to mean the process of *systematically determining the economic resource impact of alternative proposals for future courses of action*. Let us take a closer look at the words "resource" and "analysis." Here, of course, our orientation is primarily toward governmental decision-making problems, although some of the basic ideas are clearly applicable elsewhere.

Instead of "resource," we can, and often do, use the word "cost." But for present purposes, "resource" is probably the more descriptive term. Particularly in an economic sense, the word "resource" immediately gets to the heart of the matter, because economic cost implies the use of resources – manpower, raw materials, and the like – and thus the cost of something is measured by the resources used to attain it. Or, more technically, the cost of attaining a certain objective at some point in time is measured by the resources that are *not* available for use in attaining *alternative* objectives because these resources are already committed to the chosen objective. This concept of cost reflects the fact that a nation's resources are limited, which in turn explains why we must make choices – often very difficult choices – about allocating available resources among competing objectives. If, through some magic, a nation's resources were in fact unlimited, then such decisions would be essentially trivial, and we would have little occasion for discussing resource analysis – or, for that matter, most of the other subjects considered in this book.

Another reason for putting the emphasis upon resources is that to many people the word "cost" implies money cost. But money cost does not necessarily mean the same thing as economic cost. While in resource analysis we most often ultimately translate physical quantities into dollars, the real objective is to measure the probable "resource drain" on the economy that would result from various possible future actions. Dollars are used merely as a convenient common denominator for aggregating numerous heterogeneous physical quantities and activities into meaningful "packages" for purposes of analysis and decision. Occasionally, however, we need to emphasize physical quantities in addition to the dollar translations. For example, a proposal for a new weapon system may call for large numbers of highly skilled personnel, substantial quantities of a rare material or chemical, or the like. The resource analyst must give such requirements special treatment to determine if it is economically feasible to obtain them in the required time period.

Let us turn now to the word "analysis." Again, an alternative word could be used – for example, "estimating." This term, however, hardly conveys the full meaning that is intended here. To many people, "making an estimate" of the cost of something implies taking a rather detailed set of specifications and "pricing them out." While our meaning of resource analysis certainly includes such a process, a much broader frame of reference is also included.

For one thing, in decisionmaking contexts involving time horizons extended far into the future, a concrete set of specifications is usually not available. There are too many major uncertainties. Not only is there a wide range of possible alternatives, but each alternative, in turn, usually has several possible configurations. Moreover, the political, technological, and economic character of the environment in which they might operate is only dimly foreseeable. This being the case, the probable resource impact of all the relevant alternatives must be determined, with the objective of finding the really significant differences in resource requirements among them. The problem is complicated still more because decisionmakers are often interested in learning how resource requirements for any given alternative might change as key configuration characteristics are varied over their relevant ranges – a "sensitivity" type of investigation. All of this implies an *analytical* type of activity rather than just "cost estimating" *per se*; hence our preference for the word "analysis."

Another reason for preferring the word is related to the fact that in dealing with future courses of action, we are very often concerned with new equipment proposals and new methods of operating such equipment. New systems typically have components that have never been produced before, and methods for using them may differ radically from past or current methods, however successful. Thus, the cost of proposed activities involving new systems cannot be determined from a readily available "catalogue" of resource requirements. Information and data on past and current equipment and operations must be obtained, and these data must be analyzed with a view to discovering *relationships* between resource requirements and the characteristics of the equipment and the key operational variables. If meaningful analytical relationships are discovered, they can then be used to determine the resource impact of proposed courses of action. Of course, such relationships must not be applied mechanically; they must be used with discretion and informed judgment. But, again, the main point is that a significant amount of *analytical* activity is required.

Context

Central to this analytical activity is an understanding of the context, or setting, of the problem. This is so because the context determines in large part the character of the analysis – the tools that will be used, the type of results, the manner in which the results will be presented, and, indeed, the limits of useful inquiry. As such, context is worth exploring from various points of view. Here, to illustrate what we mean, we shall consider only three aspects of context: (1) the time horizon, (2) the kind of decision to be made, and (3) the scope of the problem. As we shall see, these three are certainly not mutually exclusive; there are many common threads running through all of them. As we shall also see, there are good reasons for trying to keep them separate.

Time Horizon

Time horizon is undoubtedly the most important consideration. From it stem the factors that are probably more decisive than any others in selecting the concepts, methods, and specific techniques for tackling a given resource analysis problem. Let us consider, for example, two situations that illustrate the extremes of the time horizon spectrum:

Case (a), in which the analytic problem is to examine the range of alternative weapon system possibilities that might be used to perform a certain military mission some 10 to 15 years from now.

Case (b), in which the problem is to prepare the operating portion of next fiscal year's military budget.

In (a) we are looking about as far into the future as is usually feasible, and in (b) the time horizon is essentially "tomorrow." Clearly there are numerous marked differences between the two cases. Some of the more important, from a resource analysis point of view, may be summarized briefly as follows (not necessarily in order of relative importance, nor without some overlapping):

Case (a)	*Case (b)*
(1) Wide range of alternatives (both for hardware and proposed operational concepts)	(1) Few alternatives (hardware essentially "given")
(2) Great uncertainty	(2) Small degree of uncertainty
(3) Specifications and descriptions of alternatives may be sketchy; general paucity of information	(3) Detailed descriptions; relatively good information
(4) High degree of accuracy in cost estimates is not possible; emphasis must fall on treating the alternatives consistently	(4) High degree of accuracy is required, and, in general, can be attained
(5) Emphasis on comparative or relative costs, on looking for *major* differences in cost among the alternatives to do the job	(5) Emphasis on absolute values
(6) Emphasis on presenting results of resource analysis in terms of interest to the long-range planner: "end product" or mission-oriented incremental costs	(6) Emphasis on developing and presenting estimates in terms of categories that are administrative and implementation oriented
(7) Because of wide range of alternatives and high degree of uncertainty, emphasis on developing a range of estimates: "cost-sensitivity analysis"	(7) Emphasis on developing "point estimates": limited use of sensitivity analysis
(8) Emphasis on using generalized estimating relationships	(8) Emphasis on costing out a detailed set of specifications

Between such extremes fall many kinds of resource analysis problems involving various mixtures of the characteristics listed under (a) and (b).

Probably the most significant generalization one can make is that as the time horizon extends into the future, the range of possible alternatives increases and uncertainty becomes greater (at an increasing rate). The complexity of the analyst's task changes accordingly.

Type of Decision

Generally speaking, there are three types of decisions to which the resource analyst contributes, each of which can, in a sense, be understood as merely another way of expressing the time horizon:

1. *Development*: Deciding which of a wide range of future possibilities to develop for possible operational use. (Long-range time horizon.)
2. *Initiation into service* (*investment*): Deciding which of the alternatives under development to introduce into the active inventory of the future to perform a specified mission or task. (Mid-range time horizon.)
3. *Operating*: Deciding how to operate systems that are on hand in the operational inventory. (Short time horizon.)

Because these three sorts of decisions are, in effect, sequential, they overlap in part and interact. For example, operational considerations often influence investment and even development decisions. Thus, when a resource analysis is primarily concerned with a development decision, an effort will usually be made to assess the investment and operating costs of the alternatives as well.

Since each type of decision is essentially an expression of a slice of the time horizon, the characteristics listed a moment ago are also appropriate here. For example, the characteristics of Case (a) apply to development decisions, and those of Case (b) apply to operating decisions. Investment decisions fall somewhere between (a) and (b).

Scope of the Problem

"Scope of the problem" can have several meanings. Here we want to focus on one in particular: whether the problem being investigated calls for a *total force* analysis, or, more specifically, an analysis of an *individual weapon system*. While the two are in some sense related, they nevertheless pose somewhat different problems for resource analysis.

If we undertake a total force analysis, we immediately have the problem of the magnitude of the task, since a force consists of numerous weapon and support systems, as well as various "non-system" activities. We have to estimate the *time-phased* resource impact of all of them. In addition to the sheer size of this task, we also have a problem that arises because the many components of a total force are often interrelated. If the analytical methods and techniques used to determine resource impact are not designed to take account of these interactions, the results can be seriously in error. Thus, both of these factors – size and interrelations among the components – tend to force the development of a total force model which can (at least in part) be automated. Without the use of automatic data processing equipment, it would be very difficult, if not impossible, to take the interactions into account, or to compute rapidly the time-phased resource impact of *alternative* force structure proposals.

If we undertake the analysis of an individual system, the magnitude of the job is much less, although even here the workload can be substantial, since a proposed system may have many possible configurations. The "interaction" problem is still present. The fact is that the resource impact of a new individual system can vary considerably, depending upon the projected total force into which the system is assumed to be introduced. This is especially true of facilities and personnel cost. Obviously, if a system that is being introduced into the force inherits facilities and personnel from one that is being phased out, the incremental resource impact of the new system will be less than it would be without such a carry-over.

In principle, one can conclude that to assess realistically the probable incremental resource impact of a new system, the way to proceed is to (1) determine the resource impact of the total force *without* the new system; (2) determine the resource impact of the total force with the new system included; and (3) take the difference between (1) and (2). The difference represents the incremental cost of the new system. In practice, however, it is not always possible to take this approach, and less formal methods must be used. But, in any case, the "interaction" problem should not be ignored in resource analysis of individual systems.

Impact of context on concepts and techniques used in resource analysis

As we indicated earlier, the crux of the long-range military planning problem is the systematic examination of alternative system or force proposals, with a view to finding that alternative (or combination of alternatives) which seems preferable to others according to some designated criterion of choice. The resource analysis procedure required to support such a planning activity must be geared to the concepts and methods used in the over-all analytical process. But as we have just seen, the context of the problem is vitally important in determining the specific concepts and analytical techniques used in resource analysis.

Let us consider this matter in more detail by assuming that we have been asked to analyze either a total force or an individual system to support a development decision involving a time horizon of ten years. What would be the major characteristics of a resource analysis capability designed to serve such a long-range planning activity? Here, we shall list and discuss briefly some of the more important ones:

1. "*End product*" *orientation*. Since the long-range planner is typically interested in examining alternative proposals for attaining future military capabilities, the resource analysis process must be structured to express resource requirements in terms of "major programs" that are meaningful from a planning point of view. For example, a weapon or a support system must be associated, in some sense, with military capabilities.

2. *Life cycle identification*. Within the structure of an "end product" orientation, it is desirable to identify resource requirements in terms of the major "life cycle" phases of a new military capability: development, initiation into the active inventory (investment), and operation over a period of years. This type of identification is significant analytically, because very often we will want to vary the possible force size and the number of years a new system might be in the operational force. The development/investment/operation segregation facilitates such manipulations. (Figure 7.1 gives an indication of what the life cycle identification looks like when plotted against time.)

3. *Resource and functional categories*. Within the structure of (1) and (2) above, we must set up *resource* categories (equipment, facilities, manpower, etc.) or *functional* categories (maintenance, training, etc.). These categories, which help insure completeness in identifying all required resources, should be useful from a data source and computational standpoint, and from the standpoint of serving to indicate significant areas of resource impact – special requirements for equipment, manpower skills, and so on. Regardless of what particular set of categories is established (Table 7.1 provides an example), it is vitally important to define carefully what is included in each category. This is a fundamental prerequisite to

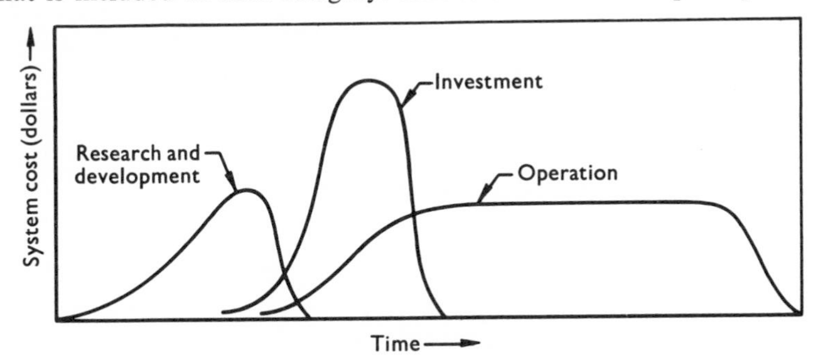

Fig. 7.1 – System "life cycle" identification plotted against time (idealized curves)

the development of estimating relationships (to be discussed later) and to help assure consistency in working out the resource impact of alternative system or force proposals.

TABLE 7.1

Resource analysis categories for individual systems

- I. *Research and development costs*
  - A. System development
  - B. System test and evaluation
  - C. Other system costs
- II. *Investment costs*
  - A. Installations
  - B. Equipment
    1. Primary mission
    2. Specialized
    3. Other
  - C. Stocks
    1. Initial stock levels
    2. Equipment spares and spare parts (initial)
  - D. Initial training
  - E. Miscellaneous
    1. Initial transportation
    2. Initial travel
    3. Intermediate and support major command
- III. *Operating costs*
  - A. Equipment and installations replacement
    1. Primary mission equipment
    2. Specialized equipment
    3. Other equipment
    4. Installations
  - B. Maintenance
    1. Primary mission equipment
    2. Specialized equipment
    3. Other equipment
    4. Installations
  - C. Pay and allowances
  - D. Training
  - E. Fuels, lubricants and propellants
    1. Primary mission equipment
    2. Other
  - F. Services and miscellaneous
    1. Transportation
    2. Travel
    3. Other (including maintenance of organizational equipment)
  - G. Intermediate and support major command operating cost (included only in exceptional cases in cost analysis of individual systems)

4. *Appropriate level of detail.* Subsidiary to point (3) is the question of the appropriate level of detail. Obviously, in a long-range planning context, the attempt to structure problems in great detail is undesirable – indeed, impossible. It is important, however, to break the problem down into

elements which will help us to distinguish the really new aspects of a system proposal from those which are not. Even the most advanced system proposals contain many elements that are not significantly new. These should be separated from those which *are* new, so that the analytical effort can be concentrated on the latter. This is a very important principle for structuring problems in resource analysis. (An example of a component structure for a ballistic missile vehicle is shown in Table 7.2.)

TABLE 7.2

Component structure for a ballistic missile

(or similar aerospace vehicle)

- *Airframe*
  - Structural
    - Leading edges
    - Body skin (including tankage)
    - Structural members (frame)
  - Sub-systems (electrical)
  - Controls (electromechanical)
- *Power Plant*
  - Liquid rocket
    - Pump drive assembly
      - Turbo-pump
      - Gas generator
    - Thrust chamber
    - Propellant lines and fittings
    - Vernier and exhaust system
    - Frame or mounting structure
    - Accessory power supply
  - Solid rocket
    - Casing
    - Nozzle
    - Propellant
- *Guidance*
  - Inertial
    - Inertial measurement unit
      - Platform
      - Accelerometers
      - Gyroscopes
    - Computers
    - Control central and associated electronics
  - Radio command
    - Decoder
    - Beacons
    - Antenna
- *Payloads*
  - Nose cone
    - Shell
    - Arming and fuzing
  - Warhead

5. *Explicit treatment of uncertainty.* Probably the most significant factor in long-range planning is uncertainty. Possible military capabilities in the distant future are subject to many uncertainties, the most important one being "configuration" or "requirements" uncertainty. Proposals for advanced systems may vary widely in such matters as hardware, system operation, and force size, and early in the development no one really knows which set of possible characteristics will ultimately prevail.[1] In resource analysis, these uncertainties must not be ignored. Among several possible ways of dealing with them, one of the most important is sensitivity analysis, which involves working out the resource impact of numerous sets of system configurations, rather than that of just a single set ("point estimate"). Sensitivity analysis is useful not only in dealing with the problem of uncertainty *per se*, but also in preliminary system design, since it is usually most helpful to have some idea as to how sensitive the total system cost is to changes in key system parameters as they are varied over their relevant ranges. In Fig. 7.2, for example, total system cost is relatively insensitive to increases in payload over the range shown; is relatively sensitive to

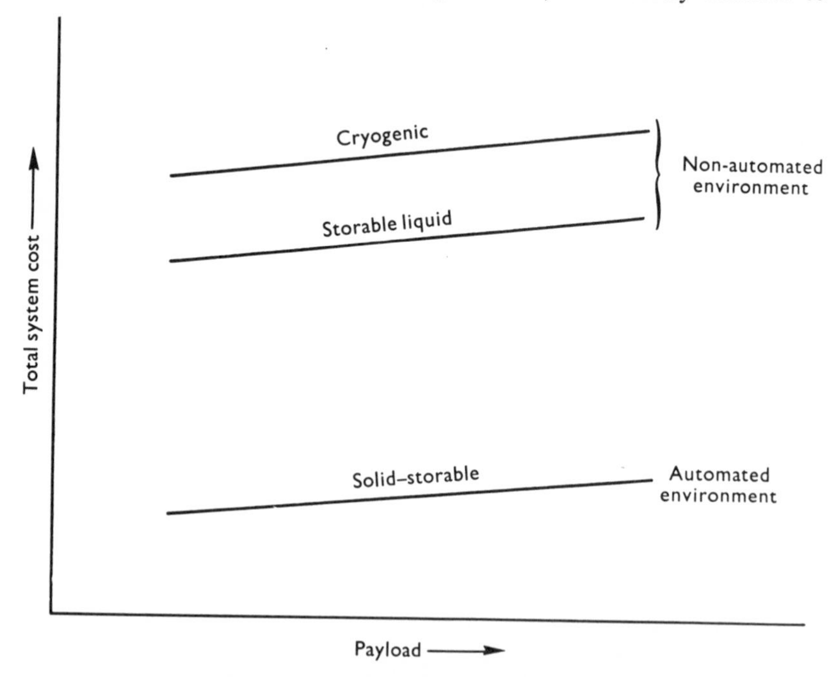

Fig. 7.2 – Missile system cost versus payload for various types of propellants and ground environments (fixed number of ready missiles)

[1] A problem discussed at length in Chapter 6.

type of propellant used; and is quite sensitive to automation of the ground environment.

6. *The principle of incremental resource impact.* In planning, it is the *incremental* or *net* resource requirements that are of interest. While it is true that "sunk costs" (reflecting resources on hand that may be used by a new proposed capability) must be taken into account in determining the economic resource impact of the new activity, these sunk costs must *not* be included in the cost for that activity. As we saw earlier, the procedure is to work out the total requirements, determine the resources that are likely to be inherited from the phase-out of other systems or activities, and subtract these amounts from the total requirements in order to arrive at the *net* resource requirements for the new system. The results are apt to be most important in the case of facilities and personnel resource categories, particularly in a "total force" analysis.

7. *Association of support activities with an "end product" package.* Related to the principle of incremental cost is the question of the *appropriate* association of support activities with end product packages – for example, weapon systems or other aggregations of activities useful to the planning process. Some people seem to suggest that the objective should be to identify *as much as possible* with "end product" activities – an accounting type of allocation or cost distribution concept. From our point of view, this is wrong. The objective should not be to associate as much as possible (often by arbitrary allocations), but rather to identify with end product packages only those support activities which are appropriate in view of the context of the problem at hand. In principle, this usually means that if a new end product activity is likely to have a significant impact on a particular support operation, the cost of that impact should be associated with the end product activity in question; otherwise, it should not be so identified. In practice – to cite an example in the case of the U.S. Air Force – the operating costs of Headquarters U.S. Air Force, the Air Academy, the Air Finance and Accounting Center, Headquarters Air Force Systems Command, Headquarters Strategic Air Command, and the like would usually not be associated with Air Force weapon systems. On the other hand, the cost of depot maintenance (in the Air Force Logistics Command) and certain course costs in the Air Training Command may be, and often are, appropriately identified with weapon systems. These principles apply to both individual system and total force resource analysis. However, it is in a total force context that we see the picture most clearly. Again taking the Air Force as an example, we find that many Air Force support activities are appropriately identifiable with systems. Others are not, but rather to a certain mission category as a whole (strategic operations, defense, and so on). Still others are not related to either missions or systems, and are

thus treated as "Air Force-wide" activities. In this latter category, we find that some activities vary with changes in the total operational force (sometimes in a discontinuous manner); others are essentially insensitive to total force size – for example, the operation of the Air Academy.

8. *The question of accuracy.* The question of accuracy in resource analysis has already been mentioned briefly. It is raised again here because it has an important bearing on the structuring of the concepts and methods of resource analysis. The key point is that in long-range planning, where uncertainty is great, a high degree of accuracy in an *absolute* sense is not attainable. This being the case, we should not waste effort in trying. Furthermore, in many of the more important long-range planning problems, comparisons among a range of alternative future courses of action are of prime interest; and in these comparisons the resource impact of the alternatives in a *relative* sense is what we want to discover. The concepts and methods in resource analysis should be oriented accordingly. This in turn means that the development and use of analytical techniques that will treat alternatives in a consistent and unbiased manner should be emphasized.

9. *Time phasing.* For many long-range planning problems, especially those involving total force analysis, explicit time phasing of resource requirements is a very important consideration. Even in individual weapon systems analyses, where the over-all analytical framework does not require explicit time-phased resource inputs, it is nevertheless often desirable to provide them. This may not only lead to better estimates, but may also provide the basis for analytical insights into the total problem that might not be readily apparent if the results are derived in a purely "static" form. In general, a resource analysis capability should provide for the generation of estimates of time-phased resource impact in terms of several "concepts of cost" – for example, obligational authority, deliveries, and expenditures. It should also provide for "equalization" ("discounting" at an appropriate rate for "time preference") of cost streams through time if the context of the problem at hand indicates that the planners are not (or should not be) indifferent about the timing of future resource impacts.

10. *Collection of information and development of estimating relationships.* To say that the results of a resource analysis are no better than the information and data that have gone into the analytical effort may seem axiomatic. Yet this *is* an important point, and we must consider the question explicitly. In fact, a really effective resource analysis capability cannot exist without systematic collection and storage of data on past, current, and projected programs. Even this is not enough. The data must be analyzed with a view to development of *estimating relationships* which may be used as a basis for determining the resource impact of future proposals. In the case of military systems, these relationships would, ideally, relate various categories

of resource impact to the system's physical characteristics, performance, and operational concept. Here are a few examples:

Initial tooling cost for turbo-jet airframes as a function of aircraft gross weight and speed.

Development cost for turbo-jet engines as a function of thrust, flight Mach number, and maximum compressor tip speed.

Ballistic missile booster cost as a function of missile weight, quantity, type of propellant, and so on. (There would be separate relationships for each of the major components of the booster.)

High-power prime radar equipment cost as a function of peak power output and antenna area.

Aircraft depot maintenance cost as a function of aircraft gross weight, speed, and activity rate.

Without an extensive and continuously updated inventory of estimating relationships, resource analysis as defined in this Chapter would be impossible. Such an inventory is particularly a prerequisite to a "sensitivity analysis" approach to the resource analysis problem.

Summary remarks

The purpose of this discussion has been to provide a basis for understanding the meaning of "resource analysis." We have pointed out that the specific meaning is heavily dependent upon the context of the particular problem at hand, and we have sketched briefly the main features of several aspects of the context.

We then attempted to outline the major characteristics of a resource analysis capability designed for use primarily in problems of long-range planning. While these characteristics are fundamental to the development and use of resource analysis, they alone will not assure good analytical studies. In the final analysis, the results will be heavily dependent upon experience, good judgment, ingenuity in creating and using analytical methods and techniques, and, above all, just plain hard work.

Chapter 8

COST-SENSITIVITY ANALYSIS: AN EXAMPLE

R. L. PETRUSCHELL
A. J. TENZER

This Chapter illustrates the application of cost-sensitivity analysis to various ways of defending the United States against a submarine-launched ballistic missile (SLBM) threat.

THE CONTEXT OF COST ANALYSIS

Cost-sensitivity analysis is one of the more interesting of the kinds of cost analysis activities carried on at RAND. As we have seen,[1] cost analyses are designed to provide information to a military decisionmaker. The analyst's frame of reference is that of the long-range planner, where "long range" means five, ten, and even fifteen years in the future. Accompanying the process, as we have also seen, is a tremendous amount of uncertainty. This uncertainty typically takes three forms, just as there are typically three steps in the planning process: (1) the uncertainty associated with the ability of the planner to predict a threat, (2) the uncertainty associated with the design of a system to counter that threat, and (3) the uncertainty associated with the ability of the cost analyst to translate the design into a statement of resource requirements. In this chapter, we will be concerned primarily with the type of analysis in which the emphasis falls on the problem of examining the resource impact of alternatives.

DEFINITION OF COST-SENSITIVITY ANALYSIS

Cost-sensitivity analysis may be defined as the process of determining how variations in the specifications of a particular system, either in design or operation, affect the requirements of that system for resources. Often the kinds of information that are presented by the cost analyst using cost-sensitivity analysis are of significant value to the system designer. Frequently, a difficult engineering design problem can be shown by the cost analyst to have little or no effect on the total resource requirements of the postulated activity and, hence, the engineer can be given guidance in allocating his effort. In providing such information, the analyst may discover that the important resource consideration is not the absolute value of the cost associated with any of the alternatives, but, rather, the relative costs

---

[1] Chapter 7.

of the alternatives and how these costs vary as the design or operational characteristics are varied over their relevant ranges.

ASSUMPTIONS OF THE EXAMPLE

This Chapter will attempt to illustrate some of the thinking that goes on in the mind of the cost analyst and some of the kinds of information that cost-sensitivity analysis allows him to present to a decisionmaker. For this purpose, we can examine the analysis of a requirement to protect the United States against the threat of a submarine-launched ballistic missile attack. Although based on an actual study, this example has been greatly simplified. Our interest here, therefore, will not be to force any conclusions from the example, but, rather, to indicate the kind of information that can be made available through cost-sensitivity analysis.

It must be made very clear that a statement of a threat, such as the one we have just named, is much less than definitive. The question of defending the continental United States against a potential submarine-launched ballistic missile (SLBM) attack does not carry with it any information as to the particular capability of the submarines or the missiles that they might launch. It does not carry with it any statement of the tactics that the submarine fleet would employ. It does not carry with it any statement of the numbers of submarines or the numbers of missiles that each submarine would carry, nor any implications about the technological capability of the United States to detect or counter such an attack. But such uncertainties are inherent in most long-range planning activities, and the cost analyst and the system designer must simply acknowledge their existence and set out to select a preferred course of action.

Figure 8.1 shows a map of the continental United States; the shaded area extends approximately 1000 miles off either shore. As an initial cut at designing a system, we will postulate that the mission is to provide defense against SLBMs (and the submarines that carry them) in this shaded area. As the study progresses, we will examine, among other things, the effect on resource requirements of either contracting or expanding this defense area.

Tactics of SLBM Interception

We want to investigate the use of manned aircraft, functioning as a continuously airborne missile-launching platform, as the central component of the system for performing this mission. The payload of the aircraft will consist of various combinations of missiles and electronics equipment. Once the radar aboard the airborne platform detects an SLBM, the air-launched missiles will intercept it during the boost phase of its trajectory (as illustrated in Fig. 8.2). Figure 8.3 presents a variation of this system. In

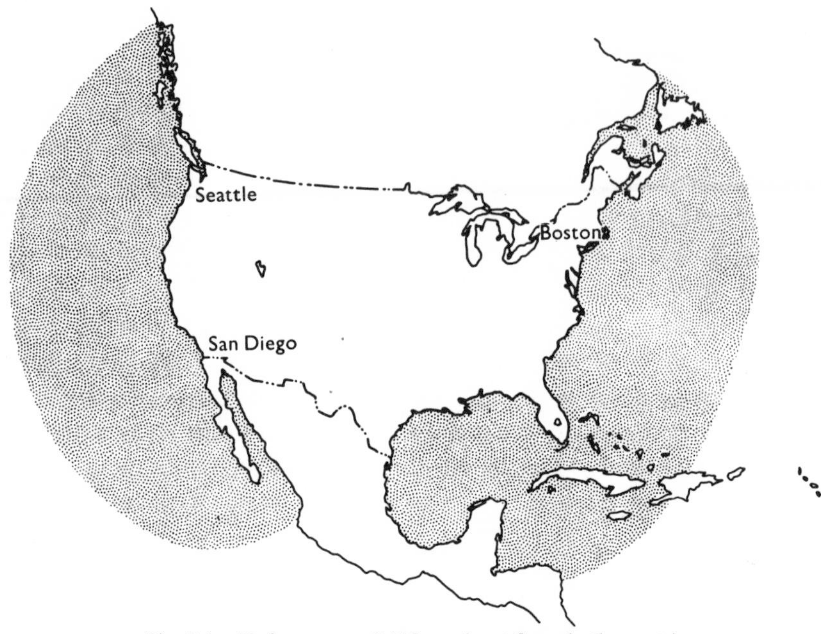

Fig. 8.1 – Defense zone (1000 n mi out from both coasts)

this case, the air-launched missile intercepts the SLBM after cut-off and during the mid-course phase of its trajectory. Figure 8.4 illustrates the addition of a counter-battery capability. Here, the fact that a missile has been launched from a submarine is observed on the aircraft, but, rather than attempting to intercept the missile itself, the aircraft directs its attack against the launching submarine itself.

A major consideration in the resource analysis of alternative systems is the deployment of their components – in this case the aircraft. There are alternative ways that aircraft can be assigned to cover a given area on a continuous basis. A "race-track" scheme might be employed, in which a succession of aircraft take off from their base, fly out over the area to be patrolled, and continue on to return to base. The pattern we will use here, however, will be to assign each aircraft a fixed area to patrol, so that it will take off, fly out to the assigned location, orbit until its fuel supply is sufficiently depleted, and then return to home base. In either case, the distance possible between the airborne aircraft is determined by the range of the surveillance radar and the speed of the antimissile missile.

The location of the bases from which the aircraft are operated is important in determining the length of time the aircraft will spend going to and from its station, and this, in conjunction with the requirement for ground

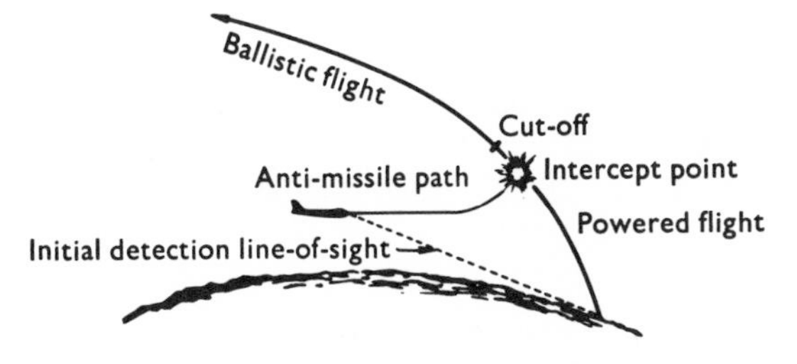

Fig. 8.2 – Boost-intercept

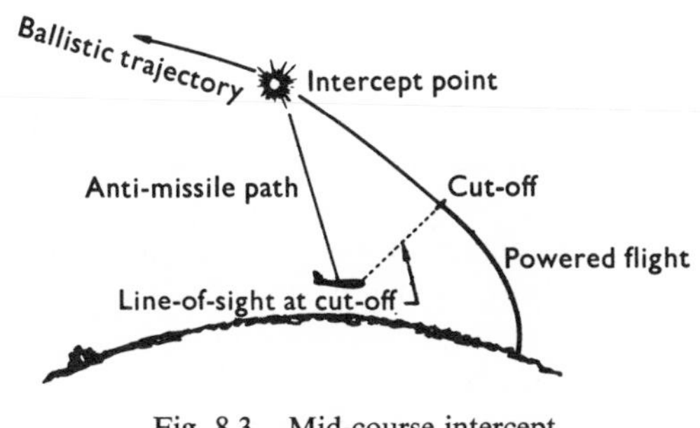

Fig. 8.3 – Mid-course intercept

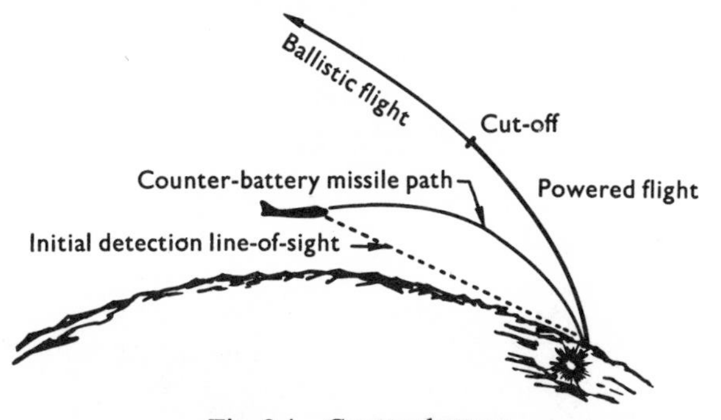

Fig. 8.4 – Counter-battery

time, will largely fix the number of aircraft required to perform the mission. The selection of the bases is also important in determining the requirement for support facilities, a resource that is often in short supply.

Construction of the model

The first step in performing a cost-sensitivity analysis is to prepare a model of the way in which the activities being considered relate to each other. Given the many interactions among the various activities that must be performed to maintain an operationally effective system of this sort, this process is often extremely complex. The model that was actually used reflected this complexity but, for the sake of the discussion here, we will assume that aircraft cycle time is the salient feature of the total model. Following a brief look at the cycle time, we can vary certain of the system's characteristics and then display the resultant estimates of the resource impact the system will have.

Effects of System Variations on Aircraft Cycle Time

Figure 8.5 shows aircraft cycle time in terms of the component activities that must be performed by or on each aircraft in the system. "Cycle time" means, of course, the time spent by an aircraft from the beginning of one mission to the beginning of the next. Part of the cycle time is spent in ground activities and travel to and from its airborne station, during which time it is not available for performance of the operational mission. Actually, only the effective time on station can be considered as a useful mission input.

As can be seen, the remainder of the airborne time is consumed in transit and depends directly on the distance from base to station and the speed of the aircraft. Ground time is spent primarily in two kinds of activities: (1) on-loading, off-loading, and general preparation of the aircraft for its sortie, and (2) performing the required maintenance necessary to make the aircraft airworthy. While we have listed airborne time and ground time separately, it should be borne in mind that they are significantly interrelated. The time required for maintenance depends on how much needs to be done and the resources available for doing it. The amount of

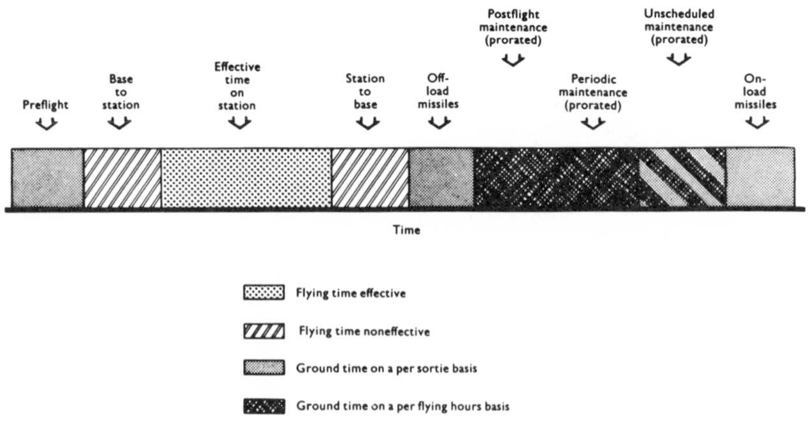

Fig. 8.5 – Aircraft cycle time

maintenance required per cycle is related both to the fact that there has been a sortie and to the number of hours flown. This would coincide, roughly, with the break between the maintenance scheduled on a per-sortie basis and that scheduled on a flying-hour basis. As far as the resources available for maintenance are concerned, such things as the number of shifts worked by the maintenance personnel are of major importance.

With this simplified model, we can begin to examine a number of aspects of the system.

Number of Aircraft vs. Endurance of the Aircraft

For one thing, we can see that the longer the effective time on station becomes in relation to total cycle time, the more efficient the system is. In this light, let us consider the effect of variations in the system parameters on this particular ratio. In Fig. 8.6 the number of aircraft required to perform the mission is plotted against the endurance of the aircraft. The curve shown represents essentially an envelope of curves for a number of different aircraft. Perhaps the most interesting aspect of this curve is that, when aircraft endurance ranges from approximately 8 to 24 hours, the number of aircraft required by the system is extremely sensitive to whatever endurance is assumed. On the other hand, as endurance goes beyond 24 hours, the net gain from additional hours diminishes. The mission of the aircraft shown in this chart is to patrol the 1000 miles off either shore; the payload carried is 50,000 lb.

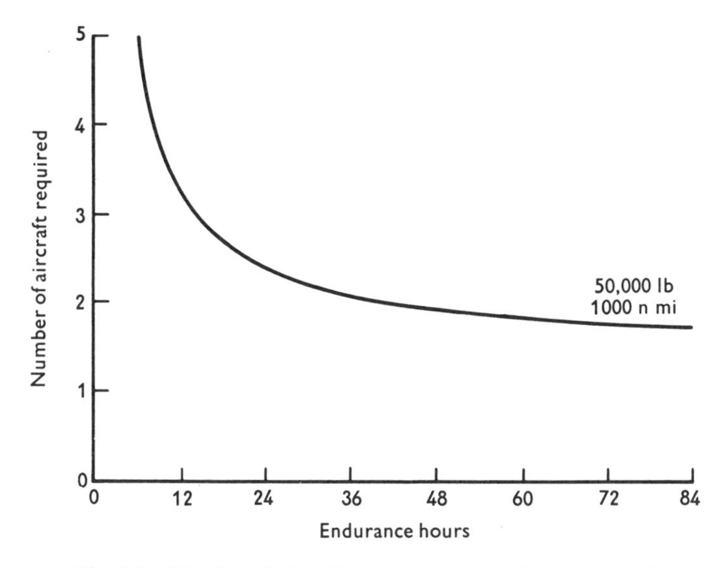

Fig. 8.6 – Number of aircraft versus endurance hours (case A)

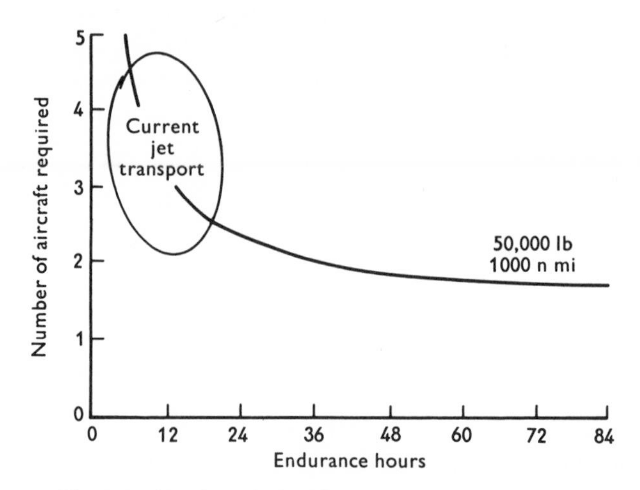

Fig. 8.7 – Number of aircraft versus endurance hours (case B)

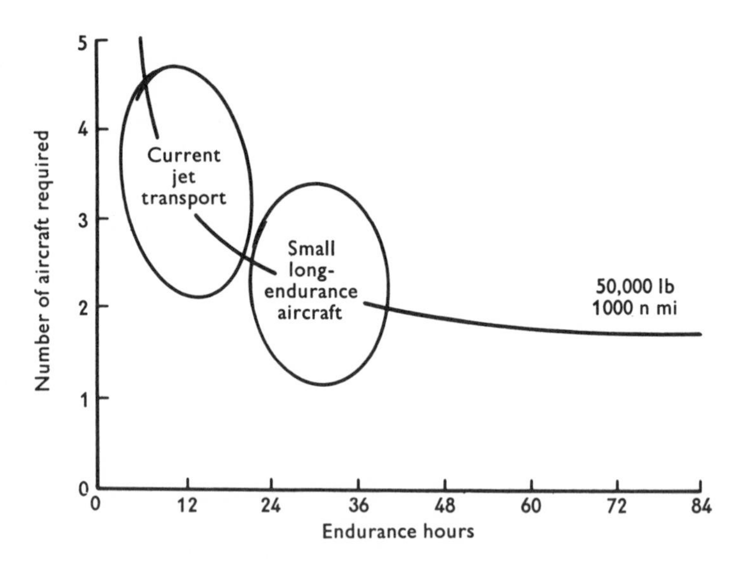

Fig. 8.8 – Number of aircraft versus endurance hours (case C)

Endurance vs. State-of-the-Art

Figure 8.7 takes the same material shown in Fig. 8.6 but relates endurance to state-of-the-art. As can be seen, our current jet transport fleet falls between the range of endurance hours from 8 to approximately 18 or 20 hours. Of particular importance is the fact that, with this capability, minimal gains in endurance again have a substantial effect on the number of aircraft required by the system.

Figure 8.8 indicates that the increment of endurance between 24 and approximately 40 hours can be achieved by going from the current jet transport to a small long-endurance aircraft (on the order of 300,000 lb gross weight). Such an aircraft is essentially within the current state-of-the-art and would involve no significant R&D effort.

Moving from about 40 hours of endurance to 60 (Fig. 8.9) would require the construction of a larger long-endurance aircraft (on the order of 600,000 lb gross weight), again one essentially within the state-of-the-art. If, however, one wished to increase the endurance beyond 60 hours (Fig. 8.10), further development, involving such matters as regenerative engines and laminar flow control, would be required.

Examining these Figures might therefore lead one to conclude that the most significant gain in the number of aircraft available for SLBM defense would be achieved by going to the small, long-endurance aircraft. Only marginal benefits would be associated with pushing the state-of-the-art beyond this point.

Notice that in the analysis we have not yet dealt directly in terms of cost. As will be seen later, however, the number of aircraft is in fact a major indicator of the total cost that would accrue to any of these systems.

Maintenance Policy vs. Number of Aircraft

Figure 8.11, which shows the number of aircraft once again as a function

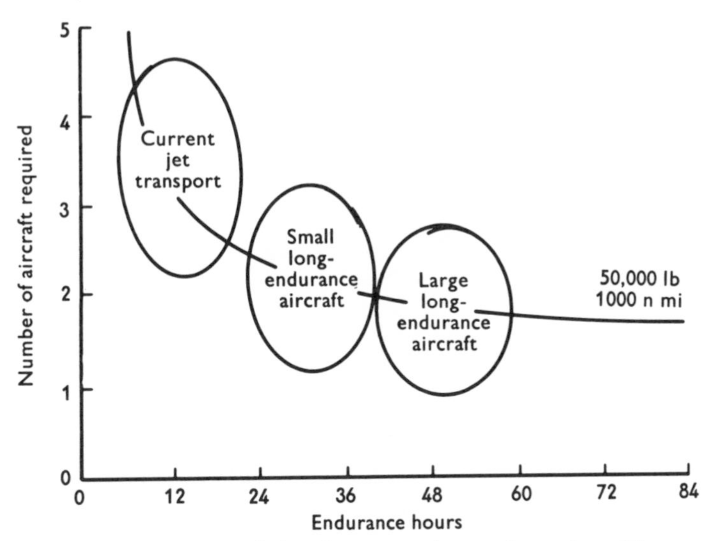

Fig. 8.9 – Number of aircraft versus endurance hours (case D)

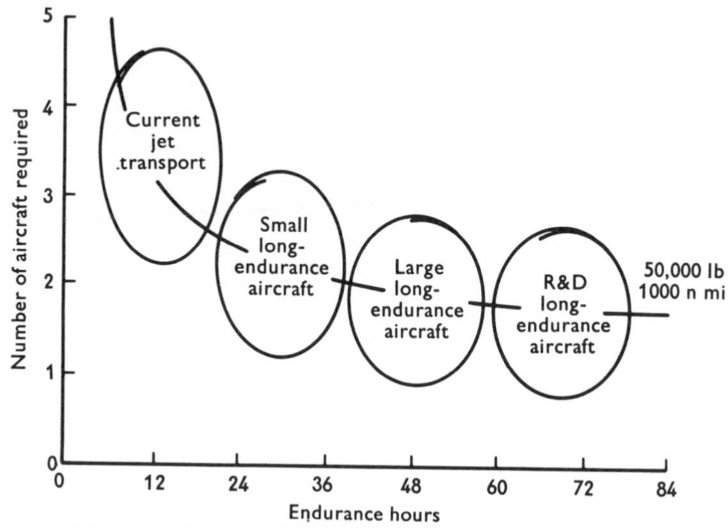

Fig. 8.10 – Number of aircraft versus endurance hours (case E)

of endurance hours, attempts to illustrate the savings that various improvements in maintenance policy can yield in reducing the number of aircraft required by the system. The curves that we have examined thus far were based on a single shift maintenance operation. In Fig. 8.11 we see the effect of adding a second and a third shift. In absolute number of aircraft required, the greatest reductions resulting from these changes occur in the area of 12 to 24 hours' endurance. When we progress to a greater endurance capability, the relative savings are essentially the same, but the absolute

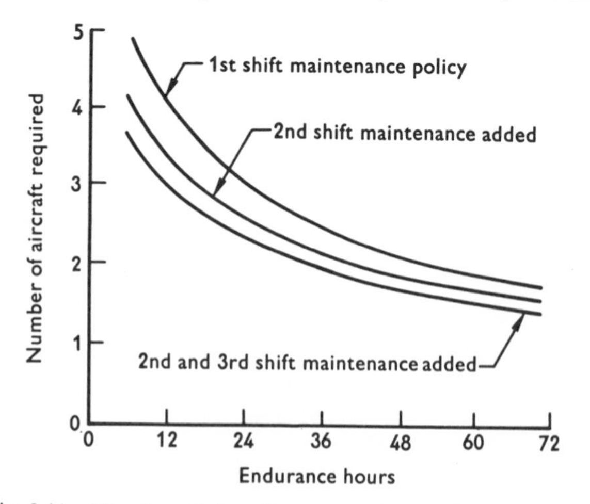

Fig. 8.11 – Number of aircraft versus endurance hours (case F)

increment of aircraft saved becomes less. As before, it appears that if we are in the position of having to rely on relatively short-endurance aircraft, the number required is extremely sensitive to some of the basic design and operational parameters.

Loading Time vs. Endurance of the Planes Airborne

Figure 8.12 shows the percentage of the fleet airborne as a function of endurance hours. Since this percentage is identical to the percentage of time out of the total cycle that each aircraft spends airborne, we have another way of getting at the number of aircraft required by the system. In this particular illustration, we have attempted to display the effect on the percentage of the fleet airborne of changes in the time required to on- and off-load the missiles. Where endurance is limited, the savings from reducing the time required to perform these loading operations are significant. As the endurance hours are increased, the absolute value of the gains accruing to the system as a result of less time spent on these tasks is much less.

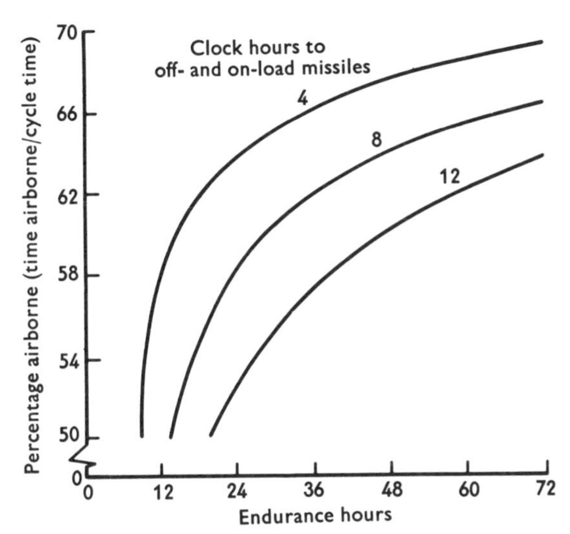

Fig. 8.12 – Percentage airborne versus endurance hours

SYSTEM COSTS

Figure 8.13 introduces the concept of system costs for the first time. Total system costs are here defined as the sum of those for research and development, the initial investment, and an arbitrarily selected five years of operation. The center curve, which relates total system costs to the endurance capability of the aircraft, is similar to the curve that we saw in Fig. 8.6. It can be seen at this point that the number of aircraft required by the system is directly related to the system's cost. Figure 8.13 attempts to suggest the effect or the potential effect on the resource requirements of the system of the uncertainty attached to the range capability of the SLBM threat. Here we have introduced a requirement, on the one hand, for extending our area of coverage to 1500 n mi and, on the other, reducing it to 500 n mi. Notice that as the area coverage is extended, the requirement for longer endurance becomes more severe. The upper curve becomes asymptotic to an endurance of approximately 10 hours, which is the same thing as saying that an aircraft with that endurance would spend all of its time going to and from the patrol area and, consequently, contribute nothing to the performance of the mission.

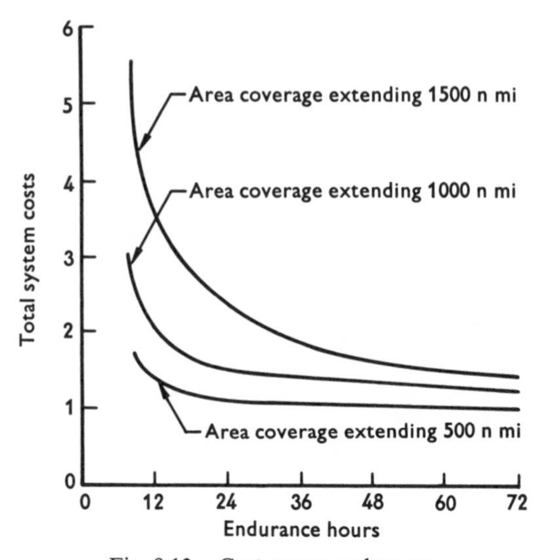

Fig. 8.13 – Cost versus endurance

If we were to suppose that the enemy, instead of randomly or uniformly distributing his forces, were to group them in an attempt to saturate the defenses, then we might have substantially different requirements for a payload in our system, assuming that it is not possible to counter such action by effective intelligence and redeployment.

Figure 8.14, which examines the total system costs as a function of the size of the defense zone, also presents the effect on the system's cost of changes that might be required in the payload. These changes are identified in this Figure in terms of single, double, and triple power. (Let us assume that these adjectives provide an accurate description of certain regular increases in offensive capability, even though their meaning is somewhat more complex.) Notice that a move by the enemy like grouping his forces

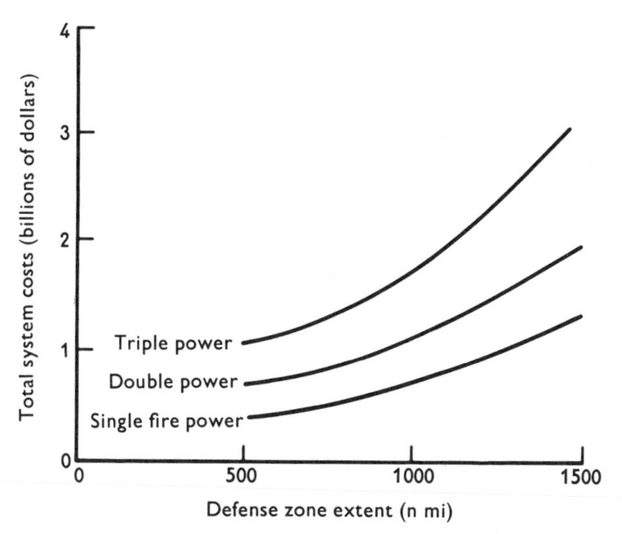

Fig. 8.14 – Cost versus defense zone extent

for a saturation attack can result in significant increases in the resource requirements of the system designed to counter it.

Figure 8.15 gives an indication of the system's costs by showing that the number of air stations required depends on the size of the defense zone, and relates the requirement to the capability of the airborne missiles that we are considering for the job of interception. The upper curve represents the capability that has been considered in the illustrations so far. The lower curve shows the reduction in the number of airborne stations that would be required if the range of the airborne interceptor missile were extended. It is clear that as the extent of the defense zone increases to upward of 1500 n mi, extending the range of the interceptor missile can result in significant savings in the number of stations.

Figure 8.16 presents the total system costs for each pound of payload on station as a function of the pounds of payload carried by each aircraft. This chart thus represents an attempt to examine the costs of each aircraft system in light of its payload capability. Notice that for the conventional jet aircraft, the costs rise rapidly on either side of a minimum cost point which occurs at something less than 50,000 lb of payload. As we go to the small, long-endurance aircraft, we find that the costs are lowest at about 75,000 or 80,000 lb of payload and much less sensitive to the particular payload than they are for the conventional jet aircraft. As we move to the large, long-endurance aircraft, the significant thing to notice is that the costs become much less sensitive to a particular loading or payload weight.

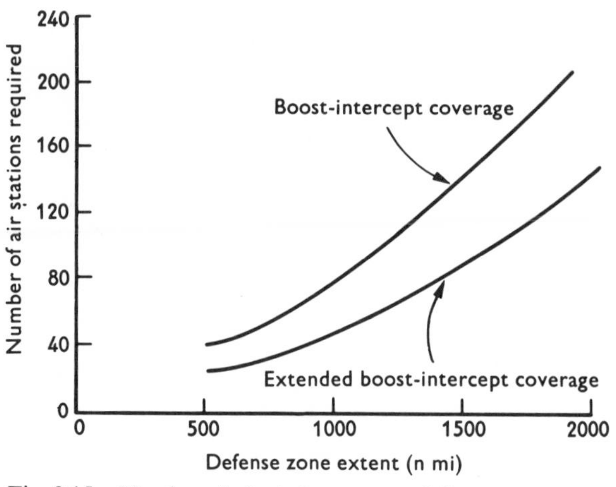

Fig. 8.15 – Number of air stations versus defense zone extent

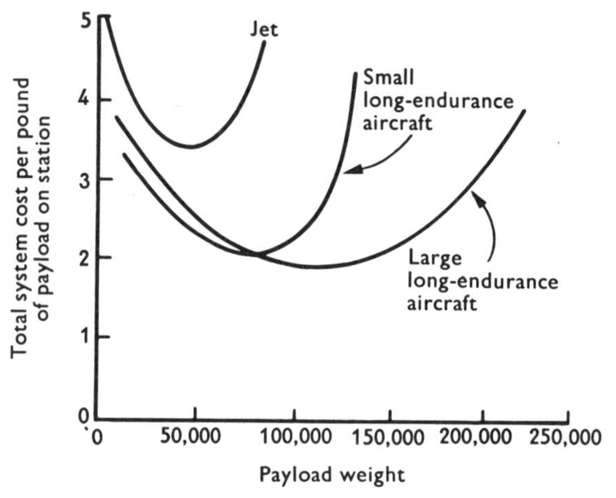

Fig. 8.16 – Cost per pound versus payload weight

Thus, a conclusion that might be drawn from this illustration is that if the size of the payload is highly uncertain, then flexibility can be achieved by going to the large, long-endurance aircraft. The cost analyst, however, would make no suggestion as a result of this analysis that either the small, long-endurance aircraft or the large should be preferred. That kind of decision can only be made after consideration of many things not dealt with in this example. On the other hand, even the kind of relative cost information that has been presented thus far, when considered together with measures of the effectiveness of each of these alternatives, might provide useful information to the ultimate decisionmaker.

In addition to this cost information, the resource analyst would generally provide the decisionmaker with some indication of the time impact of the resources necessary for the development, acquisition, and operation of each of the alternatives being considered. The military decisionmaker is not always most concerned about the total cost of particular alternatives, but, rather, about the time at which the various quantities of resources will be required. This is so primarily because the military planner, whether he likes it or not, is constrained on an annual basis through the congressional appropriation and military budget system. Figure 8.17 illustrates how time-phased costs can be presented for one of the alternatives dealt with in this analysis. Examination of this Figure should also illustrate the fact that fully operational systems do not come into being instantaneously.

USEFULNESS OF SENSITIVITY ANALYSIS

Sensitivity analysis can provide military planners with much more than dollar estimates of the costs of the alternatives that are being considered. In fact, given the uncertainty inherent in the long-range planning process, one might conclude that the absolute dollar estimates of these costs are only of minimal importance. The real benefit of the kind of analyses illustrated in

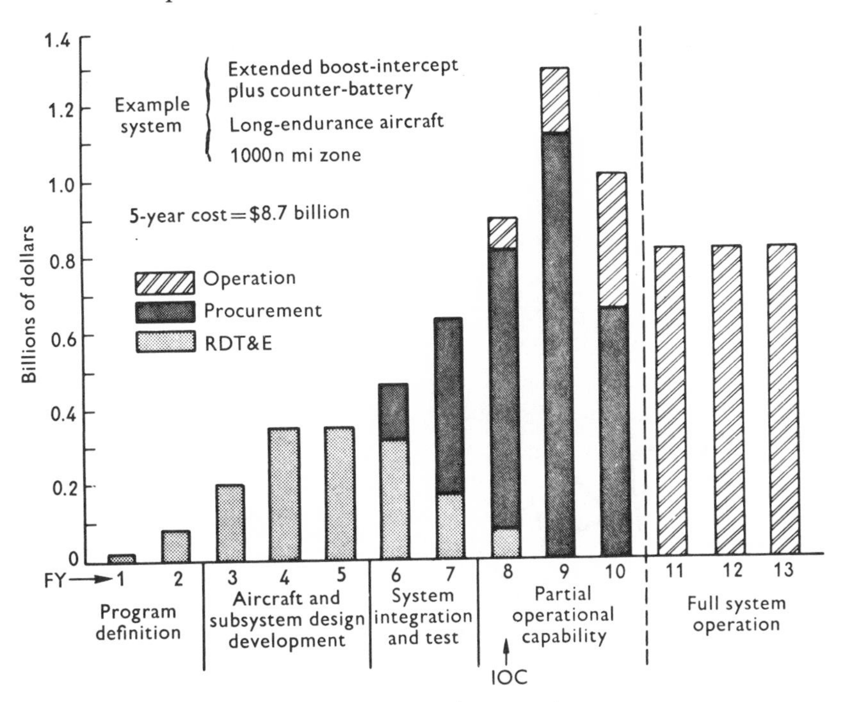

Fig. 8.17 – Time-phased resource impact

this Chapter accrues from the fact that they help to identify technical areas of potential high pay-off, either in design or in alternative operational schemes. They are also useful in dealing with the problem of uncertainty. While it is usually impossible to prepare a meaningful estimate of the absolute cost of a system, we can provide a reasonable estimate of how the costs about which we are uncertain may change as a result of changes in one or more parameters of the system. Indeed, one of the more interesting situations results when it can be pointed out that over a particular range of the values associated with a parameter, costs remain relatively insensitive, while, on the other hand, just beyond it, they may in fact become infinite.

Chapter 9

THE B-x: A HYPOTHETICAL BOMBER COST STUDY

W. E. MOOZ

This Chapter introduces a hypothetical bomber system to illustrate, step by step, the techniques used by resource analysts to cost new individual weapon systems – that is, to estimate their personnel requirements and determine the costs of their R&D, initial investment, and annual operation under various activity rates. The discussion also illustrates the role of sensitivity analysis in identifying the effects that errors in the estimating process or variations in the design or operating assumptions of such systems can have on their costs.

INTRODUCTION

G. H. Fisher discussed theoretical and practical questions that arise in any resource analysis;[1] R. L. Petruschell and A. J. Tenzer gave an example to show how cost-sensitivity analysis can help to answer some of them.[2] This chapter combines both types of analysis in a single illustration, which involves costing the resource requirements of a future weapon system – specifically, a hypothetical successor to the B-52, here called the B-x. The analysis of hypothetical systems poses special difficulties, of course, primarily because many possibilities in design and operation can be considered. The data for any of them will, at best, be uncertain; absolute accuracy will always be elusive. Intuition, caution, and informed judgment will certainly be no less necessary here than they are in analyses of other sorts of problems. For obvious reasons, however, cost analysts frequently do investigate hypothetical systems, and it is surprising how much useful information they can obtain from a few basic facts and some well-grounded assumptions. To indicate how the techniques of resource analysis and cost-sensitivity analysis provide this information is the major purpose of this chapter.

There are often several ways of approaching a problem, of isolating the question to be answered, and of answering it. Techniques will vary with cases and contexts – a point that this analysis of the B-x system should make

---

[1] Chapter 7.
[2] Chapter 8.

clear. On the other hand, many of the techniques used in the B-x study are equally valid for other systems. Where this is so, this Chapter extends the demonstration and discussion of their use beyond what might have been necessary for the bomber system alone, in order to emphasize their range and versatility. It should also be noted that, for certain problems, new techniques must be invented, or old ones refined. Costing the B-x is a case in point. The analysis led to the development of several new techniques and estimating relationships, particularly for determining the attrition of Primary Mission Equipment and the requirements for operating personnel. The discussion of these innovations is somewhat more extensive than might otherwise be necessary, so that their accuracy and rationale can be more easily examined.

If, at times, the methodology illustrated in this Chapter appears weak, or ill-defined, it is because there are parts of the analysis for which it has not been possible to gather sufficient meaningful data or develop more meaningful estimating relationships. This failing is apt to be true of any analysis, and it is important that both the analyst and the user of analysis remember that all techniques have limitations. Estimating relationships, for example, represent summations of experience; they are, therefore, severely constrained *by* experience, no matter how consistent they might be in methodology or statistical approach. If this fact is not understood, then the danger exists that an analysis of a hypothetical weapon system, such as the B-x, will be compared with an analysis of an existing system – or, worse still, one that has already been phased out of the force. Similarly, it lies in the nature of the techniques that the numbers which result from the analysis of a hypothetical system cannot appropriately be used for budget planning or any other purpose where a high degree of accuracy is important.

In any case, it is also worth noting that a real or apparent weakness in technique may or may not be significant in the final result. This is not to say that some portions of an analysis can be totally ignored, but rather that, in all things, a certain sense of proportion is important. Where a result is found to have an almost negligible effect upon the total outcome, it is not critical that the methodology used to derive it be highly developed.

This Chapter divides the analysis of the B-x into two parts. The first details, in cookbook fashion, the steps of the resource analysis. It describes the physical and operational characteristics of the B-x and then provides estimates of its personnel requirements and the costs of its R&D, initial investment, and annual operation under one activity rate. The second part illustrates the role of sensitivity analysis in identifying how the earlier cost estimates can be affected by errors in the estimating process or variations in the design or operating assumptions of the system.

I. THE B-x SYSTEM AND ITS ESTIMATED COST

INITIAL ASSUMPTIONS

Postulating the B-x as a replacement for the B-52 permits us to make certain assumptions that will both simplify the cost analysis and emphasize its comparative nature. Of the possible design configurations of the B-x vehicle, we have chosen one for which the costs of R&D and production have already been studied. This design, which incorporates variable wing geometry, provides a long-range capability and allows both high-speed flight at high altitudes and a sea-level dash at a speed close to Mach 1. There are three crew members – two pilots and one navigator-bombardier. While the types of weapons that such a bomber might carry could vary considerably, we will assume that the B-x has the same armament as the B-52. These and other characteristics of the B-x vehicle are summarized in Table 9.1.

Since no detailed information is available concerning the organizational structure of B-x units or how the system would be phased into the force, these are presumed to be consistent with present USAF practice. The same is true of B-x deployment and operation. Thus, as B-52 wings are phased out of the force, they are replaced, wing by wing, by the B-x. We assume 15 aircraft in a wing, with an additional 10 per cent in command support aircraft. No refueling squadrons are included. The organization of the new wings remains essentially the same, both administratively and functionally. The system becomes a part of SAC and each wing operates from a base within the continental United States that it has inherited from the B-52. The general operating philosophy is continued. Seven planes in each wing are on continuous ground alert. The crew has the following schedule:

Ground alert duty	130 hr/mo
Flying time (B-x)	22 hr/mo
Flying time (training and mission support aircraft)	8 hr/mo
Nonflying duty	40 hr/mo
Total	200 hr/mo

Maintenance is conducted according to standard SAC guidelines, and maintenance personnel are provided in numbers commensurate with the standard work practices for SAC bombers.

Having defined the system, the next step in the analysis is to specify the force size and the period of operation that will be considered in the cost estimate. For the B-x we will assume 10 wings and attempt to determine the costs of the first five years of their operation.

TABLE 9.1

Aircraft characteristics – the B-x

Gross take-off weight	350,000 lb
Empty weight	133,910 lb
AMPR weight	102,700 lb
Wingspan	Variable: 77 ft (swept) 145 ft (extended)
Length	182 ft
Height	31.7 ft
Engines	4
Thrust per engine, dry	16,650 lb
Thrust per engine, augmented	25,800 lb
Maximum speed at altitude	Mach 2.2 (1260 kn)
Fuel capacity	201,450 lb
Range	6300 n mi
Crew size and composition	3 (2 pilots, 1 navigator–bombardier)
Runway requirement	5000 ft to clear 50 ft
Armament	Same as B-52

The importance of being explicit about the size of the force and the time period covered by the estimate should not be minimized. Many characteristics of the system, such as the number of personnel in the force, may be simple multiples. Ten wings may require ten times as many personnel as one wing; personnel costs for five years may be five times the personnel costs for one year. But there are a few important areas where such a simple proportional relationship does not hold. One of these is the cost of the aircraft, since this cost changes with the total number purchased. Another is the number of aircraft required to replace those lost by attrition. This number changes with the degree of experience with the aircraft, and will be different for the first five years of operation than for the second five.

Even R&D costs, which are a constant for a system, may change in their relative importance to other costs as the size of the force or the length of its operating life is changed. If a small force and short operating life are chosen, then the R&D costs will appear large in comparison to the investment and operating costs. With a larger force and a longer operating period, R&D costs will diminish in comparison.

Calculation of personnel requirements

Requirements for manpower vary according to the way the weapon system is operated. Normally, however, they are calculated in four steps. The first of these entails estimating the number of operating personnel, such as those required for air crews. The second step is to estimate the number of personnel necessary to maintain, service, and repair the system under the specified conditions of operation. The third is to determine the requirements for the administrative echelon, and the fourth, those for the support organization.

This approach requires the use of an organizational chart, which will help in distinguishing the major functional divisions of the organization, as well as the smaller units in each functional area. Since we have postulated that the B-x will be the replacement for the B-52, we are probably safe in our further assumption that the functional organization of the B-x wing will be the same as that of a B-52 wing. The numbers of people may be different, but their tasks will be similar. By examining the organization of several single-wing B-52 bases, therefore, we can draw a typical organizational chart (Fig. 9.1).

Conventional methods of estimating the staffing of bomber squadrons rely heavily upon Air Force practice. This practice, as applied to strategic

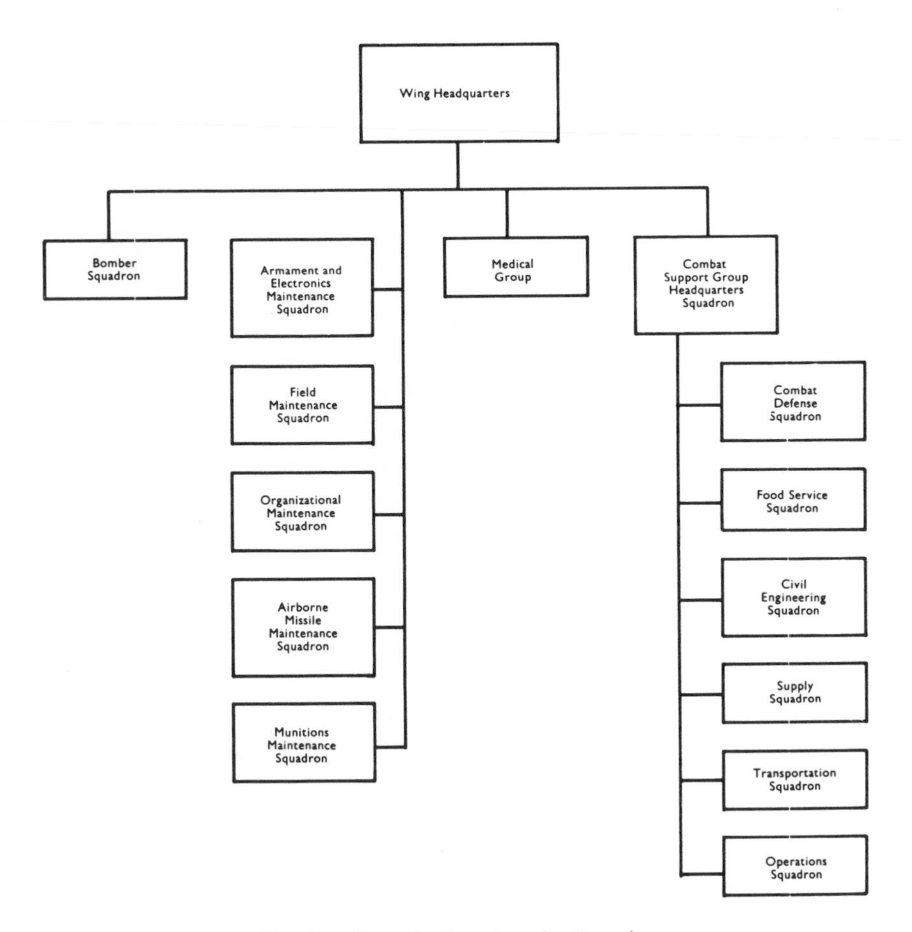

Fig. 9.1 – Organization chart for B-x wing

bomber systems, involves the use of a crew ratio, which, when multiplied by the number of planes in the squadron, yields the required number of crews. A crew ratio of 1.8 is typical. This value was no doubt derived from an analysis of the past staffing of a particular bomber system for a particular type of operation. Under the conditions of its derivation, the crews worked what the planners considered a normal work week.

But even though this crew ratio (and a number of others like it) is based on experience, and is thus one that we might use in the calculations that follow, our emphasis here is on methodology. We want to avoid ready-made answers wherever possible. Rather than use the conventional 1.8 figure, or some other less well-known ratio, we will, therefore, base our calculations of the personnel requirements of the B-x bomber squadron on the principle that there is a certain amount of work to be done and one crew can absorb only a stated amount of it. The quotient of these facts will yield the number of crews required.

Operating Personnel

By this principle, the requirements for operating personnel can be determined in a simple and straightforward manner. First, we calculate the amount of aircraft time that must be spent on ground alert duty. Since this is 24-hour duty, it entails some 730 hours each month for each aircraft:

$$\frac{24 \text{ hr} \times 365 \text{ days}}{12 \text{ mo}} = 730 \text{ hr/mo (av)}.$$

As we said earlier, seven aircraft in each wing are involved; thus, the total number of aircraft hours per month for ground alert is 7×730, or 5110.

Dividing this figure by the 130 hours per month that we assumed each crew would spend on ground alert gives us the number of crews that are required:

$$\frac{5110}{130} = 39.3 \text{ crews}.$$

Because the B-x carries two pilots and one navigator-bombardier, we can multiply by three the number of crews (which we will round to 39) to derive the requirement for crewmen:

$3 \times 39 = 117$ crewmen (or 78 pilots and 39 navigator-bombardiers).

Data on the B-52 squadron allow us to estimate the other necessary personnel in the B-x bomber squadron, and the complete staff organization then appears as follows (Table 9.2):

TABLE 9.2

Operating personnel – B-x bomber squadron

Function	Officers	Airmen	Total
Squadron Commander	1	—	1
Administrative Specialist	—	1	1
Pilots	78	—	78
Operations Officer	1	—	1
Navigator-bombardier	39	—	39
Air Operations Supervisor	—	1	1
Administrative Specialist	—	1	1
Total	119	3	122

Flying Program. The determination of the number of crews required to operate the B-x system allows us to define the flying program. As we noted earlier, each crew is scheduled to fly the B-x 22 hours per month. Since there are 15 aircraft in a wing, the flying schedule is approximately:

$$\frac{22 \text{ hr/mo} \times 39 \text{ crews}}{15 \text{ acft}} = 57 \text{ hr/mo per acft,}$$

or

$$57 \text{ hr/mo} \times 15 \text{ acft} = 855 \text{ hr/mo per wing.}$$

Maintenance Personnel

Following the organizational chart in Fig. 9.1, we can divide the maintenance personnel into five squadrons, each with a different function. The Armament and Electronics (A&E), Munitions, and Airborne Missile Maintenance Squadrons have functions that concern primarily the armament carried by the aircraft, not the aircraft itself. Because we have already specified that the armament on the B-x is the same as that carried by the B-52, we can, therefore, assume that the number of people required to staff these squadrons is also the same in both cases.

The Unit Manning Documents (UMD's) from six single-wing B-52 bases in the United States provide the figures shown in Table 9.3a for the A&E and the Munitions Maintenance Squadrons. The UMD's do not include figures for Airborne Missile Maintenance Squadrons, presumably because they were prepared before these units were activated. But such information is yielded by a review of the USAF Organizational Tables; the figures included in Table 9.3a for the Airborne Missile Maintenance Squadron are for a 15 UE wing using B-52 weapons.[3]

Estimation of personnel for the Organizational Maintenance and Field

[3] "UE" (Unit Equipment) aircraft are the operational as opposed to the command support aircraft in a wing. See p. 176.

TABLE 9.3a

Armament maintenance personnel – B-x system

Function	Officers	Airmen	Total
A&E	9	211	220
Munitions	5	61	66
Airborne Missile	3	118	121
Total	17	390	407

Maintenance Squadrons requires different treatment. We will use estimating relationships developed from an earlier analysis of data on six existing bomber organizations. This analysis showed that the number of maintenance man-hours for each flying hour is related, in both the Organizational and Field Maintenance functions, to the speed of the aircraft (see Fig. 9.2). The relations are:

Direct organizational maintenance man-hour factor
$$= 24.9 \text{ (log speed in knots)} - 55,$$

Direct field maintenance man-hour factor
$$= 47.1 \text{ (log speed in knots)} - 114.$$

The factors obtained through use of these equations must be adjusted to provide for indirect manpower and for the average working hours each month. This done, we then have the number of personnel required.

Thus, to calculate the manning for the Organizational Maintenance Squadron in the B-x system, we proceed as follows:[4]

$$\begin{aligned}\text{Factor} &= 24.9 \text{ (log 1260 kn)} - 55\\ &= 24.9\,(3.10037) - 55\\ &= 22.2\end{aligned}$$

$$\frac{22.2 \times 2 \text{ (indirect man-hour factor)} \times 855 \text{ flying hr/mo}}{140 \text{ working hr/mo}}$$
$$= 271 \text{ personnel.}$$

Since the Organizational Tables show that typical bomber Organizational Maintenance Squadrons have a complement of eight officers, we can conclude that this squadron is made up of 8 officers and 263 airmen.

The requirement for Field Maintenance personnel can be estimated similarly.

[4] AFM 26-1, *Policies, Procedures, and Criteria*, 7 September 1962, is the source of the figures used for the indirect man-hour factor and the number of working hours per month.

The completed staffing of the maintenance function, then, is as shown in Table 9.3b.

TABLE 9.3b
Total maintenance personnel – B-x system

Function	Officers	Airmen	Civilians	Total
A&E	9	211	0	220
Munitions	5	61	0	66
Airborne Missile	3	118	0	121
Organizational	8	263	0	271
Field	7	376	8	391
Total	32	1029	8	1069

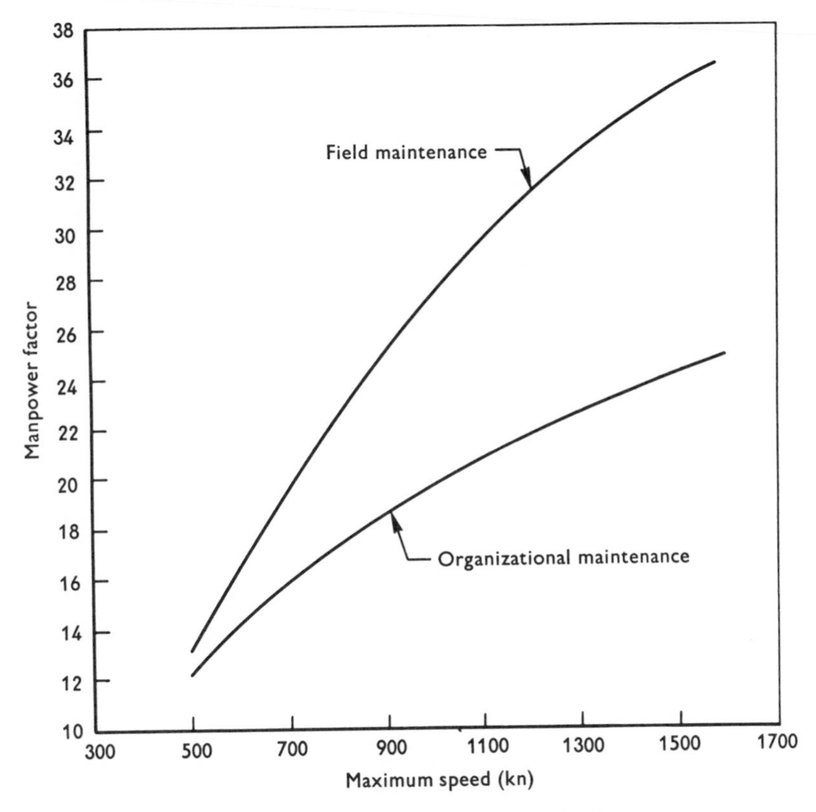

Fig. 9.2 – Estimating relationship for obtaining bomber maintenance personnel factors

While other methods of estimating the number of maintenance personnel are possible, we can have no assurance that they would give us a more credible answer. The B-x is hypothetical and differs from other aircraft which are, or have been, in the inventory. Discussions of the relative accuracy of methods of estimating maintenance personnel are, therefore, largely academic, and the analyst must simply be content to adopt a reasonable approach. There may be several of these.

At this point, the duty requirements of the aircraft and the total system personnel have been estimated. It is now necessary to determine whether or not the duty schedule will allow sufficient time for the maintenance work to be performed. We do this by calculating the monthly aircraft activity (for both ground alert and B-x flying time) and subtracting this figure from the total number of hours available in the month (that is, 730). By our earlier assumptions, we get, for ground alert time, 341 hours per month:

$$\frac{7 \text{ acft} \times 730 \text{ hr/mo}}{15 \text{ acft}} = 341 \text{ hr/mo.}$$

For B-x flying time, we get 57 hours per month:

$$\frac{39 \text{ crews} \times 22 \text{ hr flying per crew}}{15 \text{ acft}} = 57 \text{ hr/mo.}$$

Adding these two figures together, and subtracting from 730, we thus find that the time available for maintenance is 332 hours per month:

Available time	730 hr/mo
Less aircraft activity time	−398 hr/mo
Available maintenance time	332 hr/mo

Maintenance time available per sortie (assuming an 8-hr sortie) comes to 46.5 hours per month:

$$\frac{332 \text{ hr/mo} \times 8\text{-hr sortie}}{57 \text{ hr B-x flying per mo}} = 46.5 \text{ hr/mo.}$$

Since general SAC practice results in an aircraft turnaround time that is considerably shorter than this, there would seem to be more than enough time to accomplish the required maintenance.

Administrative Personnel

The administrative function depends on the number of people administered and the size and type of base, but not the characteristics of the weapon system. We can calculate the number of personnel necessary for this function through the use of an estimating relationship devised by A. J.

Tenzer, O. Hansen, and E. M. Roque.[5] This relationship, developed for the study of strategic manned bomber bases, relates the number of administrative personnel to the number of operating and maintenance personnel by the equation $Y=189+0.1203X$, where Y=the number of administrative personnel and X=the sum of the operating and maintenance personnel. Recalling that a B-x wing will require 122 people for operations and 1069 for maintenance, and applying the equation, we find that the number of administrative personnel required will be 332:

$$\begin{aligned} Y &= 189+0.1203(1191) \\ &= 332. \end{aligned}$$

Typical UMD's for strategic manned bomber bases tell us that 22 per cent of the administrative people will be officers, 75 per cent airmen, and 3 per cent civilians. Thus, our estimate for administrative personnel breaks down as follows:

Administrative Personnel

Officers	*Airmen*	*Civilians*	*Total*
73	249	10	332

Support Personnel

The services of support personnel include feeding, housing, educating, and training the base personnel; providing transportation and supplies; and protecting the base and its weapon systems. The number of support personnel is a function of the number of so-called "direct personnel" in the weapon system, that is, all of the operating, maintenance, and administrative personnel. Our estimates put this number at 1523 (see Table 9.4).

TABLE 9.4

Direct personnel – B-x system

Function	Officers	Airmen	Civilians	Total
Operations	119	3	0	122
Maintenance	32	1029	8	1069
Administration	73	249	10	332
Total	224	1281	18	1523

[5] A. J. Tenzer, O. Hansen, and E. M. Roque, *Relationships for Estimating USAF Administrative and Support Manpower Requirements*, The RAND Corporation, RM-4366-PR, January 1965. The equation used in this Chapter applies only to SAC aircraft systems located on SAC aircraft bases. A different equation would be required were the system to be located on a SAC missile base. Many other conditions can exist, and most are discussed in the reference.

On the basis of an analysis of data from 19 SAC aircraft bases, RM-4366-PR[6] defines the relationship between support personnel and direct personnel as follows:

$$\text{Support personnel} = 1089 + 0.3068 \text{ (direct personnel)}.$$

The support personnel requirement for a B-x base, then, will be as follows:

$$\begin{aligned}\text{Support personnel} &= 1089 + 0.3068(1523)\\ &= 1556.\end{aligned}$$

Applying ratios for officers, airmen, and civilians from typical SAC Unit Manning Documents for support personnel produces the breakdown shown in Table 9.5.

TABLE 9.5

Support personnel – B-x system

Squadron	%	Officers	Airmen	Civilians	Total
Combat support	15.2	23	154	59	236
Combat defense	15.0	6	228	0	234
Food service	6.2	2	96	0	98
Civil engineering	22.1	3	233	108	344
Supply	16.8	6	204	51	261
Transportation	7.6	4	100	14	118
Operations	6.7	11	88	5	104
Medical group	10.4	32	106	23	161
Total	100.0	87	1209	260	1556

Summary

Total requirements for base personnel, of all categories, may be summed up as in Table 9.6.

TABLE 9.6

Total personnel requirements – B-x system

Type	Officers	Airmen	Civilians	Total
Operations	119	3	0	122
Maintenance	32	1029	8	1069
Administration	73	249	10	332
Support	87	1209	260	1556
Total	311	2490	278	3079

At this point, the weapon system and its operation have been almost completely defined. We have established what the aircraft is and what type

[6] Tenzer *et al.*

of duty will be required of the system. We also have fixed the type of organization and derived the numbers of people required to staff such an organization. It is now necessary to determine the nature and scope of the costs to be estimated. As we said before, these will be the costs of procuring ten wings of B-x aircraft and operating them for five years.

MAJOR COST CATEGORIES

In the analysis of an individual weapon system, the costs are divided into three major categories: research, development, test, and evaluation costs; initial investment costs; and annual operating costs. *RDT&E costs* represent all outlays necessary to bring a weapon system into readiness for introduction into the active inventory. These are costs which are encountered only once, and are not related to the size of the force to be procured. *Initial investment costs* are those one-time outlays required to introduce a new capability into the operational force. These occur mainly after the RDT&E phase; and, while they are one-time expenditures, they are a function of the size of the force. *Annual operating costs* are those outlays required on a recurring basis to keep the system in operation. They are also proportional to the force size.

Chronologically, the three sorts of costs occur in the order just given, although they overlap to some extent. For the purposes of analysis, however, it is not necessary to treat them in the same order, and in some cases it is impossible. For example, fuel storage facilities on a base are a function of the fuel consumption of the weapon systems that use the base. Consequently, fuel consumption – an annual operating cost – must be calculated prior to estimating the required fuel storage facilities, which are in the category of initial investment. Analysts must therefore sometimes ignore the apparent chronology of real life. We will occasionally find it necessary to do so in what follows. But, in general, the organization of our cost analysis will be to consider RDT&E costs first, and then those of initial investment and annual operation.

Table 9.7 presents a listing of all the subcategories of costs typically investigated under RDT&E, initial investment, and annual operation. The B-x is no exception. We shall want to prepare an estimate for each element appearing in this Table. And, of course, we shall also want to outline the procedures used in preparing these estimates.

Several cost elements appear in more than one category. For example, aircraft costs appear in the category of RDT&E to cover the aircraft used for testing, in the category of initial investment to cover the actual procurement of the weapon system, and in the category of operating costs to cover the cost of aircraft procured as replacements for those destroyed in use. Now, the typical aircraft dollar cost estimate is a cost function of the

TABLE 9.7
Cost categories for aircraft

RDT&E Costs	Initial Investment Costs	Annual Operation Costs
Design and Development	Facilities	Facilities Replacement and Maintenance
Airframe	Primary Mission Equipment	PME Replacement
Initial Engineering	Airframe	Airframe
Development Support	Manufacturing Labor	Manufacturing Labor
Initial Tooling	Manufacturing Materials	Manufacturing Materials
Engines	Sustaining and Rate Tooling	Sustaining and Rate Tooling
Avionics	Sustaining Engineering	Sustaining Engineering
System Test	Other	Other
Flight Test Vehicle Production	Engines	Engines
Airframe	Avionics	Avionics
Manufacturing Labor	Unit Support Aircraft	PME Maintenance
Manufacturing Materials	AGE	PME POL
Sustaining and Rate Tooling	Miscellaneous Equipment	Unit Support Aircraft POL and Maintenance
Sustaining Engineering	Stocks	AGE Replacement and Maintenance
Other	Spares	Personnel Pay and Allowances
Engines	Personnel Training	Personnel Replacement Training
Avionics	Initial Travel	Annual Travel
Flight Test Operations	Initial Transportation	Annual Transportation
Flight Test Support		Annual Services

number of aircraft produced. Graphed and properly used, this function provides the basis for the cost inputs for RDT&E, investment, and operating costs. The level of detail required to prepare each of these estimates will obviously be the same, since all the aircraft are manufactured as part of a regular production run, and are identified within a particular estimating category only after production. It is therefore necessary to estimate aircraft costs only once. Although Table 9.7 repeats the aircraft cost element three times, this is done merely to make the point that the estimating level is the same. In practice, the aircraft cost curve is generated and then used to obtain the proper figures for the various categories.

CALCULATION OF RDT&E COSTS FOR THE B-X SYSTEM

RDT&E costs are separated into two subcategories: (1) Design and Development and (2) System Test.

Design and Development

The cost of the design and development of the airframe includes initial engineering, development support (that is, manufacturing labor and materials used to support the engineering function), and initial tooling. The design and development costs for the engines include the test equipment, prototypes, mock-ups, facilities, and the scientific and engineering manpower required to accomplish preflight rating and qualifications. The design and development costs for avionics are projected in the same level of detail as those for the engines.

Airframe. Design and development for airframes is projected in the following classes: Initial Engineering, Development Support, and Initial Tooling.

(*a*) *Initial Engineering*. The general equation for determining the initial engineering hours is:

$$\begin{aligned}\text{Log engineering hours} &= 0.90462 + 0.54716\,(\log \text{kn}) \\ &\quad + 0.8800\,(\log \text{thrust}).\end{aligned}$$

Substituting the appropriate data from Table 9.1, we derive the following:

$$\begin{aligned}\text{Log engineering hours} &= 0.90462 + 0.54716\,(\log 1260\ \text{kn}) \\ &\quad + 0.8800\,(\log 103{,}200\ \text{lb}) \\ &= 7.01117.\end{aligned}$$

Hence, the B-x will require 10,260,000 initial engineering hours. From a sample of aircraft, we can conclude that the average cost per engineering hour will be \$10.50 (in 1962 dollars).[7] The RDT&E engineering cost, therefore, will be:

[7] This dollar rate was computed from the most recent contracts available. The size of the aircraft sample varied from 15 to 25, depending on the availability of data. In this case, some 20 aircraft were studied.

$$\text{Engineering cost} = 10{,}260{,}000 \text{ hr} \times \$10.50/\text{hr} = \$107{,}730{,}000.$$

(*b*) *Development Support*. Development support costs, which include items like the static test vehicle, mock-ups, test parts, and the labor and materials costs in support of the engineering effort, are estimated as a function of the initial engineering hours. On the basis of roughly the same aircraft sample, and again using the 1962 rate, the hourly cost of this support can be calculated at $16.00.
Thus:

$$\text{Development support} = 10{,}260{,}000 \text{ eng hr} \times \$16.00/\text{hr} = \$164{,}160{,}000.$$

(*c*) *Initial Tooling*. An estimating relationship for projecting initial tooling hours is:

$$\text{Log initial tooling} = 2.79589 + 0.6637 (\log \text{kn}) + 0.46715 \{\log [(\text{gross wt} \times \text{max thrust}) 10^{-6}]\}.$$

Substituting the appropriate information from Table 9.1, we derive the following:

$$\text{Log initial tooling} = 2.79589 + 0.6637 (3.10037) + 0.46715 \{\log [(350{,}000 \text{ lb} \times 103{,}200 \text{ lb}) 10^{-6}]\} = 6.98872.$$

The B-x will thus require 9,744,000 initial tooling hours. At a 1962 rate of $9.15 per hour,[8] we have, therefore, this cost estimate:

$$\text{Initial tooling cost} = 9{,}744{,}000 \text{ hr} \times \$9.15/\text{hr} = \$89{,}157{,}600.$$

Engines. Engine costs can be calculated by using regression analysis to develop estimating relationships from data on earlier turbojet engines with afterburners. An examination of a variety of parameters indicates that the best correlation is between cost and engine thrust. Since experience suggests that RDT&E monies continue to be spent as more engines are produced, a number of equations are possible. The cost equations for 100, 1000, and 2000 engines are as follows:

$$\text{Log } y_{100} = 5.43655 + 0.71055 (\log \text{thrust}),$$
$$\text{Log } y_{1000} = 5.51180 + 0.71055 (\log \text{thrust}),$$
$$\text{Log } y_{2000} = 5.56035 + 0.71055 (\log \text{thrust}).$$

Assuming a planned production of, say, 1000 engines for the B-x, the R&D engine cost can be determined directly, as follows:

[8] This rate includes tooling material.

$$\text{Log } y_{1000} = 5.51180 + 0.71055 (\log 25{,}800 \text{ lb})$$
$$= 8.64646.$$

Thus, the initial engine cost will be $443,006,000.

Avionics. Mathematic relationships for cost-estimating the development of advanced avionics have not reached general use at this time. Devising them entails reviewing the cost of existing avionic systems, ascertaining the differences between the existing systems and the advanced ones, estimating the costs of development for individual components, and then summing these figures and adding 20 per cent for integrating the new components into the new airframe. For the B-x, we are able to escape this task, since it uses the same avionics as the B-52. RDT&E is allotted solely for integrating B-52 equipment into the B-x.

System Test

System Test includes the costs for the test vehicles; such support costs as those for test vehicle spares, ground support equipment, mobile training units, test facilities, data reduction and analysis, maintenance, and supply; and such operations costs as those for fuel, ordnance, and the engineering, tooling, manufacturing labor, material, and personnel required for the flight test program.

Flight Test Vehicle Production. The number of test vehicles to be used for the system test, and therefore chargeable to the RDT&E portion of the cost of the system, is determined by how much the development of the B-x pushes the state-of-the-art. Since, however, the B-x system obviously benefits from the B-58 development in its speed and size, and from the F-111 development in its variable wing geometry, and since no new avionics are necessary, we will assume that only 10 vehicles need to be charged against the system test. If the system were to put greater demands on the state-of-the-art, considerably more vehicles might be necessary in the test inventory.

As explained previously, the cost of the aircraft used as test vehicles in the RDT&E program is obtained from a cost-quantity relationship. The creation of this relationship is the next step in the estimating process. The level of detail used in this estimate requires the division of the aircraft production costs into discrete elements which lend themselves to consistent treatment by the use of estimating relationships. These elements are as follows:

Airframe manufacturing labor

Airframe manufacturing materials

Airframe sustaining and rate tooling

Airframe sustaining engineering
Engines
Avionics.

The estimation of the cost of each element is a function of the number of aircraft to be produced. This is because of the effect of learning upon production costs. Because the cost-quantity relation for each of the elements appears as a straight line when plotted on log-log coordinates, it is necessary to know only the *slope* of the function and one *point* on the curve in order to establish the cost-quantity relation over the entire production range. We shall see that this is so as we go through the estimating process, and the fact that various elements will be estimated at different quantity levels should not be disturbing. An estimate at *any* quantity level will suffice, since it serves merely to locate the position of the curve. Knowledge of the slope of the curve allows it to be projected for all quantities. When the curves for each of the elements listed above have been developed, they are summed to obtain the cost-quantity relation for the aircraft. It should be noted that since the slopes of the curves for the various elements may not be identical, the shape of the curve for the total aircraft costs may not be a straight line. As a result, if one desires to extrapolate the curve, he must first extrapolate the individual element curves and then sum them in order to obtain the correct position of the curve for the total aircraft costs.

At this point, we will estimate the aircraft cost-quantity curve. After this is completed, we will extract the cost of the ten flight test vehicles from it.

(*a*) *Airframe.* Of the four elements we must consider under this heading, we will estimate the costs of *manufacturing labor* first. These costs at the 100th unit of the B-x airframe, including the costs of quality control and design changes, are calculated in two steps.[9] We first determine the number of manufacturing hours:

$$\begin{aligned}\text{Log manufacturing hours} &= 0.16314 + 0.73672\,(\text{log gross wt})\\ &\quad + 0.43113\,(\text{log kn})\\ &= 5.58274.\end{aligned}$$

Thus, the manufacturing labor estimate is 383,900 hours at unit 100. The next step is to multiply this figure against three others: the cost per hour; the percentage of the direct hours spent in making design changes; and the percentage of the direct hours, plus the hours for design changes, spent

---

[9] The unit cost, if multiplied by a factor of 1.634, will yield the *cumulative* average cost at 100. A list of factors for making this sort of conversion is contained in RM-2786-PR, *Cost-Quantity Calculator*, by J. W. Noah and R. W. Smith, The RAND Corporation, January 1962.

for quality control. From an analysis of a sample of 27 aircraft, we know that the figures for these last three items are \$8.26 (1962 dollars), 11 per cent, and 14 per cent, respectively. Accordingly, we do the following:

$$\begin{aligned}\text{Manufacturing cost of}\\ \text{100th unit} &= (\text{mfg hr})(\$/\text{hr})\begin{pmatrix}\%\text{ design}\\ \text{changes}\end{pmatrix}\begin{pmatrix}\%\text{ quality}\\ \text{control}\end{pmatrix}\\ &= (383{,}900)(\$8.26)(1.11)(1.14)\\ &= \$4{,}012{,}600.\end{aligned}$$

In general, the manufacturing cost at any particular quantity X can be determined by the equation Cost $= KaX^b$, where $K = 6.762$ factor for 75 per cent reduction curve, $a =$ manufacturing cost, $X =$ the quantity, and $b = -0.415$ slope for 75 per cent reduction curve.

The second of the four elements under airframe costs is *manufacturing materials*. While general estimating relationships for determining the cost of materials are incomplete, these costs have been aggregated for a number of programs, and the interim results of this research suggest that a reasonable estimate of the cumulative average cost of the B-x at unit 100 would be \$1,850,000. Costs for other quantities can be calculated by using an 89 per cent log linear unit reduction curve.

The third element is *sustaining and rate tooling*, which includes the costs of maintenance and increased production rates. These costs are estimated by using a cumulative slope of 0.138 and a factor (F_n) as a function of the rate of aircraft production per month (R_n). In the following example, where we wish to calculate the cumulative sustaining tooling costs for 100 aircraft produced at the rate of six per month, the exponent 0.4 is a constant developed through the analysis of data on varying production rates.[10] Our first step will be to calculate F_n:

$$\begin{aligned}F_n &= R_n^{0.4}\\ &= 6^{0.4}\\ &= 2.048.\end{aligned}$$

This factor we use to solve the following equation, where "Initial Tooling" refers to RDT&E costs we found earlier, $N =$ the number of aircraft to be produced, and $0.138 =$ the cumulative slope.

$$\begin{aligned}\text{Sustaining tooling} &= (\text{Initial Tooling})(F_n N^{0.138} - 1)\\ &= \$89{,}157{,}600\,(2.048 \times 100^{0.138} - 1)\\ &= \$255{,}526{,}000, \text{ or } \$2.56 \text{ million per aircraft.}\end{aligned}$$

[10] R will vary with time and the total number of aircraft to be produced. The rate used in calculating F_n is illustrative only.

The last element under airframe costs is *sustaining engineering*. The number of hours required for sustaining engineering is estimated by using the initial RDT&E engineering hours (which we calculated earlier at 10,260,000) and a cumulative slope with a value of 0.200. The general equation and its specific application to the B-x system are as follows,[11] where N = number of aircraft produced:

Cumulative average sustaining engineering hours

$$= [(\text{RDT\&E eng hr}) \times N^{0.200} - (\text{RDT\&E eng hr})]/N$$
$$= [(10{,}260{,}000) \times 100^{0.200} - (10{,}260{,}000)]/100.$$
$$= 154{,}926.$$

Recalling that the average cost per engineering hour is \$10.50, we can quickly determine the over-all cost:

Cumulative average sustaining engineering cost

$$= 154{,}926 \text{ hr} \times \$10.50/\text{hr}$$
$$= \$1{,}626{,}723.$$

(*b*) *Engines*. The production cost of the 2000th engine, on a cumulative average curve with a 90 per cent slope, is given directly by the following equation:

$$\text{Log } y = 1.67795 + 0.87255 (\log \text{thrust}).$$

(*c*) *Avionics*. The prime reference that we will use to estimate the production costs of the avionics for the B-x system is T.O.-00-25-30, *Unit Costs of Aircraft, Guided Missiles, and Engines*. If, on the basis of data in this reference, we take the B-52 costs shown for "Electronics" and "Others, including armament" (subtracting the missile costs indicated), and add \$100,000 for flight electronics, we get a figure of \$700,000 per unit for the first 100 units. Calculated fully, the total costs of avionics production for 10 wings are approximately \$100,000,000.

Having calculated the separate costs of the elements of the B-x test vehicles, we may now plot these costs on log-log coordinates and sum them into the curve representing total aircraft costs. These results are illustrated in Fig. 9.4 (p. 177) and Table 9.8.

From Fig. 9.4, we may extract the average cost of each of the test vehicles, the first ten aircraft to be manufactured. This figure is \$30.7 million, giving us a total of \$307 million for the ten vehicles.

Flight Test Operations. The costs of flight test operation have been aggregated for a number of programs, including fighter, bomber, and cargo

---

[11] This is a single-point illustration. Additional calculations must be made to plot the type of curve shown later in Fig. 9.4.

TABLE 9.8

Cumulative average production costs – B-x system
(In $ million)

Cost Element	Quantity				
	10	50	100	200	400
Sustaining engineering	6.25	2.58	1.63	1.02	0.63
Sustaining and rate tooling	3.38	2.98	2.56	1.48	0.85
Manufacturing labor	14.78	8.34	6.56	4.86	3.68
Material	2.63	2.06	1.85	1.65	1.47
Engines	2.62	2.05	1.85	1.67	1.51
Avionics	1.00	0.78	0.70	0.62	0.56
Total	30.66	18.79	15.15	11.30	8.70

aircraft, but no one has yet succeeded in relating these costs to aircraft performance or hours of test. We can find some general guidance in past experience, however. Thus, while the B-x is a large supersonic bomber, its test requirements probably fall somewhere between those of the B-58 and the B-52, both of which required a larger fleet of test vehicles. Since their operating costs were between $105 and $185 million, we can make a rough estimate that $150 million will be necessary for the B-x.

Flight Test Support. A review of previous system tests indicates that support costs will average approximately 25 per cent of the cost of the test vehicles. Thus, for the B-x system,

$$\text{Support} = 0.25 \times \$307{,}000{,}000$$
$$= \$77{,}000{,}000.$$

MAJOR COST CATEGORIES FOR INITIAL INVESTMENT

We turn now to the matter of estimating the costs of initial investment, the second of the three main divisions of any cost analysis of a new individual weapon system. For an aircraft system, these costs are usually divided among ten major categories: facilities, primary mission equipment (PME), unit support aircraft, aerospace ground equipment (AGE), miscellaneous equipment, stocks, spares, personnel training, initial travel, and initial transportation. We shall take up these categories in the order given, characterizing them briefly and then providing an estimate of the cost the B-x would involve in each.

CALCULATION OF INITIAL INVESTMENT COSTS

In determining these costs, we benefit again from the fact that the B-x is assumed to be the replacement for the B-52. Indeed, as we shall see in the course of this discussion, the costs for two of the ten categories remain essentially what they were for the B-52, and in several other cases they change only in part.

Facilities

To determine whether or not additional investment in facilities is necessary at the bases the B-x will inherit from the B-52, we must compare what the B-x requires with what already exists on these bases. We need to discover if enough land is available, if utilities and ground improvements are required, and if there are adequate facilities for fuel storage, operations and training, maintenance, supply, medical care, administration, housing, and community activities.

Using AFM 86–4, *Standard Facility Requirements*, as a guide, we find that, since the B-x is smaller than the B-52, requires less runway, has the same armament, and entails only a small increase in personnel, the B-52 bases are generally quite satisfactory. The one significant exception concerns fuel storage. An examination of the *Air Force Inventory of Military Real Property* indicates that the storage capacity of a typical single-wing B-52 base is 2.54 million gallons. In light of the fuel consumption and flying program scheduled for the B-x, this capacity will fall some 1.95 million gallons short of providing for the 60-day fuel reserve specified in AFM 172–3, *Peacetime Planning Factors*. We can demonstrate this fact, and calculate the cost of building the necessary additional storage facilities, as follows. First, we need to determine the annual fuel consumption. We saw earlier[12] that the monthly flying schedule for each B-x wing is 855 hours; we will show later[13] that the cost for fuel is about \$266 per flying hour, assuming a fuel cost of \$0.10 a gallon. Thus,

$$\text{Annual fuel consumption} = \frac{\$266/\text{hr} \times 855\ \text{hr/mo} \times 12\ \text{mo}}{\$0.10/\text{gal}}$$
$$= 27.3\ \text{million gal.}$$

The following calculation gives us the gallons of storage required:

$$\text{Gallons of required storage} = 27.3\left(\frac{60\text{-day reserve}}{365\ \text{days}}\right)$$
$$= 4.49\ \text{million gal.}$$

The difference between this 4.49 million figure and the corresponding 2.54 million figure for the typical B-52 base is 1.95 million. According to data provided in AFP 88–008–1, *USAF Construction Pricing Guide*, and

[12] P. 159.
[13] P. 184.

shown in Fig. 9.3, it can be estimated that the cost per gallon of underground storage will be \$0.65. Therefore, the initial investment cost for additional fuel storage facilities for each B-x base will be

$$1.95 \text{ million } (\$0.65) = \$1.3 \text{ million.}$$

But this \$1.3 million is not the only initial investment cost that should be estimated for facilities. It is only the most obvious one. Experience suggests that it would be naïve to believe that introducing the B-x would necessitate no other changes. This is particularly true of maintenance facilities, where access ramps and certain equipment and tools are designed for a specific aircraft. Something on the order of \$1 million for each base should probably be provided, therefore, to handle unexpected requirements. Together with the estimate for new fuel storage, we thus have a total of \$2.3 million for initial investment in facilities – a reasonable figure, considering that it amounts to less than 5 per cent of the value of the average base.

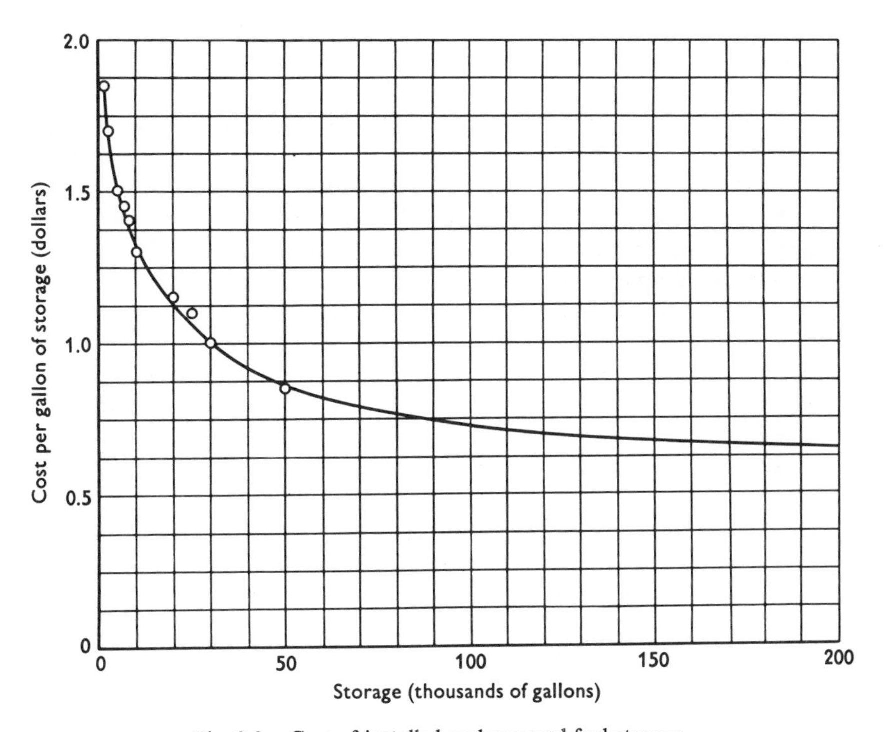

Fig. 9.3 – Cost of installed underground fuel storage

Primary Mission Equipment

Under the conditions that have already been outlined for the organization and operation of the B-x system, we know that ten wings will require a total purchase of 207 aircraft:[14]

R&D aircraft	10
UE aircraft	150
Command support aircraft	15
Replacement (5-yr operation)	32
Total	207

For the purposes of cost analysis, it is the second two sorts of aircraft – the UE and the command support aircraft – that constitute the primary mission equipment.

But to estimate the cost of initial investment in PME, we must also bring into the calculation the cost of the aircraft that will be bought to replace the PME lost through attrition. This is necessary even though attrition is an annual operating expense, since we not only have no way of knowing when attrition will occur, but also have every reason to believe that it will not wait until we have purchased all of our PME. The most plausible way to proceed, then, is simply to find the cost of all the aircraft we will buy in five years of operation (other than for R&D), determine the average cost of these aircraft, and multiply that cost against the number of PME aircraft the B-x system will require.

Now it is evident that the cumulative average cost for the aircraft purchased varies with the total number of planes produced. This is due to learning, or improvement in methods and techniques, as well as the fact that tools, production fixtures, and other capital items may be amortized over a larger number of units. Estimating and expressing this cost-quantity relationship is an extremely important part of any weapon system cost analysis, and poses problems that have received a great deal of attention. The cost-quantity curve that we will use here is known technically as a cumulative average cost curve; it is a type of learning curve in which the plot of points is generated by dividing the sum of the costs of the first n items by n for each nth observation. For the B-x, this simply means that a summation of the costs for the airframe, engines, and avionics, all estimated at various total production levels, will yield the necessary curve. This curve is shown in Fig. 9.4.

Since the R&D program has been allocated, and will pay for, the first ten aircraft, we wish to find the average unit cost from unit 11 through unit 207. Using the curve in Fig. 9.4, we can calculate this figure as follows:

[14] The source of the figure for attrition is discussed on pp. 181–182.

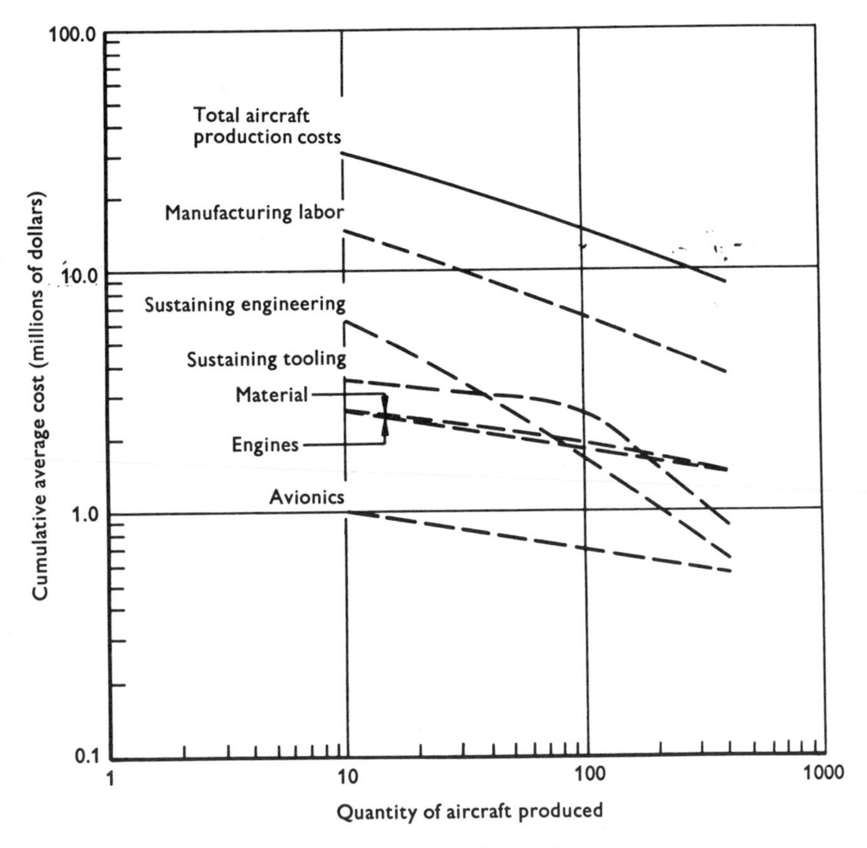

Fig. 9.4 – Cost-quantity relation

Cumulative average cost through unit 207 = \$11.3 million,
Cumulative average cost at unit 10 = \$30.7 million,

$207 \times \$11.3 \text{ million} = \$2339.1 \text{ million},$

$10 \times \$30.7 \text{ million} = \$307.0 \text{ million}.$

The difference between these last two figures gives us the cost of units 11 through 207: \$2032.1 million. Thus, the average cost of these 197 aircraft is

$$\frac{\$2032.1 \text{ million}}{197 \text{ acft}} = \$10.3 \text{ million/acft}.$$

Therefore, the initial investment cost of the 150 UE aircraft and the 15 command support aircraft that constitute the PME in ten wings of the B-x system is

$$(150+15)(\$10.3 \text{ million}) = \$1699.5 \text{ million}.$$

Unit Support Aircraft

This category includes the C-123, U-3, and T-29 aircraft assigned to the base for administrative and training support. These aircraft will be inherited, with the base, from the B-52 system, and, therefore, no investment costs are incurred.

Aerospace Ground Equipment

We include under AGE three types of equipment, and the spares for each: common equipment, such as that used to refuel, service, and tow the aircraft; specialized equipment, which is designed along with the aircraft; and all training devices.

An estimating relationship for determining the cost of AGE for the B-x can be found by examining the Air Force *Force and Financial Program* for figures on aircraft which are currently in production, noting planned expenditures wherever it is possible to do so. Such an investigation suggests that all AGE items, for all years and all aircraft examined, will equal about 7 per cent of the cost of the aircraft themselves. Consequently, we can use this figure for the B-x:

$$0.07\ (\$1699.5 \text{ million}) = \$119.0 \text{ million.}$$

Miscellaneous Equipment

This category includes general purpose vehicles, construction equipment, materiel handling equipment, general purpose communications equipment, mess hall equipment, and general purpose maintenance equipment. On bases such as those the B-x will use, this type of equipment will already exist, at an estimated average value of $1500 per military man. As Table 9.6 indicates, the number of military personnel needed in the B-x system will be 2801, just 23 more than the average of 2778 that staff the B-52 bases the B-x will inherit. But a small increase in personnel does not automatically mean an immediate $1500 increase per man in such organizational equipment. The mess halls, for example, should be put under no great strain. Therefore, we can say that essentially no initial investment is required in this cost category.

Stocks

Initial stocks are largely inherited with the B-52 base. They include such items as personnel supplies; facilities maintenance supplies; organizational equipment supplies; and petroleum fuel, oil, and lubricants (POL). Only the latter, as we saw earlier,[15] will increase. The amount of the increase

[15] On p. 174.

will be equal to the value of the extra 1.95 million gallons of fuel required, which, at about \$0.10 per gallon, is \$0.2 million.

Spares

Spares for AGE and training equipment are included in the cost of that equipment itself. Initial spares needed elsewhere in the B-x system can be calculated – as experience suggests – at 20 per cent of the cost of the aircraft:

$$0.20\,(\$1699.5 \text{ million}) = \$339.9 \text{ million.}$$

Personnel Training

Some crew retraining will be required to make the transition from the B-52 to the B-x. One of the future B-x bases will temporarily become a training base; all of the crews in the ten wings will there receive instruction by SAC personnel. From historical experience, we can assume that this training will take 10 weeks and involve some 50 hours of flying for checkout in the B-x. The direct costs can, therefore, be estimated at \$41,350 per crew, by making the relevant calculations for the costs of POL,[16] PME maintenance,[17] and TDY and miscellaneous expenses.[18] Thus:

POL	= \$266/flying hr × 50 hr =	\$13,300
PME maintenance	= \$501/flying hr × 50 hr =	25,050
TDY and miscellaneous	= \$1000/man × 3 crewmembers =	3,000
	Total =	\$41,350

Both the POL and the maintenance costs, however, represent annual operating costs. Hence, only the TDY and miscellaneous costs can be charged to personnel training. Since we have 39 crews in a wing, and we are calculating these costs for ten wings, our estimated initial investment in training will be

$$39 \times 10 \times \$3000 = \$1.2 \text{ million.}$$

Initial Travel

The costs of initial travel – the costs of bringing new military personnel into the B-x system – are actually a good deal higher than the fact that only 23 men are being added per wing would suggest. This is so because the system gains these 23 men by dropping 76 officers from the B-52 organi-

---

[16] The derivation of POL cost (\$266) is shown on p. 184.

[17] The total for PME maintenance is the sum of the cost per flying hour for base materials (\$68.60) and depot maintenance (\$432.09). The derivation of these figures is shown on p. 183.

[18] We assume \$1000 per man, primarily to cover the costs of travel to and from the training base. ("TDY" stands for temporary duty.)

zation and adding 99 airmen. Since the initial travel costs of an airman averages \$128, and 99 airmen are involved, the cost here for ten wings will be

$$\$128 \times 99 \times 10 = \$0.1 \text{ million.}$$

Initial Transportation

Initial transportation costs are those incurred in transporting all new equipment to the base, with the exception of the PME and POL. (The PME are flown in as part of the flying program, and POL is quoted on a delivered basis.) Items that would normally have to be transported are in the categories of Miscellaneous Equipment, Spares, and AGE; for the B-x system, as we have seen, nothing will be spent under Miscellaneous Equipment. In this case, therefore, the total transportation cost for ten wings is calculated by taking 1 1/2 per cent of the cost in each of the remaining categories and adding:

AGE	=	0.015 × \$119.0 million	=	\$1.79 million
Spares	=	0.015 × \$339.9 million	=	\$5.10 million
		Total	=	\$6.89 million

MAJOR COST CATEGORIES FOR ANNUAL OPERATION

Annual Operating Costs is the last of the three main categories of any cost analysis. When dealing with an aircraft system, the analyst typically estimates these costs in terms of eleven subcategories: facilities replacement and maintenance (R&M), PME replacement, PME maintenance, PME POL, unit support aircraft POL and maintenance, AGE R&M, personnel pay and allowances, personnel replacement training, annual travel, annual transportation, and annual services. As before, we will follow this order in our discussion, defining each category briefly as we encounter it. To avoid repeating a fact that applies to each of them, let us note once again that we are concerned in this analysis with the costs of operating 10 wings of the B-x for five years.

CALCULATION OF ANNUAL OPERATING COSTS

Facilities Replacement and Maintenance

This category includes the costs of all normal base maintenance, which we can calculate by means of an estimating relationship that has been developed from historical data. It indicates that facilities R&M will equal 5 per cent of the base value plus \$450 for each airman and officer on the base. The part of this relationship that depends on personnel strength covers utilities, office equipment, and military personnel supplies, while the part that depends on base value covers the maintenance of runways, roads, buildings, equipment, and so on.

The value of the average B-52 base is difficult to obtain. Many of the bases were acquired years ago at prices which would have to be adjusted to today's levels. In addition, most of the bases have been used for various weapon systems, and as a result have been modified many times, at widely varying costs, which tends to compound the difficulty. From the facts at hand, however, and from the values listed in the *Air Force Inventory of Military Real Property*, it can be estimated that the average base has a current value of about $75 million. The annual facilities R&M is then

$$0.05 \begin{pmatrix} \text{\$75 million} \\ \text{base value} \end{pmatrix} + \$450 \begin{pmatrix} 2801 \\ \text{military} \\ \text{personnel} \end{pmatrix} = \text{\$5.01 million per base.}$$

Thus:

$$\text{\$5.01 million} \times 10 \text{ bases} \times 5 \text{ yr} = \text{\$250.5 million.}$$

PME Replacement

In a normal flying program, provision must be made for the replacement of aircraft which are destroyed or lost. This attrition varies between aircraft types, and also with experience with the aircraft. Traditionally, the introduction of a new aircraft into the inventory is accompanied by a high attrition rate. As experience is gained by flying the plane, this rate falls rapidly, and then approaches stability. The attrition rate is usually higher for high-performance, many-sortie aircraft than for low-performance or long-endurance aircraft.

In practice, the total number of aircraft purchased for a weapon system includes those allocated for replacement because of attrition. To forecast the number necessary, information is required about (a) the projected life of the weapon system (or the number of years of operation for which replacement aircraft will be provided), (b) the number of flying hours each year the weapon system will operate, and (c) the relationship between the number of aircraft destroyed and the number of hours flown by the aircraft. The total number of flying hours for the system may be calculated by multiplying (a) by (b). Replacement aircraft must be purchased for this life span. From (c), the cumulative number of aircraft destroyed for this period may be obtained.

The first two facts are easily established for any particular system, and may be ascertained for various operational levels. In the present case, and on the basis of information we have already derived, we can expect a five-year program of 513,000 flying hours.

To determine the relationship indicated in (c), we can turn to copies of the *Air Force Statistical Digest* for 1949 through 1962, which will provide attrition data for jet fighters and bomber systems that have accumulated between one and six million flying hours. If we gather data on both the flying hours and the number of aircraft destroyed, analyze it on a quarterly basis, and plot the results on log-log paper, we have the graph shown in Fig. 9.5. It is apparent from this graph that attrition bands exist which portray the attrition history of fighters and bombers and that these bands have different slopes.

The B-x is an aircraft about which little is known in general, and even less about its attrition rate. It is a bomber, but incorporates some of the characteristics of a fighter, particularly in its speed and variable wing geometry (Table 9.1). However, the B-x does not push the state-of-the-art. We saw before that the B-x requires no new avionics and benefits from the B-58 and F-111 development programs.[19] Considering these facts and the rapid advance in technology, we can safely choose an attrition curve in the middle of the bomber band in Fig. 9.5. Reading directly from this Figure for 513,000 flying hours, therefore, we can estimate that 32 B-x aircraft will be attrited in 5 years of operation. At an average cost of $10.3 million,[20] PME replacement aircraft will thus cost

$$32 \text{ acft} \times \$10.3 \text{ million/acft} = \$329.6 \text{ million.}$$

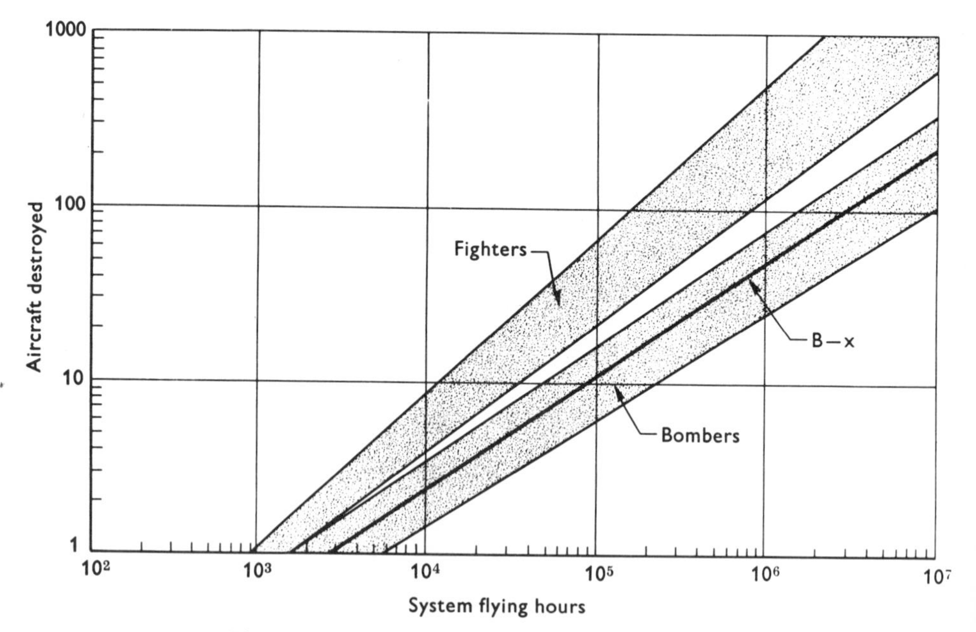

Fig. 9.5 – Aircraft destroyed versus system flying hours

[19] See p. 169.
[20] See p. 177.

PME Maintenance

The maintenance requirements of a postulated aircraft system can be forecast only generally. From historical maintenance data on seven bombers which are, or have been, in the force, we can, however, derive two workable estimating relationships. These relationships require (as inputs) the cost of the aircraft and the combat speed. They yield (as outputs) estimates of the base material cost per flying hour, and the total depot maintenance cost per flying hour. (The cost of labor for base maintenance is estimated under the category of maintenance personnel.)

The equations for these two relationships are as follows:

a) Base material costs (in dollars) per flying hour $= 31.81 + 0.00584X_1$,

where

$X_1 =$ level-off cost of aircraft in thousands of dollars;

b) Depot maintenance cost (labor and material) per flying hour (in dollars) $= 14.526 + 0.0498X_1 + 0.0824X_2$,

where

$X_1 =$ level-off cost of aircraft in thousands of dollars,
$X_2 =$ aircraft combat speed in knots.

For the B-x, the cost-quantity curve in Fig. 9.4 indicates a level-off cost of about \$6.3 million.[21] Table 9.1 indicates a combat speed of approximately 1260 knots. Substituting these values in the above equations yields \$68.60 per flying hour for base material costs, and \$432.09 per flying hour for depot maintenance:

a) Base material costs (in dollars) per flying hour

$$\begin{aligned} &= 31.81 + 0.00584X_1 \\ &= 31.81 + 0.00584(\$6300) \\ &= 31.81 + \$36.79 \\ &= \$68.60. \end{aligned}$$

b) Depot maintenance cost (in dollars) per flying hour

$$\begin{aligned} &= 14.526 + 0.0498X_1 \\ &\quad + 0.0824X_2 \\ &= 14.526 + 0.0498(\$6300) \\ &\quad + 0.0824(1260 \text{ kn}) \\ &= 14.526 + \$313.74 + 103.82 \\ &= \$432.09. \end{aligned}$$

[21] Taken at the 900th level.

With a five-year flying hour program of 513,000 hours, these figures become:

For base materials	\$ 35.2 million
For depot maintenance	222.0 million
Total	\$257.2 million

PME POL

The fuel consumption of the B-x may be inferred from the design of the aircraft and its mission. Table 9.1 states that the range of the B-x is 6300 nautical miles, and general design practice is such that 90 per cent of the fuel carried by the aircraft will be consumed flying this distance. The fuel capacity, also given in Table 9.1, is 201,450 pounds (31,000 gallons), and 90 per cent of this is 27,900 gallons. Since fuel consumption is usually stated in terms of gallons per hour, the distance must be converted into hours of flying time. This can be done by assuming that the average speed of the B-x is about 600 knots. The flying time is then $\frac{6300}{600} = 10.5$ hours. Thus, 27,900 gallons of fuel are consumed in 10.5 hours, or 2660 gallons are consumed per hour. At \$0.10 per gallon, therefore, the cost for fuel is about \$266 per flying hour.

It follows that the cost for POL for ten wings flying a total of 513,000 hours is

$$\$266 \times 513{,}000 \text{ hr} = \$136.5 \text{ million.}$$

Unit Support Aircraft POL and Maintenance

POL. We assumed earlier that each crew would spend eight hours a month flying trainer and cargo aircraft.[22] With 39 crews per wing, this comes to 312 hours a month; for a ten-wing force operating for five years, it comes to 187,200 hours. From data in AFM 172–3, *Peacetime Planning Factors*, we can estimate the cost of POL for the unit support aircraft at \$40 per flying hour, and thereby arrive at a five-year cost of \$7.5 million.

Maintenance. Again from data in AFM 172–3, we can estimate the cost of base maintenance materials and depot maintenance at \$30 per flying hour, and thus calculate a five-year cost of \$5.6 million.

The sum of the costs for POL and maintenance gives us a total in this category of \$13.1 million.

AGE R&M

From historical data, we can estimate the cost of AGE R&M as the sum of two figures: 15 per cent of the investment in AGE and 11 per

[22] See p. 155.

cent of the investment in organizational (miscellaneous) equipment. We saw earlier that AGE will cost about \$119.0 million.[23] Thus, the first total we require is

$$0.15 \times \$119.0 \text{ million} = \$17.9 \text{ million.}$$

We have also seen that each wing in the B-x system will require 2801 military personnel, and that, historically, the average value of the organizational equipment will be \$1500 per man.[24] Thus, for the second total we have

$$0.11\,[10 \text{ wings } (\$1500 \times 2801)] = \$4.6 \text{ million.}$$

The five-year cost of AGE R&M will thus be \$112.5 million.

Personnel Pay and Allowances

This category includes pay, personnel benefits, food, and other allowances. According to recent estimates made by O. Hansen,[25] these costs amount to \$11,550 for rated officers, \$9190 for nonrated officers, \$3910 for airmen, and \$7350 for USAF civilian personnel. Since rated officers[26] in nonflying positions draw flying pay, and since the number of officers in such positions appears to equal 10 per cent of the flying officers, we can restructure the data given in Table 9.6 and calculate as follows for one B-x wing:

Rated officers	215	at	\$11,500	=	\$ 2.5 million
Nonrated officers	96		9,190	=	0.9 million
Airmen	2490		3,910	=	9.7 million
Civilians	278		7,350	=	2.0 million
Total	3079				\$15.1 million

Thus, for ten wings and five years,

$$\$15.1 \text{ million/wing} \times 10 \text{ wings} \times 5 \text{ yr} = \$755 \text{ million.}$$

Personnel Replacement Training

Normal personnel turnover requires that replacement personnel be trained to maintain the required organizational proficiency. Turnover rates vary between different categories of personnel; according to O. Hansen's calculations,[27] they average 4.64 per cent per year for pilots, 3.85 per cent

---

[23] See p. 178.
[24] See Table 9.6 and p. 178.
[25] Unpublished studies of USAF civilian and military personnel cost factors.
[26] The 1963 edition of the *Air Force Statistical Digest* shows that 69.2 per cent of SAC officers are rated.
[27] Unpublished study of USAF personnel turnover rates.

per year for other crew officers, 8.22 per cent per year for nonrated officers, and 14.61 per cent per year for airmen.

AFM 172–3, *Peacetime Planning Factors*, provides training costs as follows:

Pilots:	$ 4,220	Officer basic training
	75,000	Pilot undergraduate training
	1,000	Upgrade training
	$80,220	Total
Navigator-Bombardiers:	$ 4,220	Officer basic training
	45,000	Navigator-bombardier training
	1,000	Upgrade training
	$50,220	Total
Nonrated Officers:	$ 4,220	Officer basic training
	5,473	Average cost of all nonflying courses
	$ 9,693	Total
Airmen:	$ 650	Basic training
	3,223	Average cost of all courses
	$ 3,873	Total

Assuming that two-thirds of the rated officers will have received pilot training, and the other third, navigator-bombardier training, the turnover costs for one wing can therefore be calculated as follows:

Pilots = 0.0464 × 0.667 × 215 = 6.654 = 7	
7 × $80,220 =	$561,540
Other Crew Officers = 0.0385 × 0.333 × 215 = 2.756 = 3	
3 × $50,220 =	150,660
Nonrated Officers = 0.0822 × 96 = 7.891 = 8	
8 × $9,693 =	77,544
Airmen = 0.1461 × 2490 = 363.789 = 364	
364 × $3,873 =	1,409,772
Total Turnover Cost =	$2,199,516

Thus,

$2.2 million × 10 wings × 5 yr = $110.0 million.

Annual Travel

The costs of annual travel and initial travel are analogous in that both cover the cost of bringing personnel and their dependents to the base. Annual travel, however, involves replacement personnel, and therefore the expense is a recurring operating cost. It can be calculated, from historical data, at $128 for an airman and $660 for an officer. Since we saw, in the previous category, that the annual turnover could be estimated at 18 officers and 364 airmen, we can thus calculate the cost of annual travel to be

$$
\begin{aligned}
18 \text{ officers} \times \$660 &= \$11{,}880 \\
364 \text{ airmen} \times \$128 &= \$46{,}592 \\
\text{Total} &= \$58{,}472
\end{aligned}
$$

Thus,

$$\$58{,}472 \times 10 \text{ wings} \times 5 \text{ yr} = \$2.9 \text{ million.}$$

Annual Transportation

Initial transportation was classed as an investment cost, and was estimated to be 1 1/2 per cent of the value of those items which were transported to the base as part of the B-x weapon system. The materials which are consumed in the operation of the weapon system also require transportation, and the same estimating relationship can be used to determine their cost. The items used in this relationship are AGE replacement, organizational equipment replacement, and base level PME maintenance materials. The five-year cost of the last of these items we calculated earlier at $35.2 million.[28] The first two we can determine as follows, using numbers that we have already derived:

$$
\begin{aligned}
&\text{Organizational Equipment} \\
&\qquad = \$1500/\text{man} \times 2801 \text{ personnel} \\
&\qquad = \$4{,}201{,}500; \\
&0.11 \times \$4{,}201{,}500 = \$462{,}165 \times 10 \text{ wings} = \$4{,}621{,}650 \\
&\text{AGE Replacement} = 0.15 \times \$118{,}965{,}000 = 17{,}844{,}750 \\
&\qquad\qquad\qquad\qquad\qquad\qquad \text{Total} = \$22{,}466{,}400
\end{aligned}
$$

The total in these categories for five years will thus be $112,332,000. Adding in the cost of base PME maintenance materials and taking 1 1/2 per cent of the total will give us a total five-year cost estimate for annual transportation of $2.2 million:

$$0.015 \times \$147{,}523{,}800 = \$2{,}212{,}857.$$

Annual Services

The category of Annual Services includes the costs of materials, supplies, and contractual services for such functions as base administration, flight service, supply operations, food and medical services, and operations and maintenance of organizational equipment. Historically, this cost can be

[28] See pp. 183–184.

estimated at $400 per military man. For the B-x, therefore, the estimated cost of annual services will be

$$\$400/\text{man} \times 2801 \text{ personnel} \times 10 \text{ wings} \times 5 \text{ yr} = \$56.0 \text{ million.}$$

SUMMARY AND DISCUSSION OF RESULTS

The results of this cost analysis of the B-x system may be presented in a variety of ways, and we will illustrate each of them. First, and most obviously, we can provide the sort of summary shown in Table 9.9, which lists the costs of development and procurement of the ten wings of B-x, and of their operation for five years.

TABLE 9.9

Summary of costs – B-x system

Cost Category	Per Cent of Total	Cost (In $ Million)	
Research and Development			
Design and Development	60.7	824.0	
System Test	39.3	534.0	
Total	100.0		$1358.0
Initial Investment (10 Wings)			
Facilities	1.1	23.0	
PME	77.5	1699.5	
Unit Support Aircraft	—	0.0	
AGE	5.4	119.0	
Miscellaneous Equipment	—	0.0	
Stocks	0.1	2.0	
Spares	15.5	339.9	
Personnel Training	0.1	1.2	
Initial Travel	—	0.1	
Initial Transportation	0.3	6.9	
Total	100.0		2191.6
Operation (10 Wings, 5 Years)			
Facilities R&M	12.4	250.5	
PME Replacement	16.3	329.6	
PME Maintenance	12.7	257.2	
PME POL	6.7	136.5	
Unit Support Aircraft POL and Maintenance	0.6	13.1	
AGE R&M	5.6	112.5	
Personnel Pay and Allowances	37.3	755.0	
Personnel Replacement Training	5.4	110.0	
Annual Travel	0.1	2.9	
Annual Transportation	0.1	2.2	
Annual Services	2.8	56.0	
Total	100.0		2025.5
Grand Total			$5575.1

Such a tabular presentation may be rearranged, however, to place emphasis upon the relative magnitudes of the various costs. Thus, for the 5-year operation of ten wings of B-x aircraft, Table 9.10 lists those items which, together, comprise 85 per cent of the total cost of the system.

TABLE 9.10

Those costs comprising 85 per cent of the B-x system cost

Item	Cost (In $ Million)	Per Cent
R&D	1358.0	24.4
PME	1699.5	30.5
Personnel Pay and Allowances	755.0	13.6
PME Replacement	329.6	5.9
Spares	339.9	6.1
PME Maintenance	257.2	4.6
Total	4739.2	85.1

Table 9.10 illustrates the fact that large amounts of dollar resources are committed to a few categories. There are particular dangers for the analyst when this sort of distribution occurs. Errors in estimating key items will have a marked effect upon the total cost estimate – something which will be quantitatively treated later.[29] It is necessary to be extremely attentive to these key items if the analysis is to be credible. It is equally necessary to review continually the status of the key cost items as the weapon system passes from one stage to the next on the road from conception to realization.

The concentration of costs in a few categories may lull the analyst into the belief that the most costly items are the only ones of importance. It is true that their size causes them to be important *per se*, but it does not follow that those which cost less are necessarily less worthy of consideration. For example, of the costs shown in Table 9.9 for the B-x, only 1.97 per cent of the total system cost is allocated to Personnel Replacement Training. The cost of this item would hardly be noticed if it were eliminated or increased ten times, yet its importance is fundamental: The system requires people to operate it, and if they are unavailable, the system will not work. The analyst must always remember that the estimated cost of something may not be indicative of its availability, and if the item is critical to the system, its degree of availability may eclipse major cost considerations.

Table 9.10 also reveals that replacement due to attrition of the aircraft will be a major expense. As we have seen, the attrition rate will decrease as

[29] See pp. 192 ff.

experience with the aircraft increases, and the cost of attrition may therefore be expected to fall somewhat in the years of operation beyond the five-year period we have used as the basis of this study. Despite this decrease, the magnitude of the cost involved is such that attrition will probably always be a major expense in the B-x system.

The over-all results of the analysis may also be presented graphically in several forms. One such form is a "static" analysis, shown in Fig. 9.6. The static analysis shows R&D, initial investment, and annual operating costs for forces of various sizes. Since it is oriented to force size, it illus-

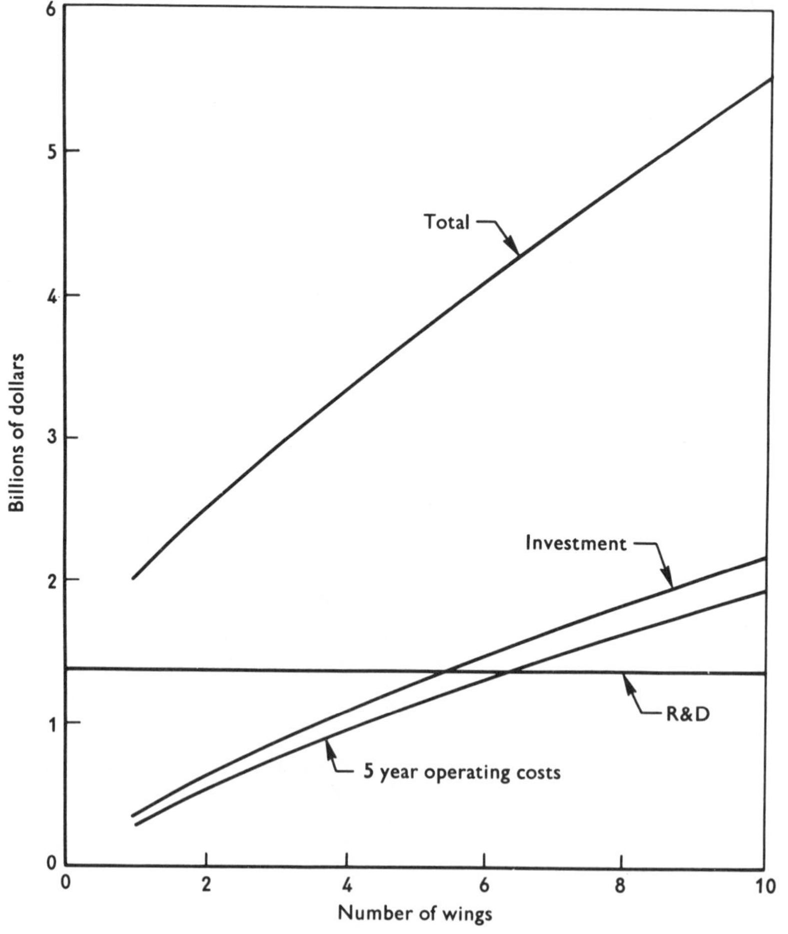

Fig. 9.6 – Static analysis of B-x system costs
(5-year operation)

trates the effect of the PME cost-quantity curve, the change in attrition rate with increased flying hours, and other costs that are affected by force size. It provides a convenient picture of the relative sizes of R&D, investment, and operating costs for a given operating period, and may be used in the comparison of alternate weapon systems. It has one major disadvantage, however, which is that it is based upon the artificial premise that a force of any size could be obtained in exactly the same amount of time as any other force. Time phasing is not considered, and therefore the picture is artificial and distorted. But if this qualification is not forgotten, the static analysis can be a helpful tool.

A static analysis is done by first making cost estimates at the various force levels, and then plotting the results. In deriving the individual estimates, care must be taken with the cost of PME, the attrition rate, and incremental costs. The latter, in particular, require attention, since inherited assets may be large enough to produce no increment in cost at low force levels, but turn out to be insufficient at higher levels, thereby forcing an abrupt need for additional outlays.

Complementing the static analysis is the time-phased analysis. While this sort of analysis overcomes the major disadvantage of the static analysis – its absence of time phasing – it cannot account for variations in force size, as is possible in the static analysis. The two work together. By itself, however, the time-phased analysis is helpful in comparing different weapon systems and pointing out differences in "concentrations" of expenditures. Such results of an analysis for the B-x system are shown in Fig. 9.7.

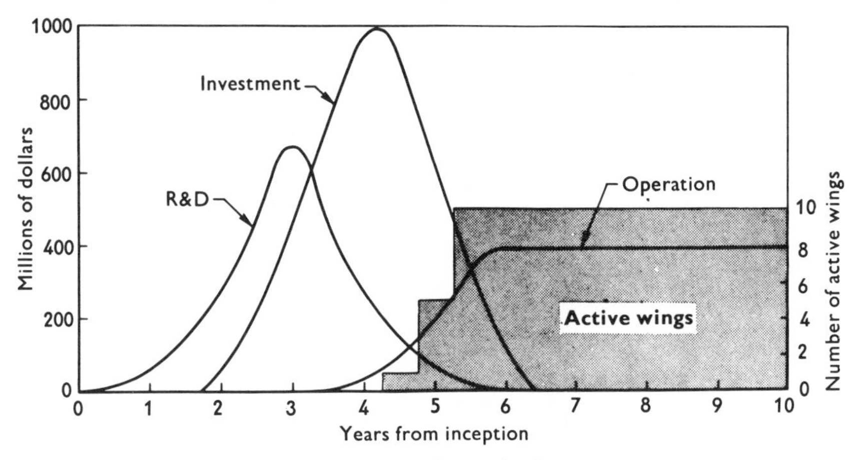

Fig. 9.7 – Time-phased costs for B-x system

The time-phased analysis requires information about the amount of time required for R&D, the rate at which the force is to be built, and the method of paying for the weapon system. For the B-x system, we can estimate that the R&D program would require about five years. Payment for R&D would be concurrent with the effort, and would be made as follows:

Year	Per Cent of Payment	Millions of Dollars
1	5	67.9
2	20	271.6
3	50	679.0
4	20	271.6
5	5	67.9
	100	1358.0

We can schedule the acquisition of the system as follows: one wing in the fourth year, four wings in the fifth year, and five wings in the sixth year. Payment for the acquired wings of aircraft thus requires, in this example, two years' lead time, with 40 per cent of the total system costs being paid in year N–2 and 60 per cent being paid in year N–1, where N is the year of acquisition.

Since operation of the system begins as the wings are phased into the force, the operating costs will increase until all the wings are activated, and then stabilize.

II. SENSITIVITY ANALYSIS OF THE B-x SYSTEM

While we have now estimated and analyzed the various costs associated with the B-x and compared the relative weight of some of them, in many respects the analytic task has only begun. It is undoubtedly important, of course, to produce a set of figures that represents the resource demands of a weapon system. But it is equally important to understand that the figures produced may vary if the conditions change from those specified. The study of the variation of costs in relation to changing assumptions is known as sensitivity analysis.

Two general types of sensitivity analysis are common. The first and simpler type examines the possible effects of errors in the estimating processes, and the importance of such errors within the context of the total estimate. It is a method of carrying through the effect of errors in individual parts of the analysis to the effect on the total estimate. Where errors might exist, sensitivity analysis provides the tool for indicating their quantitative importance, thereby providing the analyst with a measure of confidence in his work.

An example of the application of this type of analysis is easily drawn from the B-x study. The numbers of Organizational Maintenance and Field Maintenance personnel required for the B-x system were obtained through the use of an estimating relationship developed from experience with current aircraft. This was necessary because there is no positive method of reliably estimating the maintenance personnel requirements of a hypothetical system, when the system is no further defined than the B-x. The effects of errors in this estimating process could be explored via sensitivity analysis.

The second type of sensitivity analysis is concerned with possible variations in the basic characteristics of the system or its operating assumptions, and the effect of such variations upon system costs.

Any one of the aircraft's physical characteristics, for example, could be varied over a selected range, and an analysis made of the change that could be expected in the total system costs. Examples of such characteristics that could be (and have been) used in sensitivity analysis are aircraft weight, speed, runway requirements, fuel capacity or range, and crew composition. Similarly, deployment may be altered in varying degrees and the operating concepts tested accordingly.

Sensitivity analysis concerned with basic system parameters has a significance that goes beyond the mere determination of the effect upon costs of variations in the system parameters. Every system has rigid, albeit sometimes obscure, limitations which cannot be transcended. Examination of a system for sensitivity can point out these limitations. A few obvious examples: With a constant wing area, an aircraft can weigh only so much and still maintain the ability to fly; fuel capacity can be decreased only to the point beyond which the plane has too little fuel to take off and land; a plane may not remain airborne for an infinite length of time because of crew, maintenance, and fuel limitations. On every side of the system are constraints which restrict the system to a particular set of operational configurations. To transcend these limitations requires modification of the given data or assumptions, and often this is accompanied by steeply rising costs.

The B-x system can be examined for both types of sensitivity. We will consider first the effect of errors in the estimating process, and then test the effect of modifying the operating program.[30]

Possible errors in estimating

In the analysis of the B-x, some resources were estimated using relationships derived from the operation of existing equipment. While these rela-

[30] A sensitivity analysis of variations in the basic characteristics of the B-x – weight, speed, range, and so on – has been made, but the specific procedures and results are not demonstrated in this Chapter.

tionships are the best at hand, their validity is questionable for an aircraft of such advanced design as the B-x. This is particularly true in the areas of PME maintenance, PME POL, AGE replacement and maintenance, portions of the maintenance personnel estimate, and PME replacement. But a fairly simple sensitivity analysis may be performed for each of these. Let us take them in turn, and attempt to determine what effect a 50 per cent error in our estimate for each might have on the five-year operating costs of the B-x system.

PME Maintenance, PME POL, and AGE R&M

Beginning with our estimate of PME maintenance cost, we can calculate as follows:

$$\begin{aligned} 50\%\ \text{PME Maintenance} &= 0.50 \times \$257.2 \text{ million} \\ &= \$128.6 \text{ million}, \\ \text{Per Cent of Operating Costs} &= \frac{\$128.6 \text{ million}}{\$2025.5 \text{ million}} \\ &= 6.3\%. \end{aligned}$$

We may infer from this result that significant errors in estimating maintenance costs will not produce concomitantly significant errors in the final cost estimate. Similar calculations for PME POL and AGE R&M yield figures of 3.4 per cent and 2.7 per cent, respectively, thus indicating that the total operating costs also show low sensitivity to variations in these categories.

Maintenance Personnel

Determining the sensitivity of the operating costs to the maintenance personnel estimate requires a slightly different approach. The maintenance personnel estimating relationships that should concern us particularly are those involving the Organizational Maintenance and Field Maintenance Squadrons. For we find that varying the number of personnel in these squadrons by $\pm$ 50 per cent will affect the number of administrative and support personnel as well, and the sum of these personnel changes will be reflected in other related personnel costs. The investment costs affected will be Facilities, Miscellaneous Equipment, and Initial Travel; and the operating costs affected will be Facilities Replacement and Maintenance, Personnel Pay and Allowances, Personnel Replacement Training, and Annual Travel.

To calculate these modified costs, we must first determine the number of people involved. From Table 9.3b we note that the Organizational Maintenance Squadron will contain 271 men, and that the Field Maintenance Squadron will contain 391, for a total of 662. If we are 50 per cent off in our estimate, we must therefore be concerned with the consequences of adding or subtracting 331 men from the total. But we know two other things from estimating relationships discussed earlier:[31] first, that there will be 40 administrative personnel associated with these 331 operating and maintenance personnel, thus giving a total of direct personnel of 371; and, second, that 114 support personnel will be associated with 371 direct personnel. Thus, the total personnel variation will be 485, of whom 13 will be officers, 17 will be civilians, and 455 will be airmen. In other words, we must actually consider the effect of adding or subtracting not 331, but 485 personnel.

Now while the addition of this number of people will obviously result in additional incremental costs, the opposite is not true. Except for $0.1 million in savings in the category of Initial Travel, subtracting 485 people will not affect the estimates of initial investment costs. The reasons are clear: Essentially the same number of people man the B-x system as manned the B-52, and the B-x inherits equipped bases. We can easily add the additional resources necessary because of increased personnel, but we cannot subtract resources which already are present. If, then, we work through the calculations for each of the relevant categories of initial investment costs to see what effect would follow the addition or subtraction of 485 personnel, we derive these results:

Facilities	= +	$6.0 million or	−$0.0 million
Other Equipment	= +	4.5 million or	− 0.0 million
Initial Travel	= +	0.7 million or	− 0.1 million
Total	= +	$11.2 million or	−$0.1 million

On the plus side, this is 0.6 per cent; on the negative side, 0.0 per cent.

The effect upon annual operating costs can be calculated in a similar manner, and will be found to be $\pm$ 5.6 per cent.

PME Replacement

PME replacement requires still different treatment. It has been expedient so far to explore the effects of a percentage error on an estimate, and in these examples we have used $\pm$50 per cent. The attrition rate, however, was estimated from data shown in Fig. 9.5. Were we to re-examine these data, to explore the possibility that the B-x attrition curve might vary from the one we postulated, we might conclude that a reasonable range of variation would be between the upper and lower lines of the bomber band. Another way of saying the same thing is that the attrition rate of the B-x might not

[31] See pp. 162–164.

be an average, but could vary anywhere within the limits provided by our experience with bomber attrition.

Varying the attrition rate changes the number of aircraft which must be purchased for a specific system and specific period of operation, which in turn varies the amount of money required for PME replacement. When the quantity of aircraft purchased changes, the cumulative average cost of the aircraft also changes. As a result, the PME investment costs change, as do the costs of AGE and Spares, both of which are PME dependent items in the investment category. Similarly, in the category of operating costs, PME Replacement and AGE R&M will vary.[32]

TABLE 9.11

Effect of varying attrition rate

	Upper Line Bomber Band	Attrition Curve Used for B-x	Lower Line Bomber Band
Number of Aircraft Destroyed at 513,000 Flying Hours	48	32	16
UE and Command Support Aircraft	165	165	165
R&D Aircraft	10	10	10
Total Aircraft Buy	223	207	191
Cumulative Average Aircraft Cost[a]	10.9	11.3	11.7
Total Aircraft Cost[a]	2430.7	2339.1	2234.7
R&D Aircraft Cost[a]	307.0	307.0	307.0
Total Cost of Operational Aircraft (units 11-207)[a]	2123.7	2032.1	1927.7
Average Cost of Operational Aircraft[a]	10.0	10.3	10.7

[a]All costs in millions of dollars.

[32] It might also appear that both Initial Transportation and Annual Transportation would vary, since they are calculated as a percentage of the cost of PME dependent items. This, however, is not true, since transportation costs do not change with changes in the price of the item transported. This observation highlights one of the dangers in a generalized estimating relationship of this type and illustrates the need for the analyst to understand each such relation and its limitations.

Using, then, the number of aircraft destroyed represented by the upper and lower lines of the bomber band, we can derive for our cost-sensitivity analysis the data shown in Tables 9.11, 9.12, and 9.13.

These Tables show that the B-x system is sensitive to any change in attrition rate. It has already been shown that attrition cost can be expected to be approximately 16 per cent of the total five-year operating costs.[33] Now it is apparent from Tables 9.12 and 9.13 that, even if the B-x were a substantially safer aircraft, its operating costs would drop only by about 8 per cent, and this would be offset by an increase in the investment cost of about 4 per cent. Conversely, if the attrition rate were higher, the total operating costs could increase by about 7 per cent, which again would be partially offset by a smaller change in the costs of initial investment.

Summary

The results of the cost-sensitivity analysis of each of the categories we have considered are illustrated graphically in Figs. 9.8 and 9.9. These Figures also show the results of a similar cost-sensitivity analysis of PME cost, R&D, and Personnel Pay and Allowances.

TABLE 9.12

Investment cost comparison with different attrition rates based upon ten wings

(In $ million)

Category	Upper Line Bomber Band	Attrition Curve Used for B-x	Lower Line Bomber Band
Facilities	23.0	23.0	23.0
PME	1650.0	1699.5	1765.5
Unit Support Aircraft	0.0	0.0	0.0
AGE	115.5	119.0	123.0
Miscellaneous Equipment	0.0	0.0	0.0
Stocks	2.0	2.0	2.0
Spares	330.0	339.9	353.1
Personnel Training	1.2	1.2	1.2
Initial Travel	0.1	0.1	0.1
Initial Transportation	6.9	6.9	6.9
Total	2128.7	2191.6	2274.8
% Difference from Method Chosen	−2.4%	—	+3.8%

[33] See Table 9.9.

TABLE 9.13

Operating cost comparison with different attrition rates based upon ten wings
(In $ million)

Category	Upper Line Bomber Band	Attrition Curve Used for B-x	Lower Line Bomber Band
Facilities R&M	250.5	250.5	250.5
PME Replacement	480.0	329.6	171.2
PME Maintenance	257.2	257.2	257.2
PME POL	136.5	136.5	136.5
Unit Support Aircraft POL and Maintenance	13.1	13.1	13.1
AGE R&M	110.0	112.5	116.0
Personnel Pay and Allowances	755.0	755.0	755.0
Personnel Replacement Training	110.0	110.0	110.0
Annual Travel	2.9	2.9	2.9
Annual Transportation	2.2	2.2	2.2
Annual Services	56.0	56.0	56.0
Total	2173.4	2025.5	1870.6
% Difference from Method Chosen	+7.3%	—	−7.6%

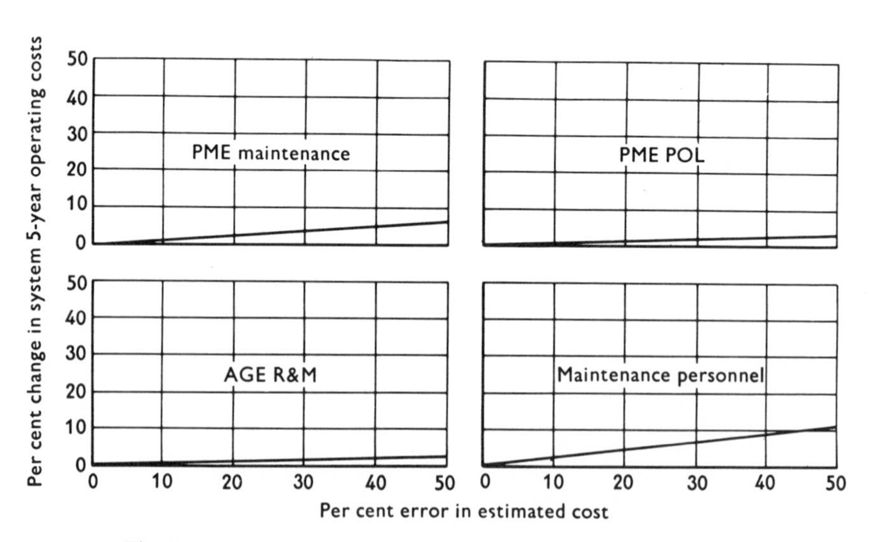

Fig. 9.8 – Effect of estimating errors upon system operating costs

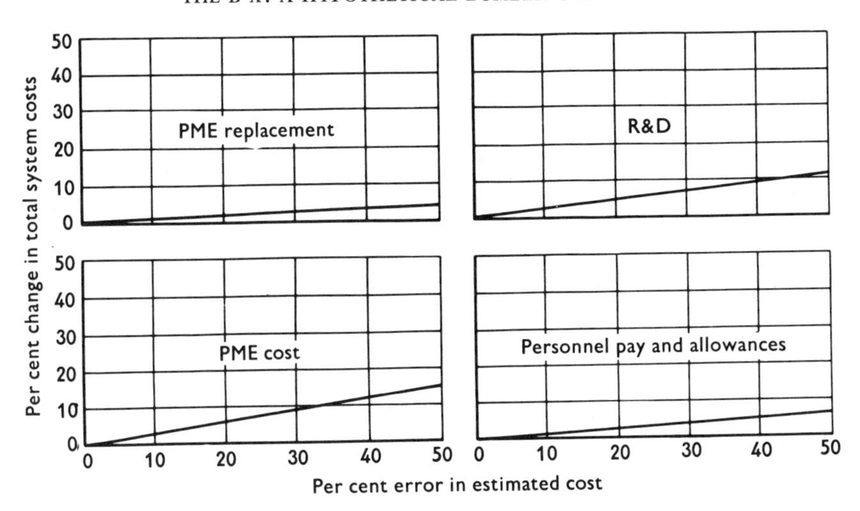

Fig. 9.9 – Effect of estimating errors upon total system cost estimate

POSSIBLE VARIATIONS IN OPERATING PARAMETERS

The preceding discussion of sensitivity analysis displayed a tool which gives the analyst some idea of the accuracy and impact of his estimating methods. The second type of sensitivity analysis is perhaps of more interest. Through its use, the analyst may probe the parameters of the system, and define its operational limits under various conditions. He may also learn a great deal about the costs of procuring and operating the system under different operating configurations, and thereby be able to identify the operating configurations of minimum and maximum cost.

To illustrate these points, we will explore the possible operating configurations of the B-x in detail, and define some of the factors which limit the operation of the system. We will then examine the system to locate the "operating envelope" within which it could function without constraint. We will then translate this envelope into a "cost envelope" which describes the resource impact of the various operating configurations. And, finally, we will illustrate a method of presenting this information graphically, which allows us to portray the universe of costs in which all the possible operational configurations of the system are contained.

Aircraft Constraints

An aircraft spends its time either on ground duty or flying. When the aircraft is fully utilized, the sum of these two activities equals 730 hours (in the average month). This may be expressed mathematically as

$$F_T + G_T = 730, \tag{1}$$

where

F_T = flying and related activities,

G_T = ground duty and related activities.

Figure 9.10 shows a graph of this relation. By using this graph, it is possible to find the amount of time spent by the aircraft on one activity when the other is known.

Other activities are necessarily associated with flying and with ground duty, and they absorb time. Maintenance is required, and it will probably differ in type and quantity from one type of aircraft activity to another. Equation (1) may then be rewritten

$$F + M_F + G + M_G = 730, \tag{2}$$

where

F = flying duty,

M_F = maintenance caused by flying,

G = ground alert duty,

M_G = maintenance caused by ground alert duty.

The amount of maintenance required will be in proportion to the amount of each type of duty. For the maintenance caused by flying, we can express the relation as

$$M_F = k_F F, \tag{3}$$

and for the maintenance caused by ground alert duty, as follows:

$$M_G = k_G G, \tag{4}$$

where

M_F = maintenance caused by flying,

M_G = maintenance caused by ground alert duty,

k_F = ratio of hours of maintenance per hour of flying,

k_G = ratio of hours of maintenance per hour of ground alert duty.

Equations (2), (3), and (4) may then be combined as follows:

$$F(1 + k_F) + G(1 + k_G) = 730. \tag{5}$$

Equation (5) states that the aircraft spends all of its available time either flying or on ground alert duty, or in maintenance caused by these duties. But we can rewrite equation (5) in a form which may be used to solve for F in terms of G:

$$F = \frac{730}{1 + k_F} - G\left(\frac{1 + k_G}{1 + k_F}\right). \tag{6}$$

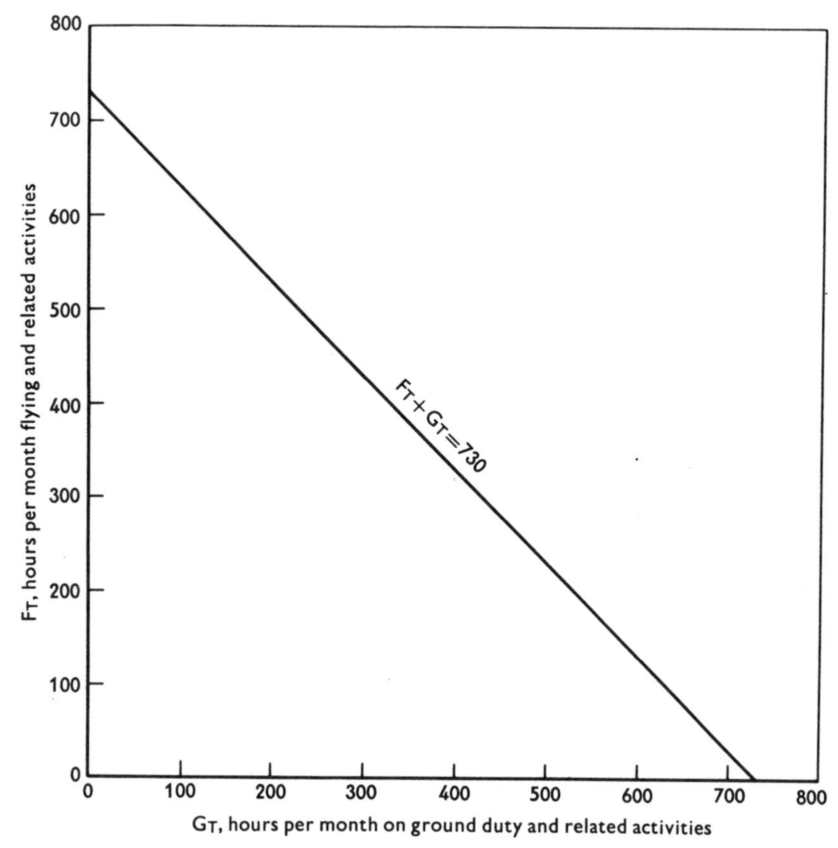

Fig. 9.10 – Monthly time distribution between flying activities and ground duty activities

The effect of the requirements for maintenance may be added to Fig. 9.10, as we have done in Fig. 9.11. In this Figure, the maintenance constraints due to equations (3) and (4) are shown. The values which have been used for k_F and k_G are 4 and 0.25, respectively, and these conform to those selected for the B-x system. The ordinate shows flying and the maintenance associated with flying; the abscissa shows ground alert duty plus the maintenance associated with this duty.

Figure 9.12, which shows maintenance and flying hours as a function of ground alert duty, has been plotted by solving equation (2) for F, M_F, and M_G in terms of G. An example will illustrate the use of this Figure. Thus, if we specify that the aircraft is to spend 300 hours per month on ground alert duty, then we can tell from the graph that the aircraft can spend 71 hours per month flying. We also find that of the 360 hours remaining in the month, 75 hours are spent in maintenance resulting from the ground alert duty, and 284 hours are spent in maintenance caused by flying. In this way, the maximum operation of the aircraft can be defined in terms of the maintenance constraints.

Crew Constraints

The operation of an aircraft (and an aircraft system) is an interaction between the aircraft and the crews which man them. Just as we have defined and illustrated the constraints on the aircraft's equipment, we must also reckon with the constraints imposed by considerations of the air crew. The combination of these two sorts of constraints will then define the overall operating constraints.

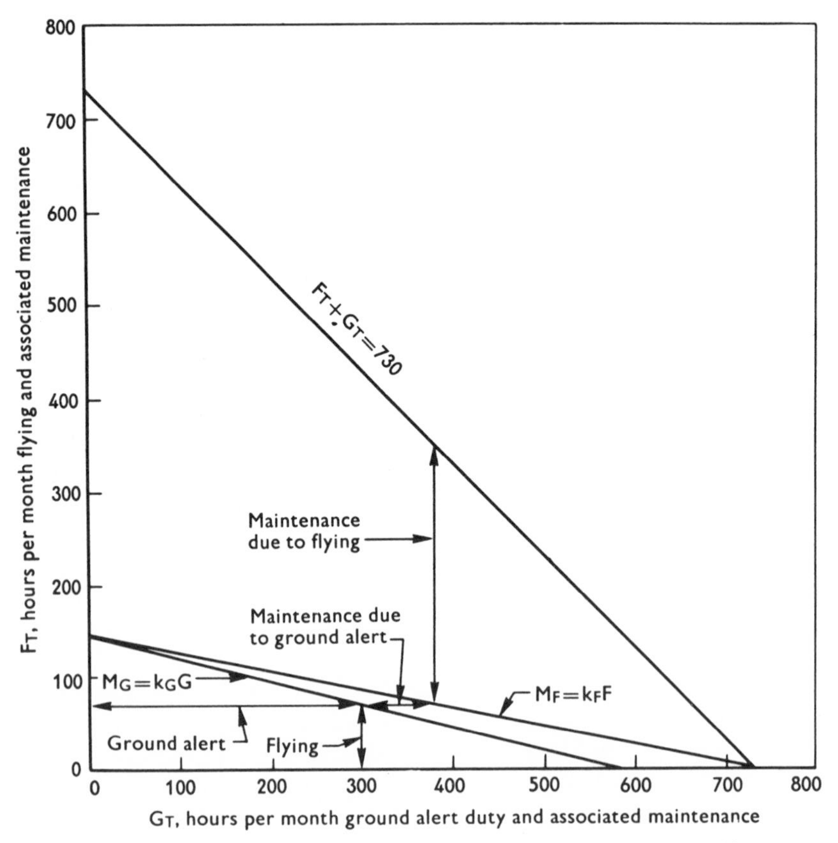

Fig. 9.11 – Effect of maintenance requirements upon aircraft activities

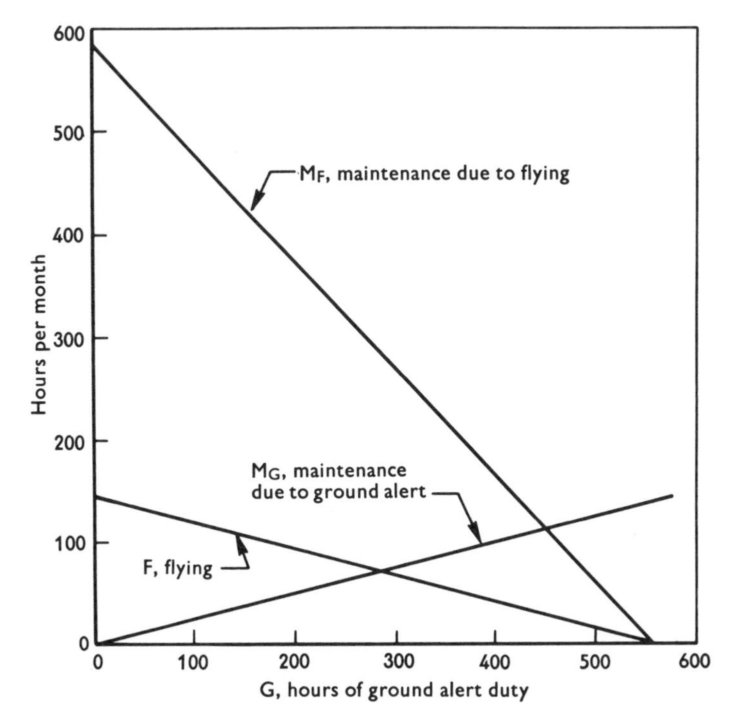

Fig. 9.12 – Maintenance and flying hours as a function of ground alert duty

Crews can spend their time flying or on ground alert duty, and the relation expressed for aircraft in equation (1) and Fig. 9.10 holds true for air crews.

The time which the crews spend on the ground will be used in many ways. Ground alert duty has been mentioned, but there are also activities that are unrelated to the aircraft – ground training, briefings, and, of course, various off-duty activities. Each reduces the amount of time that the crews have available for flying and ground alert duty, a fact we can express mathematically by

$$F + G = 730 - D, \tag{7}$$

where

F = flying time,

G = ground alert duty time,

D = the total time spent on activities other than flying and ground alert duty.

Since, as we have seen,[34] crews in the B-x system spend 152 hours per month on flying and ground alert duties, D = 578 hours. Equation (7) thus becomes

$$F + G = 152. \tag{8}$$

If the aircraft required more than 152 hours per month of crew time, an obvious solution would be to add more crews. Figure 9.13, which shows the flying versus ground alert graph for the crews, incorporates the limitations of equation (8) and illustrates the increased capacity that would result if crews were added.

While there is only one constraint upon the ground alert activity of the crews, there are two on their flying duties. The first is the time they must

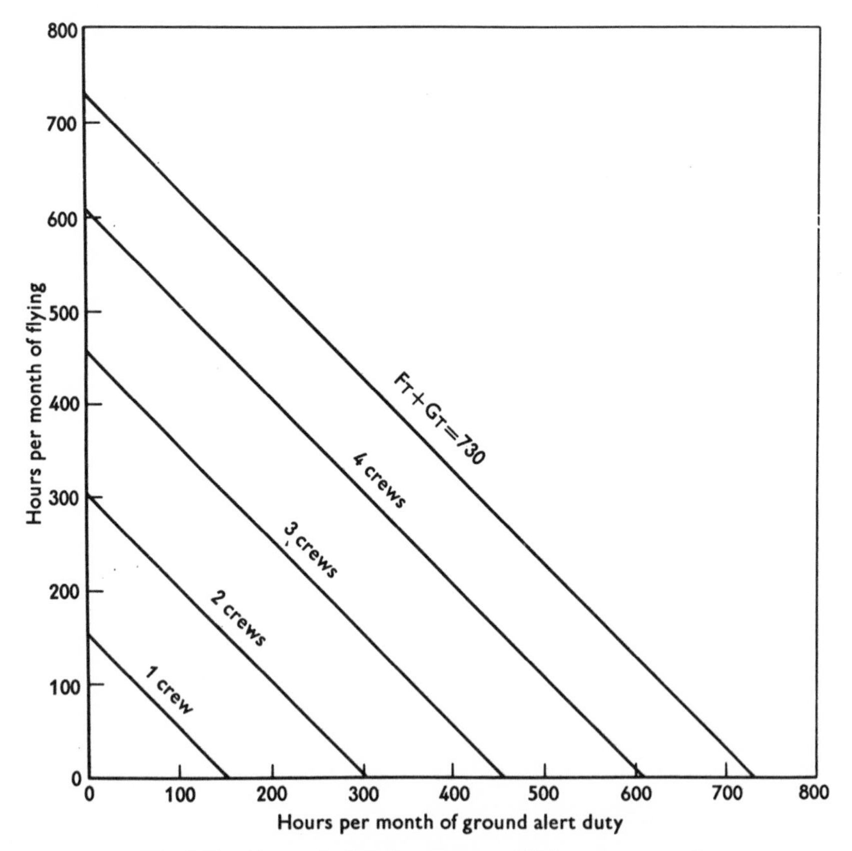

Fig. 9.13 – Duty schedule for crews, at 152 hours per month flying and ground alert duty

---

[34] See p. 155.

spend each month flying for proficiency training. In the case of the B-x, we assumed this to be 22 hours per month.[35] The second constraint is the number of hours a crew can fly without excessive fatigue. The policy is normally to establish this limit at 100 hours per month. Thus, although equation (8) indicated that the crews would spend 152 hours per month on flying and ground alert duty, and that they could freely interchange hours of either sort of duty, we now see that it must be understood with these two limitations in mind:

$$22 \leq F \leq 100.$$

Equation (8), with the limitations taken into account, is graphically illustrated in Fig. 9.14, again plotted for various numbers of crews, so as to cover an entire 730-hour month. Included in Fig. 9.14 is the area within

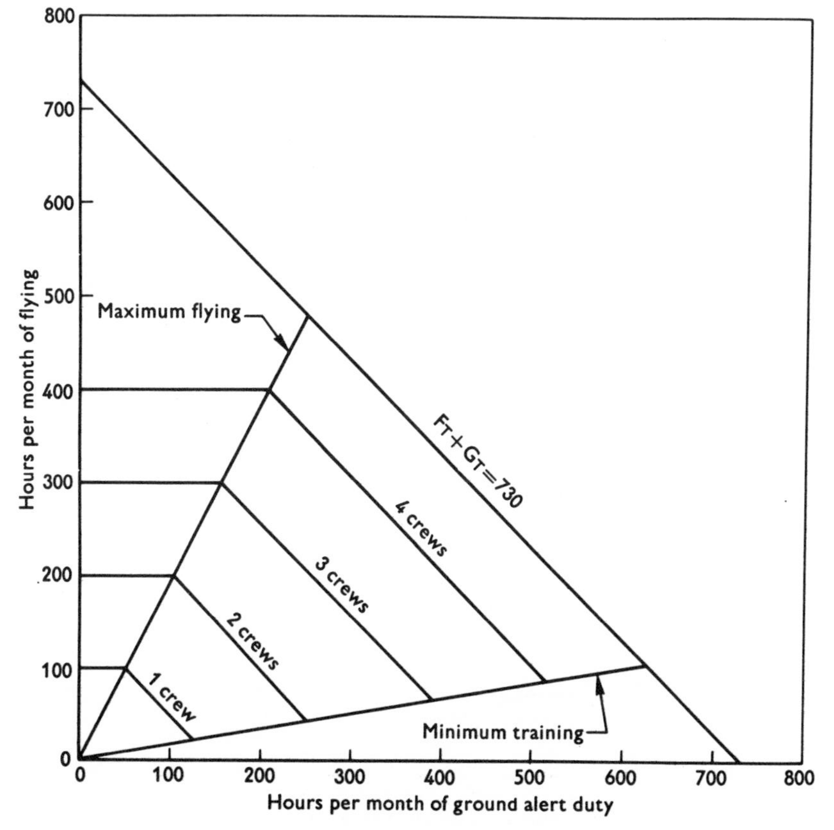

Fig. 9.14 – Effect of flying and training constraints upon crew duty

[35] See p. 155.

which a free interchange may be made between flying and ground alert duty. Of the two lines bounding this area, the lower one represents the monthly 22-hour training requirement for each crew, and the upper, the maximum permissible flying duty of 100 hours per month. The area to the left of this 100-hour line represents operations in inefficient systems, since crews may fly and may perform ground alert duty, but no longer can trade flying hours for ground alert hours. The graph shows, for example, that while the manpower requirement for 300 hours of flying and 100 hours of ground alert is three crews, the same number could also accomplish 156 hours of ground alert, without any loss of flying time. It should perhaps be noted that no operation is possible below the line representing minimum training. That the crews must be trained is an axiom of the system, and we have already seen that $F \geqq 22$.

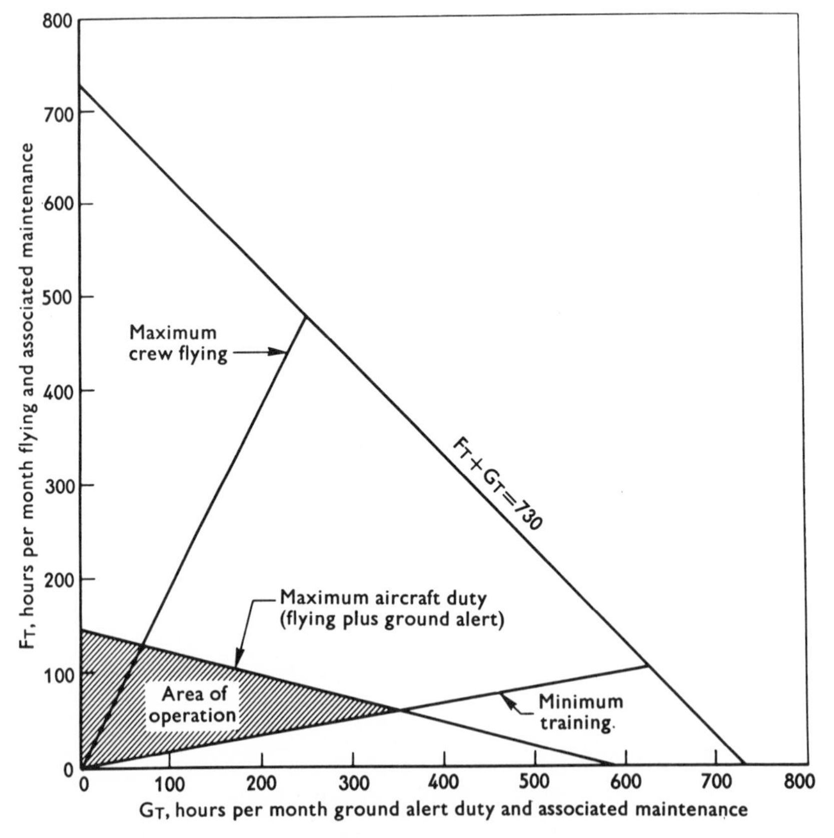

Fig. 9.15 – Possible ground alert and flying duty in terms of aircraft and personnel constraints

The B-x Operating Envelope

As we stated previously, the combination of aircraft and crew constraints will define the over-all possible area of operation of the B-x system. We can illustrate these general limits, therefore, simply by superimposing Fig. 9.12 on Fig. 9.14, as we have done in Fig. 9.15. With the criteria that have been established for the crew duty and for maintenance, no operation outside the delineated area is possible. A few examples may illustrate the utility of Fig. 9.15.

EXAMPLE 1

Problem: Determine the maximum number of flying hours possible for the B-x system.

Solution: This number is found at the intercept of the "maximum aircraft duty" line with the ordinate. At this point, 146 hours per month per plane may be flown. Since there are 730 hours in a month, this is equivalent to a 20 per cent alert $\left(\frac{146}{730}\right)$; for a wing of 15 aircraft, it is the same as maintaining three of them on continuous air alert. No ground alert duty is possible with this flying configuration, and the remaining 584 hours of the aircraft's time each month are spent on maintenance caused by flying. The crews will fly only 100 hours a month, which is the maximum that they may fly.

EXAMPLE 2

Problem: Determine the maximum number of hours that it is possible for the B-x to be on ground alert.

Solution: This number is found at the intercept of the "maximum aircraft duty" line and the "minimum training" line. For the B-x this is 350 hours per plane per month. This is equivalent to a 48 per cent ground alert $\left(\frac{350}{730}\right)$, or to 7.2 planes on continuous ground alert in a wing of 15. From Fig. 9.12, or from equations (2), (3), and (4), we find that this amount of ground alert implies a flying schedule of 58.7 hours per plane per month, and a maintenance schedule of 234.8 hours per month required by the flying and 87.5 hours per month required by the ground alert.

EXAMPLE 3

Problem: Determine the maximum and minimum numbers of flying hours that may be scheduled when there is a 30 per cent ground alert.

Solution: Thirty per cent of 730 equals 219 hours per plane per month. At this point on Fig. 9.15, the minimum flying is 37 hours per plane, and the maximum is 91 hours. No less than 37 hours may be flown because of crew training; no more than 91 hours may be flown because the time will be required for maintenance. From Fig. 9.11, we see that, if the aircraft is scheduled for 91 hours of flying per month and 219 hours of ground alert, then 364 hours are required for maintenance as a result of flying, and 55 hours are required for ground alert maintenance. If, on the other hand, the minimum monthly flying schedule of 37 hours per month is used, then the aircraft schedule is as follows:

Ground alert	219 hours
Ground alert maintenance	55
Flying	37
Maintenance due to flying	148
Total	459 hours

The balance of time in the month (730 – 459=271 hours) is spent by the aircraft doing nothing; that is, waiting. Under either configuration, the crews will work a full 152 hours.

Example 4

Problem: Determine the maximum and minimum numbers of ground alert hours that may be scheduled with a flying schedule of 100 hours per plane per month.

Solution: The maximum number—180 hours—is found on the "maximum aircraft duty" line. The minimum number is zero. At the maximum ground alert point, the plane will be fully utilized (no waiting) and the crews will work 152 hours. At the minimum of zero ground alert, the plane will have excess time, and will encounter periods of waiting. Under this configuration, the crews will fly 100 hours per month, and will, therefore, have 52 hours of time available for other duties.

Example 5

Problem: Assuming that the aircraft is scheduled for 100 hours of flying per month, as in example 4, determine how much ground alert may be scheduled without altering the number of crews.

Solution: The crews may add ground alert duty to the 100 hours per month that they are allowed to fly until they have reached a total of 152 hours of scheduled duty. This figure can be found by reading at the appropriate point on the "maximum crew flying" line; in this case, they can add 52 hours per month of ground alert.

Information of this sort may be used in many ways. Its most apparent use is that it allows us to make sure that a postulated operating configuration is in fact possible. It also allows the analyst to examine the weapon system for cost sensitivity to the various possible configurations, and thereby describe the estimated cost "universe" in which the system will operate, given its basic assumptions. Such a description appears in Fig. 9.16, which shows the five-year system operating costs for all of the operating configurations of the B-x that are consistent with the data and assumptions we have introduced in the course of this discussion.

The B-x Cost Envelope

In effect, then, what Fig. 9.16 presents is analogous to the "area of operation" shown in Fig. 9.15. It defines the limits of operation of the system in terms of ground alert and flying, and indicates the estimated operating costs for any selected operating configuration. As might be expected, the operating costs are shown to be heavily dependent upon the flying schedule, and affected to only a minor degree by the addition of ground alert duty.

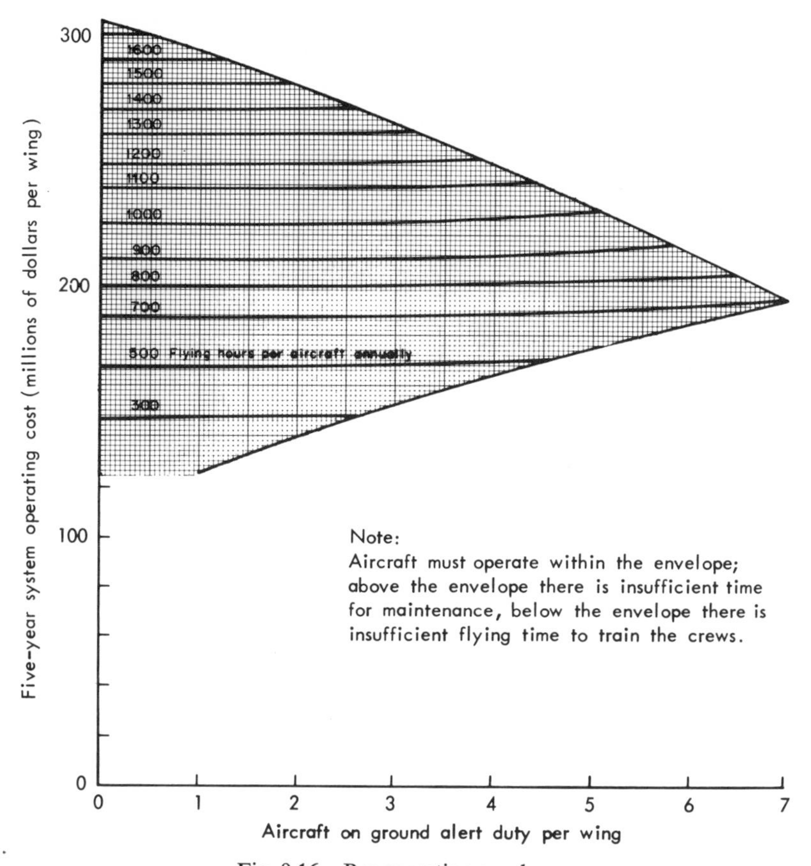

Fig. 9.16 – B-x operating envelope

The lower line of this graph represents those costs incurred when the system is operated as a "minimum system," that is, when the only flying done is for minimum crew training. The upper line represents the costs due to maximum use of the system, that situation in which the total time available to the aircraft is completely consumed by flying duty, ground alert, and associated maintenance. Between the upper and lower boundaries lie other possible configurations. Costs for any of these may be found from the plot.

Chapter 10

THE NATURE OF MODELS

R. D. SPECHT

This Chapter classifies the various types of models discussed elsewhere in this book and explains their nature – what they are; how they are designed; what forms they may take; how and to what extent they embody mathematics; how they relate to the techniques of analysis, to systems analysis as a whole, and to the real world; and how they can and do assist decisionmakers. Among its other goals, this discussion attempts to present an accurate idea of the sometimes indirect and tenuous connection between systems analyses and the real world. It indicates how much significance can properly be attached to this relationship, and what procedures the analyst, in constructing models, has at his disposal to keep the connection as close as possible and ensure that his conclusions accurately reflect the connection as it exists.

WHAT IS A MODEL?

If this were a psychological test in which I give you a word and you respond by writing down your free association – the first thing that comes to your mind – and if I were to say "model," you might react by writing down "36-24-36."

Now you may think that this is irrelevant – that this is not a model in the sense that concerns us in this book. But if you think so, you are wrong. Our definition of "model" will include your "36-24-36." Our definition of "model" will be broad enough to cover even an uncovered broad.

In this Chapter and those that follow, you will meet a surprising variety of things that we shall classify as "models" – a collection of mathematical equations, a scenario, a program for a high-speed computer, a war game. And the list of creatures that we could include in our model zoo is much longer yet. We could add an organization chart, a map, a set of questionnaires, a copy of Plato's *Republic*, a Link trainer, and a group of people and machines acting as if they were an air defense direction center.

What is it that all these have in common? Each is an idealization, an abstraction of a part of the real world. Each is an incomplete representation of the real thing. Each is an analog, an imitation of reality.

But why settle for an imitation? Because the real thing is not available for study or is too expensive to experiment with. Some of the things we contemplate are too expensive to allow to happen even once.

A model, then, is an analog of reality. It is made up of those factors that are relevant to a particular situation and the relations among them. We ask questions of the model and from the answers we get, hopefully, some clues, some hints, to guide us in dealing with that part of the real world to which the model corresponds.

We must not object that a model does not look like the real thing or that it does not represent all aspects of reality. It seldom does. The important thing is whether or not the outputs of the model, the answers it gives to our questions, are reasonably appropriate and valid.

We would like to test the results of our analysis of a model and determine the correctness and relevance of these results for real-world decisions. Perhaps we could make this test if we lived in the best of all possible worlds. But, unfortunately, we live next door. We can never be certain, in this sinful world, that we have been wise. Perhaps the best that we can hope for is to be honest.

We must not object if the analyst changes models on us, if he produces different models for the same reality. The model depends not only on the thing being modeled, on the part of the real world with which we are concerned. The model also depends on the questions to be asked of it, the decisions to be affected by its results.

A trivial example: If you are driving from Santa Monica to San Francisco and have not yet decided on a route, then an adequate model of this part of California may be a road map. If you are a trucker concerned about maintaining a schedule between here and San Francisco, then an adequate model may be a timetable that tells you, among other things, when you are due to pass Pismo. If you are a highway planner who must recommend a freeway route between the two cities, then quite a different model or set of models is necessary – road maps, topographic maps, maps of land use and value, traffic charts showing origin and destination, and a model, implicit and subjective, of the behavior of a population surfeited with taxes, attached to their real estate, and not altogether enchanted with freeways. Each is unrealistic in its own way, but each is useful when shaken well and taken as directed.

In Chapter 1, and again in Chapter 3, E. S. Quade defined a model in something like the following terms. Given a set of alternatives (including ones that the analyst may have invented in the course of studying the problem), a model is a black box. The analyst has designed the particular box to deal with his particular problem, and he has constructed it to reflect the state of the world of which the alternatives are a part. Into this box as inputs the analyst feeds information about the alternatives, and from the box as outputs comes information about the effectivenesses, plural, and the costs of each of the alternatives. With the help of a criterion, the analyst or the decisionmaker can then rank the alternatives in order of desirability and can select the optimum.

The black box, of course, is simply a figure of speech to represent any device or process with which we can take into account, in a way as nearly logical as possible, the interrelations of the relevant factors. And the black box isn't really that color. If the analyst, the model builder, has done his work satisfactorily, the walls of the box will not be black; they will be transparent. The spectator and the user of the model will be able to see inside, will be able to understand and evaluate the structure of the model.

Now Quade's definition of a model – a means of producing measures of the costs and effectivenesses of various alternatives – is a handsome definition, and I can't improve on it. But you will soon discover, if you haven't already done so, that, unfortunately, the world is not this tidy.

For one thing, instead of a single model that produces information about both cost and effectiveness, it will often be convenient to have separate models – a cost model, for example, or a campaign effectiveness model. Indeed, more often than not, the analyst will use a collection of models for various parts of his problem, knitting these submodels together by means of verbal arguments.

There are other problems. For example, we said that a model interrelates the relevant factors; but, at least when he begins his study, the analyst probably does not know just which factors are the crucial ones, which may safely be neglected. Part of his job is to discover what is important, what is trivial. This means that the analyst may go several times around the cycle of building a model, experimenting with it, deducing its implications, building a better model, and so on.

But the major difficulty with this definition comes only if you misuse it, if you let it suggest that an analyst armed with model and criterion can arrive at an optimal course of action to recommend to the decisionmaker. In his wonderful address in May 1960 as retiring president of the Operations Research Society Charles Hitch laid this ghost to rest. The operations researcher, he said, is

> . . . faced by his fundamental difficulty. The future is uncertain. Nature is unpredictable, and the enemies and allies are even more so. He has no good general-purpose technique, neither maximizing expected somethings, nor *max-min*ing, nor gaming it, to reveal the preferred strategy. How can he find the optimal course of action to recommend to his decisionmaker?
>
> The simple answer is that he probably cannot. The same answer is also the beginning of wisdom in this business. There has been altogether too much obsession with optimizing on the part of operations researchers, and I include both grand optimizing and sub-optimizing. Most of our relations are so unpredictable that we do well to get the right sign and order of magnitude of first differentials. In most of our at-

> tempted optimizations we are kidding our customers or ourselves or both. If we can show our customer how to make a better decision than he would otherwise have made, we are doing well, and all that can reasonably be expected of us.[1]

And this much, said Hitch, we frequently can do.

If the analyst with his models is not computing optimal solutions, what is he about? Computation is not his most important business. His functions are to define alternative objectives, to design alternative solutions, to discover the critical uncertainties, to recommend ways of reducing them, and to explore the implications of alternative courses of action. And computations help do these things.

Let us leave generalities for a spell and look at a model, a real one.

Design and use of models: an example

This is a model a RAND analyst, T. F. Burke, devised to help him think about the problems of hard point defense of missile sites. To keep the explanation brief, I have simplified his model, but only slightly.

The problem, simply put, is whether or not to buy active defense for a land-based ICBM force. Should we buy an undefended missile force, or should we spend the same budget for a smaller defended missile force? It is obvious without either model or analysis that if ICBMs are expensive and defense is cheap, and if there is appreciable danger of attack, then we buy the defense system. But how expensive? How cheap? We set up a model that will quantify, even if crudely, some aspects of this problem.

We first need some definitions. Let

M = number of undefended missiles we can buy with a given budget,

A = number of shots fired by an attacker at our missiles,

p = probability that 1 shot kills an undefended missile,

p_D = probability that 1 shot kills a defended missile,

$\$_M$ = cost of 1 undefended missile,

$\$_D$ = cost of defense for 1 missile.

These are the factors that we have chosen as relevant – at least for our first cut at the problem. Later we shall call the roll of factors that have been omitted from this model.

Consider first the attack upon the undefended missile force. There are A attacking shots against M targets or A/M shots per target. In order to simplify the arithmetic we assume that the number A/M is an integer, that the number of attacking shots fired at each missile is a whole number.

[1] C. J. Hitch, "Uncertainties in Operations Research," *Operations Research*, Vol. 8, July-August 1960, pp. 443–444.

(Burke did not make this simplifying assumption.) Then:

p = the probability that a missile is killed by 1 shot,
$1-p$ = the probability that a missile survives 1 shot,
$(1-p)^{A/M}$ = the probability that a missile survives A/M shots,

and the expected number of our missiles that survive the attack is this probability multiplied by the number of missiles in the force, all undefended:

$$M(1-p)^{A/M}.$$

Now the attack upon the defended missile force. First we calculate the size of the defended missile force:

$$\text{Total budget} = M\$_M = (\text{number of defended missiles})\,(\$_D+\$_M)$$

or

$$\text{number of defended missiles} = \frac{M\$_M}{\$_D+\$_M} = \frac{M}{\frac{\$_D}{\$_M}+1}.$$

There are again A attacking shots, but this time their kill probability P_D is lower and distributed over the smaller force just calculated, $M\$_M/(\$_D+\$_M)$ defended missiles. The number of shots per missile is then

$$A \div \frac{M\$_M}{\$_D+\$_M} = \frac{A(\$_D+\$_M)}{M\$_M} = \frac{A}{M}\left(\frac{\$_D}{\$_M}+1\right).$$

Again we make the simplifying assumption (for this explanation, and not for the original analysis) that this number of attacking shots per missile is an integer. Then, as before,

p_D = the probability that a missile is killed by 1 shot,
$1-p_D$ = the probability that a missile survives 1 shot,

$(1-p_D)^{\frac{A}{M}\left(\frac{\$_D}{\$_M}+1\right)}$ = the probability that a missile survives all shots,

and the expected number of our missiles that survive the attack is this probability multiplied by the number of missiles in the force, all defended:

$$\frac{M}{\frac{\$_D}{\$_M}+1}(1-p_D)^{\frac{A}{M}\left(\frac{\$_D}{\$_M}+1\right)}$$

We now choose a criterion. We recommend buying defense for the missile force if this leads to the expectation of a greater number of missiles surviving. That is, buy defense if

$$\frac{M}{\frac{\$_D}{\$_M}+1}(1-p_D)^{\frac{A}{M}\left(\frac{\$_D}{\$_M}+1\right)} > M(1-p)^{\frac{A}{M}},$$

or, dividing both sides by M, buy defense if

$$\frac{(1-p_D)^{\frac{A}{M}\left(\frac{\$_D}{\$_M}+1\right)}}{\frac{\$_D}{\$_M}+1} > (1-p)^{\frac{A}{M}}.$$

We note, incidentally, that the missile defense model uses expected values – it is not what is called a Monte Carlo model. From the probability p that a missile target survives a single shot we computed the expected number of surviving missiles, neglecting the matter of fluctuations that may occur about this value. In some cases, the fluctuations may be important. If your employer offered to toss you double or nothing for your month's paycheck, you might develop an interest in such fluctuations, even though your expectation had not changed. But in most cases, these statistical uncertainties are not crucial when compared with such real uncertainties as future enemy capabilities and actions, not to mention our own costs and capabilities.

Where statistical fluctuations must be reckoned with, they can be handled sometimes by mathematical analysis. At other times, the situation is treated by drawing random samples from a carefully determined distribution. In a bombing campaign, for example, we may follow the airplanes by tail numbers and for each one draw random numbers to determine whether or not it aborted, made a navigation error, killed its target, and so on. Such a model is referred to as a *Monte Carlo* one, in contrast to an *expected-value* model; our missile defense model is an example of the latter kind. These ways of handling chance events are discussed further in the chapters on mathematical models and simulation.[2]

Now the analysis is not completed at this point. In fact, we have hardly begun. For example, we need to study the dependence of the kill probability p on the attacker's accuracy and the missile site hardness. Thus another model is introduced. With it we can study the worth of hardening, a

[2] Chapters 11 and 12.

competitor to active defense for promoting missile survival. Again, we can study the effect of varying the force ratio, M/A, of missile sites to attacking shots. We can fix the values of all the factors in the inequality above with the exception of the defended kill probability, p_D. We can then ask how small this inequality requires p_D to be – how effective the active defense must be before it is worth buying. Alternatively, we can fix the value of every factor except the unit cost of defense, $\$_D$, and ask how small it must be – how cheap defense must be before it is a good buy. How does this price depend on the other factors – the kill probabilities, the size of the attacking and attacked missile forces, the cost per missile?

Omissions in This Example

What have we omitted from the model? Many things, and I name a few:

> We have not considered buying defense for only part of the missile force.
>
> We have not considered the possibility of grouping several ICBMs at each defended point in order to decrease the cost of defense.
>
> We have not let the attacker use a shoot-look-shoot policy.
>
> We have not let the attacker saturate the defense by simultaneous penetration.

And you will think of other omissions. The ones I have mentioned could be taken care of by a more complex model and more costly analysis. (And Burke did this for some factors.)

This model involves implicitly some simplifying assumptions:

> All missile sites are equivalent in hardness, in cost, and in worth.
>
> Unit costs of missiles and defenses are fixed, independent of the number procured. This means, for example, that we have neglected the research and development costs of an active defense system (or, alternatively, that we have estimated the number of missiles and have prorated the cost of research and development).

And others. These simplifications, too, could be removed by more extensive analysis, if it were thought worth the doing.

And finally, there are idealizations in the model that could not be removed easily, if at all, by more complex analysis. For example, we have neglected the values, military and political, that may come from owning a larger missile force, apart from survival in the attack considered. And so on.

I have spent some time on this model not because we are concerned in this chapter with hard point defense or with this particular study; not even because we have an undue concern with this type of mathematical model. Rather, this model furnishes a concrete example around which we can make some comments that, hopefully, will apply to other instances of analysis as well. The following remarks, then, are less comments upon this particular model and more in the nature of generalities draped for convenience around it.

The Problem of Selecting Criteria

To begin with, consider the matter of criteria. In our example we chose as the criterion the maximizing of surviving missiles. We recommended the alternative that led to the largest expected number of surviving missiles. But this, by itself, is not an adequate criterion (as Burke points out). Suppose, for example, that costs, attack size, and kill probabilities lead us to the conclusion that a small defended missile force is preferable to a larger undefended force because it has more surviving missiles; but suppose, further, that in neither case does the force survive in sufficient strength to be useful. In this case, we do not buy either of the two alternatives; we look for a third and more satisfactory one.

This suggests that in any study it may be hazardous to choose the first criterion that comes to mind, reasonable as it may seem – an observation that you will encounter more than once in this book. The problem of selecting a criterion is more difficult than we have just indicated. If the question at issue is whether or not to start the development of a new weapon system – an ABM system, say – then the decisionmaker will be interested in the analysis whose beginnings we have outlined. But he will have a host of additional questions. How effective might the system be in the damage limiting role? What countermeasures can the enemy develop? How is he likely to react? What are the technical prospects for development to meet an advanced enemy threat? And so on. Some of these questions affect the choice of criterion; others determine the form of analysis that is appropriate; still others add qualitative factors that cannot be translated into elements of a mathematical analysis. It is unlikely that any quantitative model can do more here than throw light on some aspects of the problem.

The Problem of Deciding What Is Relevant

Another lesson from the missile defense model is that it is not an easy matter to decide which factors are relevant, which may be omitted. Are the results sensitive to a shoot-look-shoot capability and tactic on the part of the attacker? There is no firm guide here except the experience and intuition of the analyst as he devises his model, gains experience in working with it, and, as is likely, revises it.

The Necessity of Being Explicit

Again, the simplifying assumptions made in the model may not be readily apparent to the user of the model's results. This makes it all the more important that the analysis not be cast in the form of a black box with the user asked to take the analyst's word for it that all is well within. As E. S. Quade has said,

> All of the assumptions of a model must be made explicit. If they are not, this is a defect. A mark of a good systems analyst (or any wise person communicating with others) is that he state the basis on which he operates. This does not imply necessarily that he makes better assumptions, but only that his errors will be more evident.[3]

The Treatment of Nonquantitative Considerations

From the missile defense model we learn that most problems involve considerations that cannot be handled quantitatively – for example, any military and political values that may come from owning a larger missile force. If we were trying to assist in a decision on the initiation of a new weapon system development, and if we were trying to see whether or not the new development was justified on the basis of the damage limiting objective, then many intangible factors would become essential to the decision: What kind of war? Can we use superiority in surviving forces to coerce the enemy? And so on.

The Static Character of the Model

The missile defense model, like almost all models, is static. This is no criticism; one would always begin this way, and often it is sufficient for the problem at hand. On the other hand, force optimization studies have sometimes been carried out to assist decisions on new weapon system developments, and it is not enough in such cases merely to predict a future Soviet posture against which one then attempts to evaluate various U.S. force structures. It is necessary for the analyst to recognize the dynamic nature of the arms race. For example, development by one side of improved warheads and of an antiballistic missile system might stimulate subsequent work on mobile or hidden basing and on new warheads by the other side. Over the years each side acquires knowledge about its own major weaknesses and those of the enemy, and this knowledge is reflected in the sequence of decisions about advanced weapon systems. This does not mean

---

[3] "Methods and Procedures," in E. S. Quade (ed.), *Analysis for Military Decisions*, Rand McNally & Company, Chicago, 1964, p. 168.

that analysis is impossible. But it does suggest that simple arithmetic models have some limitations.

Other Observations

We can use the missile defense model as a hook on which to hang a few additional comments. We saw that a model may involve sub-models – like the one that relates the kill probability p to an attacker's accuracy and the hardness of missile sites. We saw that building and working with a model consitute only part of a study. We saw that there is no experimental proof that a model is correct and appropriate. The physicist can test his models in the laboratory, but the systems analyst does not have access to an experimental war. And if he did, it might not help too much. Even a war might not resolve all doubts. I observe that the Civil War is still the subject of some dispute. And in some aspects, an actual war resembles a game played only once. So the analyst cannot test his models satisfactorily. The best he can do, as E. S. Quade has said, is to determine answers to the following questions:

> Can the model describe correctly and clearly the known facts and situations?
>
> When the principal parameters involved are varied, do the results remain consistent and plausible?
>
> Can the model handle special cases in which there is some indication as to what the outcome should be?
>
> Can it assign causes to known effects?

A few words about the role of judgment and intuition. The missile defense model gives the appearance of being coldly objective and free of the foibles of human intuition. We will indeed see examples of models in which human judgment plays no explicit and integral part. But in all models, including the missile defense one, human judgment and intuition enter, if not in an explicit fashion. In the first place, man designs the model, that is, he decides what factors are relevant to the problems and what the interrelations between these factors are to be in the model. In the second place, man decides the numerical values of the input variables fed into the model. And, finally, man inspects, analyzes, and interprets the results, the outputs of the model.

This fact – that judgment and intuition and guesswork are embedded in a model – should be remembered when we examine the results that come, with high precision, from a model.

I would like to think that all RAND analysts have always understood this. We haven't. I find in a 1947 document the following statement made with calm and impressive assurance:

> In so far as practicable RAND attempts to eliminate intuitive thinking and comparisons from its evaluation work. Wherever possible, the optimum instrumentality is selected by precise mathematical methods.

The role of judgment and of objective analysis in present and future studies has been described by M. M. Lavin in an internal RAND memorandum:

> In the last 5 years, study technique has really taken a back-seat at RAND. For one reason, we've just gotten more thoughtful about the criterion problem and have admitted to its multi-dimensional and semi-qualitative nature. Good decision criteria just seem overwhelmingly more important in broad problems than do analytical models. For another reason, we've begun to face up to Air Force decision problems so broadly ramified, with so many intangible components that only the most naive analyst would attempt to deal with them by analytical models . . . I venture the following anticipation: For future broad studies, particularly those concerning Air Force posture and compositions and others involving the criterion problem in its most obtrusive form, we shall continue to use intuitive, subjective and *ad hoc* study schemes . . . No individual or organization can hope to be objective. They can, however, be honest in identifying and displaying their bias. The notion that big decisions can be an automatic consequence of the application of mathematical models, cost-effectiveness analysis, or computer simulation belongs to that dreadful era when science-fiction writers, including some on the editorial pages of the N.Y. *Times*, were heralding the advent of "push-button" warfare (in some instances, with the buttons being pushed by computers).

TYPES OF MODELS

If we were to look at very many models it would be convenient to have some scheme of classification, some set of characteristics according to which we could group them. But there are many ways in which we can slice this cake, many characteristics by which we can organize our knowledge of models. For example, we can classify models according to

Purpose – training, study, and so on

Field of application – strategic, tactical, logistic, and so on

Level – from national policy to base operations

Time character – static or dynamic

Form – two-sided or one, conflict or not

Analytical development – degree to which mathematics is used

Use of computers – how much and how

Complexity – detailed or aggregated

Formalization – the degree to which the interactions have been planned for and their results predetermined.

And so on.

The classification scheme I shall use is as unsatisfactory as any other, but it will serve to suggest some of the relations between the models described elsewhere in this book (particularly in the next few Chapters), as well as between many other models that we could describe.

We shall file our models in 10 pigeonholes (Fig. 10.1) arranged in five rows and two columns. The five rows of our filing scheme describe the form of the model: verbal, analytical, and so forth. Each of these five categories is broken into two according to whether or not an active opponent is involved and conflict is an essential part of the model.

Our five categories of model form are

I. Verbal.
II. People – as an integral part of the model.
III. People and computers interacting as a part of the model.
IV. Computer.
V. Analytical.

If conflict is an essential element, then we shall speak of model types

Ic, IIc, IIIc, IVc, Vc.

(A conflict situation does not, of course, exclude the case of opponents who also have interests in common.) If conflict plays little or no role, then we shall speak of model types

Inc, IInc, IIInc, IVnc, Vnc.

When the "conflict" or "no conflict" subscript is lacking, both kinds of models are contemplated. Thus V includes all analytical models, both the game-theoretic ones dealing with conflict situations, and the host of analytical models used in operations research.

If we wished to be complete, we would have to add a few more pigeonholes. At present our classification scheme excludes physical models – for example, wind tunnels. More important, it has no place for visual models like the organization chart or the blackboard chart that could be filled in as we describe the various categories and which would make the clumsy symbolism Vc unnecessary in this model of models.

Our five categories, from "verbal" to "analytical," constitute a scale that measures, roughly speaking, how broad or narrow a part of the real world can be satisfactorily treated. Let us take a look at each type of model, beginning at the narrow end.

Analytical Models

Our missile defense example fits into category Vnc here – as do most of the models built by operations researchers. In this pigeonhole are found the models that use the interesting techniques of linear and dynamic programming, queueing theory, network theory, and so on. A computer (or computress) may be used in category V, but as an aid and after the mathematician has finished most of his work. It is characteristic of the models in V that they deal not with specificity, but with generality; not with a single play of a situation (game or otherwise), but with all possible plays of the situation.

Pigeonhole Vc contains the models of game theory. Here the analyst is concerned not with playing tic-tac-toe – that is done in box IIc or IIIc – but with the theory of optimal play. He is concerned not only with our decisions, as in the case of the missile defense example, but also with the decisions of our opponent. You may remember the cartoon which showed a high-level conference in Washington, with the speaker saying, "The way I see it, Russia thinks we think they think we're not willing to go to war." Game theory does not solve that problem, but it does furnish a framework in which one can think more clearly about the difficult problems of conflicting interests. In Chapter 11, Melvin Dresher will discuss this type of model.

		CONFLICT	
		LITTLE OR NONE	ESSENTIAL
	I. Verbal	Inc	Ic Some scenarios
	II. People	IInc Command Post Exercise	IIc War game Crisis exercise
	III. People and computers	IIInc Logistics Systems Laboratory	IIIc TAGS
Mathematical models	IV. Computer	IVnc FLIOP SAMSOM	IVc STAGE
Mathematical models	V. Analytical	Vnc Missile defense example	Vc Game theory

Fig. 10.1 – Categories of model forms

Computer Models

In the missile defense example, the relevant factors were few enough in number, and the relations between them were simple enough, that we could trace out the interactions with pencil, paper, and a little mathematics. We arrived analytically at the relation which specified, for any value of the parameters, when the defense option was preferred. In the problems addressed by the models of category IV, the relevant factors are too numerous or their interrelations too complex to be handled analytically. Instead we must write our instructions for an electronic computer, and the model thus appears as a computer program. In contrast to category V, a particular run of a computer model deals with numerical values and hence with a specific play of a situation.

One example of the models found in pigeonhole IVnc is SAMSOM,[4] a Monte Carlo model which simulates the capability of an aircraft organization to generate sorties and turn aircraft around to support peacetime flying-training programs, meet maximum effort readiness requirements, and provide combat capabilities. In Chapter 12, N. C. Dalkey will describe the global air war model STAGE (which is used in the Air Force by the Office of the Assistant Chief of Staff for Studies and Analysis), the strategic planning tool FLIOP, and other models of category IV.[5]

Categories IV and V together constitute the class of "mathematical models."

People Models

We skip, temporarily, category III. As we observed earlier, humans are involved in all models – as designers, and experimenters, and users. But in category II humans are an integral part of the model. In category IIc we find the war game, the business game, and the military and political crisis exercise. In Chapter 14, M. G. Weiner will discuss some of these models. A command post exercise in which the opponent is either absent or plays only a *pro forma* role is an example of a category IInc model.

People and Computer Models

Here both people and computers are embedded in the model. RAND's Logistics Systems Laboratory and the air defense simulations of RAND and the System Development Corporation are examples of type IIInc. The limited war game China-5, and the tactical air and ground support game TAGS, both played at RAND some years back, go into pigeonhole IIIc.

Categories IV, III, and II together make up the class of "simulation models."

---

[4] Discussed in Chapter 13.
[5] See pp. 248–250.

Categories IVc, IIIc, and IIc together make up the class of "gaming models."

Verbal Models

As we have seen, the model-builder decides what factors are relevant to his study, determines the relations between them, and traces out their interactions and implications. This activity is, more or less, what anyone does who thinks about a problem. (We have been speaking prose all our lives without realizing it.) The model-builder merely does these things explicitly and, where possible, quantitatively – his assumptions laid out on the table for any man to inspect and criticize.

If a model has no quantitative content it goes, perforce, in category I. Note that the most common study is one that combines verbal and analytical models; it is a mixture of I and V.

The scenario, whether used alone or in conjunction with other models, is often cast in the form of a verbal model. Seyom Brown will discuss scenarios in Chapter 16.

All of us use models of various parts of the real world, though in most cases we do not make them explicit and, indeed, would probably have great difficulty in laying them out for others – or even ourselves – to see. And most of our decisions must be made on the basis of these implicit models.

When an *ad hoc* committee of experts addresses a problem around a table, it attempts to arrive at a consensus on the basis of whatever analysis may be done, together with the knowledge and intuition and the implicit models of each member of the committee. For some problems this may be a satisfactory approach in spite of the difficulties that can arise –"the hasty formulation of preconceived notions, an inclination to close one's mind to novel ideas, a tendency to defend a stand once taken, or, alternatively and sometimes alternately, a predisposition to be swayed by persuasively stated opinions of others."[6]

The search for better ways of making systematic use of expert judgment has led to various techniques, including Olaf Helmer's Delphi method, which exposes the experts' views to one another's critiques by a program of sequential individual interrogations interspersed with feedback of prior and preliminary consensus.[7]

While verbal models often remain unstated and implicit, there are exceptions. For example, in a verbal model built by Anthony Downs, a

[6] N. C. Dalkey and O. Helmer, *An Experimental Application of the Delphi Method to the Use of Experts*, The RAND Corporation, RM-727-PR (Abr.), July 1962, p. 2.
[7] The Delphi technique is discussed in Chapter 18.

bureaucracy is defined as an organization that has the following four characteristics:

1. It is large; that is, the highest-ranking members know less than half of all the members personally. This means that bureaus face substantial administrative problems.
2. A majority of its members are full-time workers who depend upon their employment in the organization for most of their income. That is, the bureau members are not dilettantes but are seriously committed to their jobs. Also, the bureau must compete for their services in the labor market.
3. Hiring, promotion, and retention of personnel are at least theoretically based upon some type of assessment of the way in which they have performed or can be expected to perform their organizational roles (that is, rather than on some characteristics such as religion, race, or social class or periodic election).
4. The major portion of its output is not directly or indirectly evaluated in any markets external to the organization by means of voluntary tit-for-tat transactions.

Some typical examples of bureaus covered by the theory are the Roman Catholic Church (except for the Pope, who is elected), the University of California, the Soviet central planning agency, the U.S. State Department, the New York Port Authority, and the Chinese Communist Army. The theory has been designed to make practical predictions about the likely behavior of real-world bureaus. The theory generates specific propositions linking certain elements of the internal structure of bureaus with certain aspects of their functions and their external environments.[8]

Conclusion

As a RAND staff member, D. Ellsberg, has observed, those critics of analysis who object that it deals with an artificial and oversimplified version of the real world often have even more artificial, more highly simplified models of the world, although implicit. For example, they sometimes appear to think that the arms race may be summarized merely by referring to budgets or by counting warheads; that weapons are either invulnerable or vulnerable, first-strike or second-strike; that postures are characterized either by "superiority" or "stability"; that reliability is either perfect or impossible; that both U.S. and SU wartime objectives are simple; that many things can be assumed as certain: escalation, or all-out war, or spread of arms, Allied response, nuclear war; or that many things can be assumed impossible: thermonuclear war, big threats, big conventional war, bigger non-nuclear forces, a U.S. first strike.

Systems analysis and the use of logical models, Ellsberg argues, will not eliminate uncertainty or insure correctness; will not foresee all major

---

[8] A. Downs, *Inside Bureaucracy*, The RAND Corporation, P-2963, August 1964. (For a full account of this theory, see A. Downs, *Inside Bureaucracy*, A RAND Corporation Research Study, Little, Brown and Company, Boston, 1967.)

problems, goals, contingencies, and alternatives; will not eliminate the necessity of judgment or the effect of bias and preconception. Hopefully, they will tend to increase the influence of the "best," most informed, judgments, both on component matters and in the final weighing of decisions; they can provide choices and a market of ideas. They can discover problems, stimulate relevant questions, and encourage people to face complexity and uncertainty explicitly and honestly.

Chapter 11

MATHEMATICAL MODELS OF CONFLICT

MELVIN DRESHER

Every systems analysis must fall back finally upon a description of a battle. It is therefore natural to make the analysis within the framework of the theory of games. This Chapter describes the game-theoretic model of conflict and presents two examples to show how the model is formulated, how the game-theoretic analysis proceeds, what the resulting optimal strategies look like, and what the implications of these strategies are for military planning. The examples concern an air-defense problem and a tactical air-war problem.

INTRODUCTION

Many military problems are concerned with the allocation of resources in space or time, or both, in a competitive or hostile environment. In these problems the game factor dominates; that is, the outcome of operations can be described only in terms of the decisions made by the participants with conflicting interests. Thus, they resemble games of strategy, such as the parlor games of chess, poker, and bridge.

We can characterize the major aspects of a game of strategy in the following manner. A participant is in a situation in which one of several possible outcomes will result. He has personal preferences about these outcomes. Though he may have some control over some of the factors which determine the outcome, he does not have full control. Some factors are manipulated by another participant who, like him, has preferences among the possible outcomes. In general, the participants do not agree in their preferences. Further, chance events may influence the final outcome.

The types of behavior which result from such situations have long been observed and recorded. But it has been a challenge to devise theories to explain the observations and to formulate principles which should guide intelligent action. Indeed, even though these games of strategy have been discussed for thousands of years, the modern mathematical approach is hardly forty years old.

The theory of games attempts to abstract a certain large class of these problems of conflict into a mathematical system by further abstracting common strategic features for study in theoretical models termed "games." In brief, one may formulate the game-theoretic model in the following manner. There are two participants, whom we may call Blue and Red. Each participant is required to make one choice from a set of possible choices, and this choice will be made without any knowledge of the choice of his opponent. Each choice is commonly referred to as a *strategy*. Now, given the choices of each of the players, there is then a certain outcome, or *utility*, for each of the participants. This outcome is a function of the strategies selected by the participants. This characterization of the game is commonly known as the *normal* form in contrast to the *extensive* form, which is described in terms of the moves of the game.

Game theory does not cover – and probably no mathematical theory could – all the problems which are included in our brief characterization of conflict of interest. Two simplifying assumptions are particularly noteworthy. First, the possible outcomes are assumed to be so well-specified that each participant is able, either directly or indirectly, to assign a numerical utility to each of them in such a way that the larger of two values always indicates the preferred outcome. Thus, an individual's desire for an outcome he prefers becomes in game theory a maximization problem involving numerical utilities defined over all outcomes.

Second, it is assumed that the factors which control the possible outcomes can also be well-specified; that is, the participants can precisely characterize all the factors and all the values which they may assume.

An air defense game

Every systems analysis must fall back finally upon a description of a battle. It is therefore natural to make the analysis within the framework of the theory of games. We shall present some examples to show the formulation of a game model, the optimal strategies of the game, and the implication of these strategies for military planning.

In order to indicate how game-theoretic analysis proceeds, let us look at an air-defense problem. Like most battle situations, the combat between air attack and air defense can be viewed as what is called a two-person zero-sum game. It is two-person because there are two sides and zero-sum because the gains of one are assumed equal to the losses of the other. Thus, we have an attacker who seeks the greatest possible gains (the destruction of targets) and a defender who attempts to make these gains as small as possible. Now, the attacker has a large number of different decisions to make. He must choose a time and target for the attack, the type and number of planes, the formation and flight tactics, the weapon yields, and the diversionary maneuvers. Since these decisions, however, are made in order to destroy targets, perhaps the most important choice that the attacker has to make is of the targets for the attack.

The defender has a more limited choice of possibilities. He can choose to locate his weapons and their tactics. His most important choice, however, concerns the distribution of his defense resources among the targets subject to attack. This distribution may be known by the attacker. It is therefore important to the defender to make the selection of this allocation as carefully as possible. (Incidentally, if the defender makes an optimal allocation, he may reveal this allocation to the attacker without the attacker being able to take advantage of it. This is a property of an optimal strategy.)

We shall consider this game in a very simplified form, in which we assume only a single possibility of choice for each participant, namely, for the attacker the choice of targets for attack, and for the defender the choice of targets for defense. We wish to answer such questions as these: Should all the targets be defended? If only some of the targets are to be defended, how shall these be selected? How should the attacker select his targets for attack?

Objections may be made that we have oversimplified the problem, but we shall try to demonstrate that the problem as stated does introduce the essential elements of air defense. Further, this simplified model will yield some general principles regarding the optimal allocation of resources useful in military planning.

Description of Payoff

We will assume equal exchange between attack and defense – that is, that one unit of defense can check one unit of attack – and also that the amount of damage to any target is proportional to the number by which the attacking units outnumber the defending units. It then follows that the payoff at the target may be expressed by $v(x-y)$ if the attack x is larger than the defense y. Otherwise the payoff is 0. The proportionality constant v depends on the target being attacked. Finally, we will assume that the total payoff to the attacker is the sum of the payoffs at each target.

This air-defense game may be summarized as follows:

There are n targets, $T_1, T_2, \ldots, T_n$, which have values $v_1, v_2, \ldots, v_n$, respectively. The defender, who has D units of defense, allocates y_1 units to T_1, y_2 units to T_2, and y_n to T_n. The attacker, who has A units of attack to allocate among the n targets, allocates x_1 units to T_1, x_2 units to T_2, and so on. The total damage done to the defender can be written as follows:

$$\sum_{i=1}^{n} v_i (x_i - y_i),$$

unless x_i is smaller than y_i, in which case the i th term is replaced by 0. This expression also represents the payoff in a game-theoretic analysis of the game between the attacker and the defender. The attacker attempts to maximize this yield by selecting the targets for the attack and the defender seeks to minimize it by appropriate choices of the defenses y_i.

This game has optimal strategies that provide some general principles of optimal allocations. One is that the defender distribute his resources among the targets so that the attacker's maximum yield is as small as possible. Any other distribution of resources by the defender would be more favorable for the attacker. The optimal strategy for the attacker is to choose his targets for attack with the aid of some randomization device.

Optimal Strategies

Some properties of the optimal allocations are as follows:

1. Only high-value targets are defended.
2. Only high-value targets are attacked.
3. The targets that are defended are those which may be attacked, depending on randomization.
4. Low-value targets are not defended, nor are they attacked.
5. The attacker and defender are indifferent to the same targets. That is, if the defender has made an optimal allocation of his resources, he presents the attacker with a set of defended targets which he will prefer equally.
6. There are no soft spots in the defenses. That is, each of the defended targets yields the same payoff to the attacker concentrating his attack at that target. An undefended target yields a smaller payoff, even for concentrated attack.

It should be pointed out that a solution of this air-defense game yields optimal strategies for both the attacker and the defender. At first glance one may think that only the defender's optimal strategy is required in air defense. But since a player may be the defender of one set of targets and the attacker of another, it is useful to know the optimal strategies for both the defender and attacker.

Generalization of the Air Defense Model

As we mentioned earlier, we are looking at an extremely simplified picture of the attack-defense air game, a picture chosen only to illustrate the principles of reasoning involved. We shall now attempt to show how, keeping these principles, we can remove a number of the limitations implicit in the model. First, we can remove the assumption of equal exchange between attack and defense, $(x-y)$, and replace it by an arbitrary exchange rate. In this way we can obtain general conclusions about the allocations as a function of the exchange rate.

So far we have assumed that the attacker has been concerned with the effect on the target without regard for his own losses. We can rectify this by changing the payoff to:

$$\sum_{i=1}^{n} v_i(x_i - y_i) - \sum_{i=1}^{n} \ell_i y_i,$$

which decreases the payoff by the losses. The formal treatment of this game, as well as its general conclusion, is exactly the same as in the previous case.

The effect of the attack, (x–y), has been assumed to have the same form for every target. If the targets are very dissimilar, however, we need to vary this attack function from target to target. When we do, it again turns out that the general form of the optimal strategies is unchanged and essentially no new principles are introduced.

This is not to say, however, that every element of air defense can be introduced into this simple game by simple modification. Frequently, fundamental changes have to be made.

In view of all the reservations made in discussing the original model, it might seem that its value and the analysis are too limited. Without doubt, this is true if the model is regarded as an attempt to arrive at a comprehensive description of air defense, but this, as we mentioned, is not our objective. On the other hand, it is possible – if one closely compares the model to what air defense actually involves – to distinguish three main types of factors in air defense:

1. Those that the model can describe immediately or after slight modification.
2. Those that do not fit the model, but can nevertheless be described quantitatively by means of other models.
3. Those that cannot be stated quantitatively.

In the study of air defense – or any other problem – by systems analysis, no distinction can be made between the first two types of factors. Whatever can be stated quantitatively will be taken into account, regardless of the formal scheme it introduces. In fact, one should regard this procedure as characteristic of systems analysis: To get a comprehensive view of the problem at hand one must set up many models, each of which takes into account only certain specific factors in the problem. The results of these model studies are then combined and evaluated in relation to the importance and accuracy of the formal assumptions on which each model rests.

In this latter activity one encounters the third sort of factor mentioned above. Even in the foreseeable future there is little reason to suppose that all the conditions significant in evaluating military situations could be described quantitatively. This is especially true for those involving human reactions. Indeed, because the task of integrating all such elements can still best be performed by the human intellect, the goal of systems analysis might be to provide such a foundation that these special qualities of intellect can be put to actual use.

A TACTICAL AIR WAR GAME

Our previous example of air defense started with a very simplified game model of air defense. After solving this game, we generalized the problem by introducing additional factors to make the air-defense game more comprehensive. But we can also start with a very detailed and comprehensive game model, simplify it, and then solve the simplified game. Further, in such cases, no loss of actual detail need be necessary. An example is the game-theoretic analysis of tactical air war.

The problem of the optimal employment of a tactical air force in various theater air tasks can be analyzed as a multimove game between two sides. Tactical forces may be used on many tasks, such as the following:

Counter Air. These operations are against the enemy's theater airbase complex and organization in order to destroy his aircraft, personnel, facilities, and so on.

Air Defense. These represent air-defense operations against the enemy's counter-air operations.

Close Air Support. The targets for close-support operations are concentrations of enemy troops or fortified positions. They are attacked in order to help the ground forces in the battle area.

Interdiction. These operations reduce the enemy's military potential by attacking his transportation facilities.

Reconnaissance. The most important function of these operations is to obtain information about enemy targets.

Airlift. In this operation the planes are used to transport troops and equipment.

If we set up three categories – attack, defense, and support – it is apparent that each of the six tasks just mentioned can be placed into one or more of these three categories. For example, counter air would go into the attack category. Air defense, of course, would be placed under defense, and air support under the support category. Reconnaissance would go into the categories of attack and support, and so forth. In this way we have performed an integration of the tactical war model from many tasks to three tasks, which we can handle analytically. Thus, the problem of tactical air war becomes the problem of employing the tactical air force in the three tasks of attack, defense, and support.

Formulation of the Tactical Air Game

We view the tactical air-war game as consisting of a series of strikes, or moves, each of which consists of simultaneous counter-air, air-defense, and close-support operations by each side, undertaken to accomplish a given theater mission or payoff. Let us assume that at the start of the air operations the stronger side (the opponent with the larger air force), say, Blue, has p planes; and the weaker side, Red, has q planes, where q is smaller than p. Let us look at a strike in the campaign, say, the initial strike. Suppose that on this strike Blue dispatches x planes on counter-air operations and u planes on air-defense operations, and the remaining amount, m, on ground-support operations. Similarly, suppose that for his first strike Red allocates y planes to counter air, w planes to air defense and the remaining number, n planes, to support his ground forces. For this initial strike and for any future strikes, these decisions are made by each side in ignorance of the allocation of the opposing side. We will assume, however, that each side knows the number of planes that he and his opponent have.

Since Red allocates w planes to air defense, we can expect a reduction in the number of Blue's planes that get through to counter-air targets. The number of interceptions by Red will be proportional to w – say, cw – unless Blue's attacking planes are saturated. The proportionality constant, or kill potential, depends on the planes' characteristics and flying altitudes, and on their weapons' characteristics. The number of Blue attacking planes that penetrate Red's defenses is x–cw, as long as cw is not larger than x. If cw is larger than x, no Blue aircraft will penetrate. Hence, the number of Blue attacking planes that penetrate Red's defenses is the larger of the two numbers x–cw and 0.

The objective of Blue's counter-air operations is to reduce the enemy's air force by dropping bombs on certain targets, and the number of aircraft destroyed will vary with the number of attacking planes that penetrate Red's defenses. Increasing the number of Blue's penetrating planes will diminish the enemy's air force, but cannot reduce it by more than q. If we assume that each of Blue's penetrating planes can destroy b planes of the enemy, and that all of Red's aircraft are at risk at the time of a strike, then Blue's initial counter-air strike will destroy the smaller of the two numbers of Red planes, b max (0, x – cw) or q. The proportionality constant b depends on the target, as well as the characteristics of the aircraft used.

Red's air force is further reduced during the strike by such factors as accidents and antiaircraft fire. Let us assume that these losses are proportional to the number of planes used by Red during the strike, or aq. Let us also assume, finally, that the planes used in air defense will survive, and the Red aircraft that fail to penetrate the Blue air defense will return to base. Subtracting the losses, we obtain Red's aircraft inventory at the conclusion of the strike.

In exactly the same manner we can analyze the effect of the initial strike on Blue's inventory.

Description of Payoff

Let us look at Blue's employment of theater air forces during the campaign. Where his objective is to assist the ground forces in the battle area, the results will vary with the number of planes he allocates to ground-support operations. We assume that it is possible to construct for Blue a payoff function, giving the payoff for each strike of the campaign, in the form of the distance advanced by the ground forces as a function of the number m of planes allocated to ground support. This function depends heavily on the characteristics of the ground-support targets – for example, on the degree of concentration of troops, vehicles, and material, and on the fortification of positions. We make no attempt to give the explicit form of this function, but merely assume that the payoff, f(m), is a positive function that increases with increasing allocations.

If Blue's ground forces now must advance while being subjected to Red's ground-support sorties, Blue's yield in ground support is no longer equal to f(m) as described above, but is reduced in accordance with the number n of planes allocated by Red to close-support missions. If g(n) is the function that measures the distance gained by Red's ground forces, then the net advance of Blue's ground forces, if he allocates m planes to ground support while Red allocates n planes to ground support, can be written as

$$Y(m, n) = f(m) - g(n).$$

This expression represents the payoff to Blue for one period or one strike. The payoff for the entire campaign of N strikes is the sum of these net yields for each of the N strikes, or

$$M = \sum^{N} [f(m) - g(n)].$$

The problem faced by each side is now apparent. For example, Blue would like to allocate a large number of planes to ground-support missions and thereby increase the value of f at a given move; yet he would like to destroy the Red air force by means of counter-air operations in order to ensure that g is small, or zero, for subsequent moves. Further, if he does not provide for air defense, he may suffer severe losses to his own air force, if Red elects to mount a large counter-air strike. Each player has to take the future and the possibilities open to his opponent into account.

For the game described here, it turns out that optimal procedures for play, or optimal strategies in the game-theoretic sense, do exist. For particular functions f and g, we shall give a qualitative description of the optimal tactics.

Further Simplifying Assumptions

To simplify the computations, we assume that Blue and Red have the same air-defense potential: Each plane allocated to defense can prevent one attacking plane from reaching its target; that is, we assume that $c=1$. We also assume that each attacking plane that penetrates the defense can destroy one plane in an airfield strike, or $b=1$, and that losses because of aborts, accidents, and antiaircraft fire are negligible. We again emphasize the fact that these simplifying assumptions have no effect on the general form of the optimal strategies. They are introduced merely to simplify the calculations.

To further simplify the computations, we assume that the yield functions $f(m)$ and $g(n)$ are linear, say, $f(m)=m$, $g(n)=n$. The payoff in the campaign, then, is

$$M(x, u; y, w) = \sum^{N} [(p-x-u) - (q-y-w)].$$

Blue wishes to make this payoff as large as possible by properly choosing the x's and u's during each of the N strikes, and Red wishes to make the payoff as small as possible by properly choosing the y's and w's.

Strategies of the Tactical Air Game

The strategies available to Blue and Red are characterized by the strength of their forces during each strike of the multistrike air campaign. Blue's strategy can be specified by the number of planes he allocates to counter-air operations, the number of planes he allocates to ground support, and the number of planes he commits to engage Red's attack. Red's strategy can be similarly described. These specifications are to be given for the first strike, the second strike, . . . , and the last strike of the campaign. Of course, the allocations during any strike will depend on the strengths of the forces of the two sides at the beginning of that strike, since we are assuming that each side knows the strength of the opponent's forces at that time.

Optimal Tactics

A complete description of the optimal employment of tactical air forces must be given in terms of the number of strikes and the relative strengths of the two sides. However, there are certain general conclusions that apply to all campaigns.

Campaign ends with ground support. The campaign always ends with a series of strikes on ground support; that is, during the closing period of the campaign both Red and Blue concentrate all their forces on ground-support missions. In this terminal period both sides have the same optimal tactics, regardless of their relative strengths.

Stronger player splits his forces. Except during the closing phase of the campaign, Red and Blue have very different optimal tactics. During any of these early strikes, the stronger side, say Blue, has a pure strategy. That is, there exists a best allocation of Blue's air force among the three air tasks. In this connection, there is a critical value of the ratio of the Blue force size to that of Red that governs Blue's allocation during the early period in the following manner. If the force ratio is less than this critical value (which is about 2.7), then the optimal allocation in the early period consists of splitting the stronger air force between two air tasks, counter air and air defense, and neglecting ground support. The size of the split depends on the relative strength of the two air forces and the number of strikes left in the campaign. However, if Blue's strength relative to Red's is greater than the critical value, then Blue should divide his force in a fixed way, regardless of his strength, among the three tasks, counter air, air defense, and ground support. The number of aircraft allocated to each mission is, however, still dependent on the number of strikes remaining.

Weaker player mixes tactics and concentrates forces. The weaker combatant cannot use a single strategy, but must bluff during all the strikes other than those of the closing phase. Unlike the strong player, the weaker player does not have a single allocation that is best. He must use a mixed strategy. He concentrates his entire force either on counter air or air defense, but which of these tasks receives the full effort is decided by some chance device.

Blue's defense decreases during campaign. Thus, prior to the closing phase of the campaign, Blue splits his forces among his air tasks. The actual value of the split is a function of the force sizes of Blue and Red and the number of strikes left in the campaign. As the campaign proceeds, the fraction of Blue's force allocated to air defense will decrease, and the fraction allocated to counter air will increase. During this time, the chance that Red will attack Blue also decreases, while the chance that he will defend himself increases.

Blue's defense increases in early stages of a long campaign. In the early stages of a relatively long campaign, the stronger side defends itself against a concentrated attack by the weak side. During this period, Blue dispatches on air defense a force of planes approximately the size of Red's entire force.

(Recall that we assumed a particular value for the effectiveness of air defense.)

Generalization of the Tactical Air Game Model

We can generalize our model by removing some of the restrictions. For example, we have assumed that each plane can destroy as many planes in the air as it can on the ground. Actually, the air-kill potential is much less than the ground-kill potential. However, the general properties of the optimal tactics as described are applicable for arbitrary values of the kill potentials; that is, the game still ends with a series of moves in which both sides concentrate on ground support, while at each move prior to this the stronger side splits his force between counter air and air defense and the weaker side randomizes among the various tasks. The magnitudes of the kill potentials determine the actual split of forces for the stronger side, and the probabilities associated with the mixed tactics determine the split for the weaker side.

In addition, we have assumed that each side has been concerned only with the effect on excess ground support and not his own plane losses in the air battle. But since the losses in the air duel are small compared to the losses on the ground, this omission cannot appreciably affect the solution.

Although the model does not contain force replacements, no difficulty arises in the analysis of the game if we assume a replacement schedule that is independent of the tactics employed. In such cases the optimal tactics are the same, though the game value is changed.

Even with such extensions as these, however, it is important to note that not all aspects of tactical air war can be handled by the model described. Essential modifications must be made in order to analyze tactical air war. We list briefly a few of the limitations of our model:

We have assumed that each plane is not only capable of performing each of the three air tasks, but is equally effective on each. Actually, of course, there are different types of planes, not all of which can perform all three tasks.

We have assumed a continuous payoff function. The realistic case where a discontinuity in the payoff occurs if the front line reaches a particular point has been excluded.

We have assumed that the counter-air strike is equally effective at all times, regardless of the defense. It would be more realistic to assume that, if a combatant employs no air defense, then strikes by an attacker will be very effective since, among other things, he may now fly at a lower altitude.

The design is static. We have assumed that the destruction of a target always has the same value to the attacker regardless of the status of the campaign.

The duration of the campaign is known when the campaign begins. Actually, the duration of the campaign may depend on the tactics employed.

In order to overcome these limitations, we must design additional models, perhaps many models, each of which takes into account only certain features of the tactical air problem.

Applications of game theory

No presentation of the theory of games would be complete without a discussion of its applications. Of those that concern military problems, John Williams, in 1954, stated in his book, *The Compleat Strategyst:*

> While there are specific applications today, despite the current limitations of the theory, perhaps its greatest contribution so far has been an intangible one: the general orientation given to people who are faced with overcomplex problems. Even though these problems are not strictly solvable – certainly at the moment and probably for the indefinite future – it helps to have a framework in which to work on them. The concept of a strategy, the distinctions among players, the role of chance events, the notion of matrix representations of the payoffs, the concepts of pure and mixed strategies, and so on give valuable orientation to persons who must think about complicated conflict situations.[1]

Ten years later at a NATO conference on the theory of games held at Toulon, France, Dr. Clayton J. Thomas of the Operations Analysis Office, Vice Chief of Staff, Department of the Air Force, reported:

> Applications of game theory have been neither non-existent on the one hand, nor yet very dangerous to sound defense planning on the other hand. The theory has been most useful at a "tactical" level, well below that of "grand strategy."[2]

Specifically, the "no soft-spot principle" discussed in the air-defense game model – the principle that each defended target yields the same payoff to the attacker and undefended targets yield less – has had much application. Again quoting from the speech by Dr. Thomas:

> The principle of "no soft spots" has been of tremendous value in the allocation of defenses. This derives in part from its appeal as a simple unifying concept. Also of importance, however, has been the recognition of the principle by key personnel and their ingenious applications of it in defense planning over a period of several years. This has given a mathematical framework within which to fit otherwise

---

[1] J. D. Williams, *The Compleat Strategyst: Being a Primer on the Theory of Games of Strategy*, A RAND Corporation Study, McGraw–Hill Book Company, Inc., New York, Revised Edition, 1966, p. 217.

[2] C. J. Thomas, "Some Past Applications of Game Theory in the United States Air Force," Paper presented at the NATO Conference on the Theory of Games and Its Military Applications, June 29–July 3, 1964, Toulon, France.

unrelated observations. It has aided in the evaluation of different weapon systems and different defense systems. The principle has served as the point of departure for other interesting investigations also, like the search for a method of dividing a budget between strategic offense forces and continental defense forces.

In light of the progress of game theory in the last decade, it might be wondered why we restricted ourselves, in this Chapter, to examples that represent such highly simplified problems of strategic and tactical air war. If, in order to solve a game analytically, we must limit our inquiry to essentially only a single factor, do we not thereby severely limit the usefulness of the results? For several reasons, this need not be the case, as was pointed out by S. Golubev-Novozhilov, in his preface to the Soviet edition of the author's *Games of Strategy: Theory and Applications:*

> We must forewarn the reader regarding errors related to the degree of approximation of the mathematical models discussed in this book as illustrative examples, to real conflict situations encountered in solving war problems. These models have been greatly simplified, to be sure. Nevertheless it has seemed to us that this is justified methodologically. The complexity of the game models would have led to an overly cumbersome book, and would not have added to its value. In addition, the complexity of a model does not always add to its usefulness in finding correct solutions.
>
> On the other hand, by adopting simplified game models and by penetrating them in essence and in substance, the reader can move on to independent construction and analysis of more complex models, as may prove necessary from practical considerations. It is certain that this must be done while bearing in mind the computational difficulties to be encountered in solving game problems.

Chapter 12

SIMULATION

NORMAN C. DALKEY

Simulation is a technique for studying complex military processes. It consists of an abstract representation of the more important features of the situation to be studied, designed to be played through in time either by hand or by computer. The basic advantages of simulation are that hypothetical future conflicts can be investigated in terms of elementary events, and precise, reproducible models can be constructed of processes for which there are no general theories or analytical descriptions. The major disadvantages are that simulation is slow and expensive and the range of cases that can be treated is highly limited. New developments in computers, simulation languages, and model structures will extend the utility of simulation for the military analyst.

INTRODUCTION

Simulation is one of the more extensively used of the tools available to the military analyst. The primary reason is that military conflict involves a complex interaction of numerous elements – kinds of weapons, patterns of deployment and employment, rapid changes over time. In many instances, simulation is the only technique by which this intricate interplay of factors can be studied in a precise and reproducible fashion.

The word "simulation," a distant relative of the term "similar," refers to a construct which resembles the process or system to be studied but which is easier to manipulate or investigate. Examples might be a model aircraft in a wind tunnel, an army field exercise, a group of subjects in a psychological laboratory where the individuals play the role of major military commands, a computer routine which describes the minute-by-minute activity of aircraft and missiles in a nuclear exchange. In these cases, the real object of interest, because of its size, expense, uncontrollability, or danger to national security, is difficult or impossible to study. A representation of the object which is similar to it in essential properties is investigated instead.

If we include all the kinds of investigative aids mentioned above, we wind up with a very diverse set of things to talk about. For the purposes of this book, it is convenient to focus on a limited few. Accordingly, study objects which physically resemble the phenomena of interest, such as the wind-tunnel model or the field exercise, will not be discussed. Constructs where human subjects play the analogue role or where human judgment and decisions influence the course of the exercise will – rather arbitrarily – be assigned to gaming, which M. G. Weiner will discuss in Chapter 14.[1] There remains the case where the construct or representation is a logical or mathematical model and where the course of a play or run of the simulation is determined by a set of formulated rules.

There is no sharp distinction between simulation and analytic models of the sort R. D. Specht discussed in Chapter 10. What difference exists lies primarily in degree of generality or abstractness, and in the way in which the models are manipulated.

The analytic model is likely to be quite abstract and deal with aggregated entities, such as number of weapons and number of targets, whereas the simulation is likely to refer to a list of specific weapons or individually named targets. The analytic model is likely to be expressed by a set of equations, whereas the simulation may be expressed by a set of rules determining what "happens" under various circumstances. Finally, the analytic model is likely to be formulated primarily for the purpose of finding a "solution" to the equations – for example, an optimal strategy, a least expensive combination of weapons, a best mix of warheads and decoys in a missile payload. The simulation, however, will be used to investigate a specific case, such as the outcome of a duel between a given type of fighter and a given type of bomber when the fighter undertakes a tail chase from a given position relative to the bomber; the relative damage to each side in a central nuclear war if each uses a specific allocation of weapons to targets; and so on. In short, the simulation is likely to be used in an experimental fashion, to generate specific case studies or instances; the analytic model will be used to compute some over-all quantity or strategy. Simulation can be used in a laboratory fashion to generate data to suggest or test hypotheses. If systems analysis is viewed as a scientific endeavor, the experimental role of simulation is probably the preferred employment. However, in many studies this process is shortened – because of exigencies of time, cost, or the simple failure of the results to suggest a useful theory – and specific results are presented for the edification of the decisionmaker.

Although a simulation can be carried through manually, using more or less traditional map exercise and hand accounting methods, the high-speed computer is ideally suited for keeping track of the many items and performing the large number of separate computations that determine the changing status of the elements. In the case of more extensive simulations,

[1] This limitation is arbitrary in the sense that the term "simulation" is frequently applied to laboratory exercises involving human subjects as representations of suborganizations.

manual computation is only a theoretical possibility. Literally millions of man-hours would be required to perform the exercise.

SIMULATION BASICS

Most military simulations exhibit a common structure which can be expressed in terms of

1. Elements,
2. Attributes,
3. Activities,
4. Plans,
5. Time.

"Elements" are all the items involved in the interaction – missiles, aircraft, missile sites, airfields, support targets, and the like. "Attributes" are the properties of the elements – location, type, speed, status (surviving or destroyed), and so on. Some of the attributes such as CEP, weapon load, or kill probability, may be designated as *parameters* on the ground that these are the factors likely to be varied during a study. "Activities" are rules prescribing what will occur under various circumstances – such as radar detection, fighter-bomber duel, bomb drop, missile interception, and so on. "Plans" are the prescriptions of how weapons are to be employed (strategy, tactics, doctrine). In many instances where plans are reducible to a simply formulated doctrine – as, for example, a simple x to 1 allocation of fighters to bombers in an air-defense model – they may be included as an aspect of "activities," but in other cases they may require elaborate instructions that indicate how each individual weapon is to be employed. In such cases, a large part of the input may consist of the plans for each side.

The role of time in a military simulation is of basic importance. Military conflict is not a static balance of forces; it is a dynamic interaction of destructive events, where the relative time of an occurrence can be crucial. A large measure of the complexity in military matters stems from the intricacies of temporal change. One of the fundamental values of simulation is that it can display a complex pattern of events in time.

There are two ways of handling time in a simulation: the *interval* and the *event* technique. In the interval technique, time is divided into a number of sections, usually equal. The conflict is examined interval by interval and, except for the order in which activities are taken up, the occurrences during a time interval are considered as simultaneous.

In the event type of model, a list of potential occurrences is compiled on the basis of planned actions and the events these can produce. This list must be edited frequently, some events being cancelled and new events added. For example, the planned arrival of a bomber over target must be cancelled if, prior to arrival, it is shot down by air defenses. This event may, on the other hand, give rise to a new one, namely, the unplanned detonation of weapons during the crash. Otherwise, the events list is processed by taking up each event at the time it is scheduled to occur and applying the appropriate activity. Theoretically, the event type of model allows as fine a division of time as might be wished; practically, the fineness of division is limited by the number of events that can be managed.

Event-type models have become more popular than interval types because of their greater freedom in dealing with highly time-dependent interactions. However, this advantage is compensated for by the need for storing and updating a large list of events. In some models, both methods of handling time are used, where event processing is resorted to only for critical interactions.

Another basic consideration relates to the treatment of chance, or probabilistic events. Most events of military interest are partially determined by chance. Abort probability, target damage probability, kill probability, CEP, and so on represent instances of basic planning factors in military analysis that are by and large statistical rather than exact. As in the case of time, there are two major techniques for handling chance events, although combinations are also possible. The two are *expected-value* and *Monte Carlo*. In an expected-value model, when a chance event arises, the expected result of that event is assumed to occur. For example, if a group of 10 aircraft is flying over a defended area, and the probability of a plane being shot down is 30 per cent, the model presumes that 3 are shot down and 7 survive. In a Monte Carlo model, on the other hand, the outcome of probabilistic events is determined by chance. In the example just given, a random number between one and a hundred might be generated for each aircraft. This random number would be compared with, say, 30. If it is greater than 30, the aircraft is assumed to survive; if less, the aircraft is presumed to be shot down. Although, in a large number of such events, the average number surviving will be 7, in any particular case the number might be larger or smaller.

In a Monte Carlo model, a single run is one sample out of a very large number of possible cases. In order to discover what the expected or average outcome of the conflict would be, it may be necessary to run many cases – take a large sample – varying only the random numbers selected. This is a drawback of the Monte Carlo method of handling chance events in large models. On the other hand, the expected-value model says nothing about the variance – how widely outcomes can differ from the average result. The Monte

Carlo technique can, with sufficient cases, give some indication of how far away from the average outcomes may be.

Both types of models have their value. The Monte Carlo type appears to have greater favor at the moment because of the advantages mentioned, but also because it lends an air of realism to a simulation, and because it allows chance to be applied to single events. If a lone aircraft is examined flying over a defense, and the probability of its being shot down is, say, 45 per cent, there is something repellent about saying that 45 per cent of the plane is shot down and 55 per cent continues on to bomb a target.[2] But while the language of the expected-value model is perhaps strange, it is not necessarily wrong. With the Monte Carlo technique the plane is either shot down or it is not, depending on the random number drawn.

Another basic consideration in constructing a simulation is level of detail. Air forces can be described in terms of individual aircraft flying separate routes to specific bomb release points, or in terms of groups of aircraft of varying size attacking groups of targets. Local defenses can be expressed in terms of individual missile launchers, missile complexes, or simply the level of local defense for a set of targets. Time can be divided into seconds, minutes, or hours. In general, there is no "right" level – the level of detail that is selected should depend on the problem being posed and the resources of the analysis team. However, it rarely is worth creating a detailed simulation for a single study unless the study is very extensive. The reason is that building a simulation is expensive in time and effort, and an extensive use of the model is necessary to make the cost worthwhile. The statement is made more than once in this book that there are no general-purpose models, that the model should be tailored to the problem.[3] This is correct, but in the case of large-scale simulations an attempt is usually made to introduce some general-purpose features – to make the model sufficiently flexible, so that a variety of problems can be dealt with. This can be done to some extent by allowing a fairly large number of parameters that can be easily changed to define new weapons characteristics, changed force structures, or different modes of employment. As a result, simulations are commonly constructed with more detail than a particular problem requires.

Finally, a basic question concerning simulation is the computation technique selected. It has already been pointed out that high-speed computers are ideally suited for the type of data manipulation involved in most simulations. For some elaborate simulations, it would be completely imprac-

---

[2] I have had an Air Force officer challenge me by asking, "If you were 65 per cent of a fighter pilot, how would you attack 15 per cent of a bomber?"

[3] Note especially Chapters 3 and 10.

tical to conduct the exercise without a high-speed computer. In other cases, however, different methods have been useful. One technique, the map exercise, has proved of continuing value, especially for simulations involving ground forces. Variants of the map exercise, in which forces are represented by poker chips, have been useful in exercises on a smaller scale. The chips are convenient for bookkeeping purposes. Furthermore, the graphic "picture" of the course of the war afforded by the map is often valuable to the analyst in furthering his understanding of what is going on.

Choice of simulation techniques

The preceding remarks lead naturally to the question, When is it reasonable to use a simulation, and how does the analyst go about setting one up? The principal reason for resorting to simulation, as was indicated in the introduction, is that the phenomena to be studied are too complex to be manageable in any other way. It is fair to say that simulation is often used because of ignorance – the analyst does not know how military events proceed in the large, and hence cannot formulate a simple, general model of the conflict. On the other hand, he can express the situation in terms of elementary events because he understands them or, equivalently, he may have data only on elementary events and not on global interactions. There have been too few modern wars to derive general relationships from direct experience. War is especially difficult in this regard, because of the two-sidedness of the conflict. Many wars would have to occur to give some indication of the effect of different war plans. Elementary events, on the other hand, are more closely related to peacetime activities. Many are subject to peacetime exercise and test. By formulating the simulation in terms of elementary events, complex interactions can be synthesized.

A problem can be complex in ways other than involving a large number of elements or intricate interactions. In general, the formulation of a simple model of a military situation requires that a simple expression for the payoff or the criterion must be available. But in many cases it is not possible to express a simple payoff. For example, in a nuclear exchange we would be interested in the damage to civilian targets on each side, in the forces remaining, in the fallout contamination, in effects on allied nations, and so on. But there is no simple trade-off among these effects that will allow us to produce a single index, and, above all, there is no simple criterion for determining which mode of attack is preferable. This is so because – among other reasons – nuclear conflict is nonzero-sum. Both sides can lose catastrophically, depending upon the modes of attack selected.

In a case where the payoff and the criterion are unclear or complex, it frequently happens that the only useful method of proceeding is to exhibit the outcomes of several cases, and let the decisionmaker "make up his feelings" about them.

A third kind of complexity is related to uncertainty. There may be many factors in the situation about which – even in elementary form – we do not have sufficient information. This is particularly true if we are examining proposed weapon systems or conflicts several years from now. One popular mode of procedure in this kind of situation is to express in the formal analysis just that part of the situation that we do have solid information about and leave the uncertainties to an informal, judgmental "discussion." For many types of problems this is a reasonable way to proceed. But for many others it is desirable to include the uncertain elements in the formal analysis. There are several ways in which this can be done. Separate cases can be run for a range of values of the uncertain items; or the cases can be examined closely to see at what stage – if at all – the uncertain items are crucial; or a set of extreme values can be tried to see what the total effect of the uncertain factors might be. None of these procedures is a completely satisfactory answer to the problem of uncertainty because it is rarely possible to do a thorough job – the number of cases required is generally prohibitive. But they are frequently more informative than the less formal exercise of judgment.

When it comes to the specifics of laying out a simulation, the familiar caveat that we are concerned with an art and not a science holds. In general, the form of the simulation and the level of detail will be determined by the problem to be tackled and by the resources at the analyst's command. Other factors will be influential – the kind of data available, deadlines, and so on.

It should be reiterated that a simulation is generally only a part of the over-all systems analysis. Usually, simulation will be concerned with the effectiveness computation. The simulation model needs to be supplemented by a cost model and an evaluation technique. This last is frequently referred to as "analysis of results," but a great deal more is involved than mere tabulation.

In setting up a simulation it is generally a good practice to formulate first a small, possibly aggregated example and play this simulation by hand. The small model provides a way of checking whether or not the simulation has a complete structure and adequately contains the factors of primary interest. Several attempts to develop simulation of ground warfare have failed at this stage. There is no problem in defining the elements, attributes, activities, and time scale for ground warfare, but the representation of *plans* – that is, the simulation of the complex decision process for ground forces – has not been satisfactorily solved. The small model also provides a way of obtaining gross estimates of the significance of various factors.

Some items can be left out of, or aggregated in, the production model if the exploratory model indicates they are not critical.

After the exploratory phase there are several directions in which one can proceed. The basic model must be translated into a detailed program and this in turn must be checked by trial runs. If the program is to be run on a computer, additional coding and debugging are involved. A large proportion of the study effort is usually absorbed by the collection of data and transformation of those data to a form suitable for the model. After the runs or exercises, analyzing the voluminous information generated is normally a major task.

Examples of the use of simulation

Some examples may give a feeling for the wide range of military situations which have been dealt with by simulations and some feeling for the variety of model structures that have been employed.

Perhaps the archetype of the large-scale military simulation is the global air war model, STAGE, used in the U.S. Air Force by the Office of the Assistant Chief of Staff for Studies and Analysis. This model, which has gone through a number of versions (it was known as the Strategic Operations Model when initiated at RAND and later as the Air Battle Model), provides a highly detailed play of a world-wide nuclear exchange. It follows the movement of individual aircraft and missiles as they leave their airfields or sites, as they travel through air or space, and as they pass through surveillance radar coverage and defenses; it computes how many are lost to enemy interception; and it determines the damage that the surviving weapons inflict on military and civilian targets. For bombers, the activity of supporting tankers is followed, and, for missiles, decoys and other penetration aids can be programmed in.

The air battle is scrutinized every few minutes of simulated time, allowing a rich interplay of operational constraints, relative timing of penetration and attacks on defenses, and so on. Because of its large size, STAGE requires several hours on an IBM 7090 computer for a single Monte Carlo run, in which many thousands of chance events are evaluated. It also requires the investment of hundreds of man-days in the preparation of inputs. STAGE is admirably suited for the detailed evaluation of war plans and for testing the effect of operational constraints on the execution of a war plan during enemy attack.

On the opposite extreme is a model called FLIOP, which was designed at RAND as an aid in strategic planning. FLIOP checks the feasibility of a single bomber mission. The input is a detailed profile, including potential refueling points, of a bomber mission from its beginning at the take-off base, through its flight over enemy territory (visiting one or more targets), to its ending at a recovery base. FLIOP is also coded for high-speed computer. It operates by "flying the bomber backwards," that is, by starting at the landing base and accumulating fuel and the weight of bombs as it backs up. Missions are judged infeasible if the fuel required exceeds the capacity of the bomber or if the over-all weight exceeds the maximum flying weight before a refueling point is reached. If these restraints do not operate, a tanker can, as it were, remove the excess fuel at refueling and the bomber can accumulate more weight on its way back to the take-off base. The routine computes the required off-load and also determines the feasibility of the tanker mission. The routine takes only a few minutes on the computer, allowing the computation of hundreds of flight plans in a day. As an aside, we might note that the peculiarity that the simulation operates in reverse has been modified in practice. It is clearly more intuitive to have the aircraft fly forward, even in a computer!

Several models exist which evaluate an anti-ballistic missile defense against combinations of incoming warheads and penetration aids. Because time is critical in ballistic missile interception, these models break time down into very fine-grained intervals. They are examples of a rather rare sort of simulation using high speed computers. It may take several hours to compute an interaction that in reality would involve only seconds.

The models follow the trajectories of incoming warheads, decoys, and other penetration aids, assess the probability of detection at various stages, schedule interceptors according to input doctrines, and assess kill probabilities. A variety of interceptor characteristics, payloads of ICBMs, and firing doctrines can be examined.

Similar models exist for evaluating a duel between a fighter plane and a bomber. An initial relative position and heading for the fighter and the bomber are selected, and the model simulates the chase, taking into account the constraints on speed and turning radius of the fighter and the effects the bomber can produce on the fighter's radar by using electronic countermeasures. The fighter's armament will be activated if it is successful in achieving a firing position (as defined by a preassigned doctrine). By running a large number of initial positions and headings, we can estimate an average kill probability for a specific fighter configuration against a given bomber type.

It should be pointed out that simulations, in the sense in which the term has been used here, can be submodels in a larger, less formal exercise. Thus, the activities of a depot can be simulated as an element in a larger logistics exercise. Attrition models, or damage models, have been used as parts of a theatre war game where much of the play is determined by the decisions of teams of players or of umpire teams, or both.

The preceding examples are only a few of the hundreds of simulations that have been developed in the last few years for military analysis. Simulation routines have been developed for evaluating damage to communications in a nuclear war, evaluating the effects of conventional bombing on troops and equipment in a non-nuclear attack, assessing the effects of terrain on the coverage of defensive radars, computing the rapidity with which a task force can deploy from the United States to a foreign theatre. There are very few areas of interest to the Air Force that have not been made the subject of a simulation.

Pros and cons

Because of the widespread use of simulation for military analyses, it is important to achieve a perspective on its good and bad features. On the positive side, and pre-eminently, a simulation may be the only feasible way to analyze a highly complex system or process. By reducing the complex process to more elementary activities, simulation provides data from peacetime experience or tests that can be used for analyzing hypothetical conflicts. In addition, by breaking a complicated situation down into a series of simple interactions, a simulation can make the evaluation of plans or weapons effectiveness more understandable to military decisionmakers. The language of a simulation is usually much closer to the language of the military officer than the language of a mathematical analysis. This aspect of simulations has a number of ramifications. Since the simulation is a completely formulated model, all concepts employed in it must be completely and sharply defined. This is especially true if the simulation is programmed for a computer. Hence, the simulation furnishes a common and precise language for a team of specialists working with it. It also can provide a common language for the specialists and the decisionmaker. The fact that the assumptions of the model are explicit and the results can be duplicated is extremely important when a sizeable community with differing interests (for example, the various agencies of the Department of Defense) is interacting on a problem.

The model also furnishes a logical structure or framework for the data involved. Frequently the simulation is itself a check on the consistency and completeness of the data. Some of the most useful data files in the Pentagon are those developed for simulation exercises.

More generally, even in those areas where solid data are lacking and the analyst must proceed by assumption, judgment, or guesstimate, a formulated model requires that the assumptions be clearly spelled out. The simulation can check the consistency of assumptions. Most important of all, the simulation can derive the consequences of assumptions in an impersonal, objective fashion. These properties, of course, are common to all completely formulated models.

On the negative side, simulation has a number of shortcomings from the point of view of an analyst conducting studies for military decisions. Above all, simulation is likely to be a slow and cumbersome method of attacking a problem. This is especially true if the simulation is coded for a computer. Despite advances in the programming art, a sizable effort is required to formulate and code a simulation of even moderate size. Furthermore, a large computer routine is difficult to modify. If some assumptions turn out to be inappropriate after the first few runs, a major reprogramming may be required. In many cases, this fact leads the analyst to build in general-purpose features ("flexibility") which overly complicate the model and extend programming time. Almost by definition, simulations pose difficulties in predicting which features will turn out to be inappropriate, or be outrun by technology.

A simulation is likely to be restrictive with respect to the range of cases that can be examined. In most situations of military interest, the number of possibilities worth looking at is enormous. There will be a range of possible weapons, a variety of possible employments, and large areas governed by chance. And these possibilities are inflated by applying to both contestants. For most simulations, the best the analyst can do is select a very few cases out of this vast spectrum of possibilities for examination.

Some of the values of simulation can easily be overplayed and turned into liabilities. The fact that simulations are couched in a language close to that of the military can give a false air of reality to the results. The fact that activities are examined in minute detail and at electronic speed can lend an air of glamour to an exercise that can turn the head of a decision-maker who is not fully aware of the guesses and approximations that went into the study. Because simulations can be set up in a fairly direct fashion, by stringing elementary processes together, it is all too frequently the lazy way out of a problem. In many instances, simulation is not the best approach, but it is the easiest.

One of the more insidious drawbacks of large simulations is that – although the elementary events are perspicuous and easy to understand – the sheer volume of occurrences is so great that the model will be treated as a black box into which data are shoveled and out of which neatly packaged results are delivered. I once sat in on a briefing on a large military model in which one "explanatory" chart showed precisely this analogy, complete with hopper and conveyor belts.

The question inevitably arises, "How do you know that a simulation actually represents what it is supposed to?" This question is usually accompanied by the query, "Why don't you try your simulation on some historical war or battle and see if it can predict the outcome?" This issue is common, of course, to all military systems analysis, but it arises more naturally with regard to simulation because of the greater air of authenticity which surrounds it.

The answer is that in most cases we can't determine how good the simulation is. The ultimate test might appear to be a war of the type being simulated. But even if such a war should occur, the outcome might look quite different from the outcome of the simulation, and yet the simulation might not be wrong. Chance, accidental features not in the model, on-the-spot decisions of commanders, and so on may strongly affect the outcome. This, in effect, is the answer to those who ask why we do not try to simulate a specific historical battle. One of the major determinants of historical battles is the specific decisions made by commanders. To be candid, we do not have a good way to simulate such decisions. In a simulation, the course of the exercise is determined by a set of decisions – plans or doctrine – usually devised by the analyst. The plans and the doctrine may be very good, but they may not represent the decisions of any real commander in the thick of battle – both may be right, but very different.

Why not limit simulation to those situations or those problem areas where there is good solid information? A rather good case can be made for the presumption that simulation has been applied where not enough is known to justify the elaboration of ignorance. On this question, the good sense of the analyst generally provides the only guide. Limiting simulation to problem areas where impeccable data exist would exclude the technique from most systems analyses. Frequently, the attempt to build a simulation in a shadowy area uncovers significant features of the problem that were overlooked or hidden in qualitative discussions. The simulation then serves to define the needed information in a sharp fashion, and to point up its importance.

New developments

Some of the drawbacks of simulation can be ameliorated, and several developments are under way that promise to increase its practicality and power.

This Chapter is not the place to discuss at any length the impact of increasing computer capabilities, but it should be pointed out that developments in faster, more capacious machines, with the capability, for example, of parallel rather than purely serial computation, will make some of the larger types of simulation easier to manage. Improvements in data handling – for example, more direct access by the machine to data files – will simplify generating inputs. The major transformation that is now under way toward maintaining military information in machine accessible form will undoubtedly make the task of simulation-building easier. Conversely, the construction of more extensive simulations will improve the structure of data files.

Of more direct interest is the rapidly expanding number of simulation languages being developed for high-speed computers. These languages, which bear interesting names like GASP, SOL, Militran, Simscript, and SOS, are an attempt to furnish the analyst with a certain basic skeleton of a simulation: list-processing structures to simplify the defining of elements and attributes and the handling of internal data; an event processor for automatically scheduling and computing events according to activities specified by the analyst; output techniques for more or less automatically generating the information the analyst wishes from his simulation. These meta-simulations save the analyst the time required to reproduce those elements which are common to a large number of simulation routines. They have proved highly useful in a number of studies in reducing the time required to set up a simulation and also in simplifying modifications in a routine after its initial formulation. Although not applicable to all types of simulation, and probably not yet of great value for the larger models, these efforts are expanding rapidly and promise to remove much of the repetitive, housekeeping part of the job of constructing simulations.

One other development we might note is the marriage of simulation with other techniques. Outstanding among the serious handicaps of simulation is its case-study quality. On the other hand, one of its strong points is its ability to deal with a richly detailed problem area. It is feasible to construct not a single model, but a family of models, at different levels of generality or aggregation. By covering the same problem at several levels, it would be possible to employ the tools of mathematical analysis – optimization, sensitivity analysis, trade-off analysis, and so on – at the higher levels of generality and then check the accuracy and feasibility of the solution against more detailed models of the simulation type. This prospect appears most inviting in the area of complex planning – for example, in strategic attack planning or in logistics planning for a large deployment – where it is desirable to survey a large number of possible plans before deciding on a preferred one, but where, at the same time, the operational feasibility of the plan is a significant criterion.

In the area of strategic attack planning, RAND has developed experimentally such a family of models. It consists of a highly aggregated, two-sided war game which plays a single nuclear exchange in a hundredth of a second, an intermediate-level simulation of a two-sided air war which computes the result of implementing a pair of plans in a matter of minutes, and finally a planning routine which takes the intermediate-level plan and unpacks it in great detail, including tanker support for bombers, relative timing of penetration, and specific geography for launch sites and targets. The smallest model can be employed to survey thousands of cases in a hasty or – so to speak – back-of-the-envelope fashion. The lower-level models can then be used to spell out the details and check the reasonableness of the outcome of the upper-level survey.[4]

This type of hierarchy of models can be used to compensate for the cumbersomeness of the simulation, and to relieve the abstractness of more analytic techniques.

---

[4] For a fuller description, see N. C. Dalkey, *Families of Models*, The RAND Corporation, P-3198, August 1965.

Chapter 13

SAMSOM: A LOGISTICS SIMULATION

CHAUNCEY F. BELL

SAMSOM, an acronym for Support-Availability Multi-System Operations Model, is a computer simulation model designed to study the influence of resource and policy changes on aircraft capabilities – that is, the interactions between logistics and operations. It considers aircraft characteristics of reliability and maintainability by subsystem, operational and logistics policies and schedules, and the manpower and aerospace ground equipment needs for the support of such policies and schedules. To demonstrate the potential usefulness of such models, SAMSOM is used here to examine a problem involving increased training flights and alert requirements for an organization typical of the Air Defense Command.

Additionally, on the basis of SAMSOM outputs or the records of an operational organization, an analytic technique for determining economic quantities of maintenance manpower and AGE is presented.

INTRODUCTION

RAND's Logistics Department has been developing and using simulations for a number of years to assist in the study and solution of weapon system support problems. These efforts have included both man-machine laboratory simulations (which sometimes approach "gaming," the subject of a later Chapter[1]) and all-computer simulations. In this Chapter, we will consider one of the latter type and then use it in answering some fairly straightforward questions. In so doing, it should be possible to gain an insight into some of the potential applications of computer simulation models.[2]

DESCRIPTION OF SAMSOM

SAMSOM, a Support-Availability Multi-System Operations Model, is a third-generation version of a computer simulation maintenance-operations model developed at RAND in the late 1950's. An earlier model has

[1] Chapter 14.

[2] For a general discussion of simulation, see Chapter 12.

been used by several defense contractors in developing similar simulation models for their own purposes.

The key features of SAMSOM are these:

An ability to simulate operations from several bases at the same time, permitting examination of dispersed operations, or study of a Military Airlift Command organization;

Ability to handle several types of aircraft in the same simulation.

Ability to simulate interactions among types of missions by the same type of aircraft, such as reconnaissance, air superiority, and interdiction missions; and

Substantial capability in the handling of resources, which include facilities, maintenance manpower by shop assignment or skill classification, and maintenance ground equipment by individual type, as well as parts shortages.

Additionally, maintenance requirements can be treated in a good deal of detail. Reliability and maintainability factors can be handled flexibly down to the subsystem (work unit code) level. Constraints on the number of people working in one area of the aircraft, conflicting maintenance, critical versus non-critical maintenance, and so on, can be introduced.

Figure 13.1 provides a simplified schematic of the model. Generators of logistics activity are listed in the box in the lower left-hand corner.

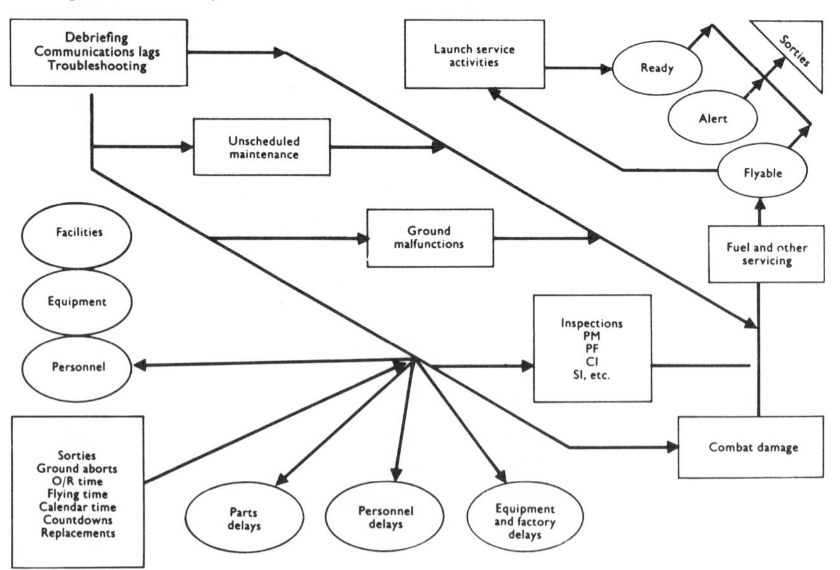

Fig. 13.1 – Model logic schematic

Two of them might be noted here. In the model, sorties and ground aborts generate servicing, unscheduled maintenance, and other support requirements in much the same way they are identified in the field. Similarly, accumulated flying time or calendar time is used as a basis for triggering inspection requirements.

Either as separate inputs (sorties) or as the results of simulated events (accumulated flying time), these generators activate the maintenance modules or routines in the model, several of which are identified by the small boxes placed on the diagonal.

The first maintenance routine represents debriefing times, lags in communication, or troubleshooting activities. If used in a given simulation, this routine is always activated first after an aircraft lands. It also may be used to represent weapon downloads, delays in recovery activities, or travel time to a missile site.

Unscheduled maintenance requirements usually are generated next. In most simulations this routine handles a major portion of all maintenance activity. It represents failures and discrepancies reported by the flight crew, diagnosed by the debriefing crew, or discovered by maintenance teams as maintenance proceeds.

The ground malfunctions routine would be especially active in the simulation of ICBM missile operations or a communications center. (While we have not run non-aircraft simulations on SAMSOM to date, we do not see major difficulties in doing so.)

The inspection module represents several different kinds of inspections. Periodics, hourly postflight, special, and calendar inspections are handled through separate but essentially similar routines. Combat damage is generated by attrition inputs which specify the probability of such damage for each different route or mission configuration. Since combat damage may represent special or unusual maintenance requirements, together with special skills and equipment, it is handled in the model as a special kind of unscheduled maintenance.

Fueling and other routine servicing operations result in a flyable aircraft, which may also receive launch service activities to simulate weapon uploads or other mission-configuration requirements, or to simulate preflights.

Each time a sortie lands or some other generator activates a routine, the model draws random numbers to identify failing systems and to determine the resources required to correct the malfunction or satisfy the requirements. The number drawn has a relationship to the real or estimated reliability and hardware characteristics of the aircraft. The model then searches through the appropriate pools for available resources to accomplish the required tasks. If personnel are off duty or busy on other jobs, or if equipment or facilities are not available, or if parts are not available, the job and the aircraft may be delayed. Such delays are depicted by the large "Q's" at the bottom of the schematic. When resources become available, additional draws of random numbers by the model determine repair times. When all malfunctions have been corrected and servicing is completed, the aircraft is added to the appropriate flyable, ready, or alert pool.

An example will illustrate how a simulation model such as this can estimate effectiveness capabilities and costs for a specific situation and explicit assumptions. Let us assume that an Air Defense Command unit possessing 23 aircraft is asked to consider an increased capability posture which would involve keeping 7 aircraft on alert at all times and flying 25 sorties a day, each averaging 1.3 hours in length. We have information concerning failure characteristics, repair time of subsystems, and so on. Figure 13.2 shows the flight schedule desired by the operations people. It calls for flying 8 aircraft each day at 0630 hours, 7 at 0930 and 7 at 1230. Two additional aircraft are launched every other day from the alert pool to rotate those aircraft, and every other day four aircraft participate in night flying at either 1800 or 2030 hours. The simulation shows that, with the existing manning and workshift policy, only 20.3 sorties can be attained on the average. This kind of answer could have been determined by actual controlled test, of course, with some difficulty. An obvious question would be raised, however: What would happen if there were no manpower problems? In the model, added manpower comes quickly and easily, but the answer is disappointing – 21.1 sorties (Fig. 13.3).

Before reluctantly deciding that 25 sorties a day cannot be met, we can

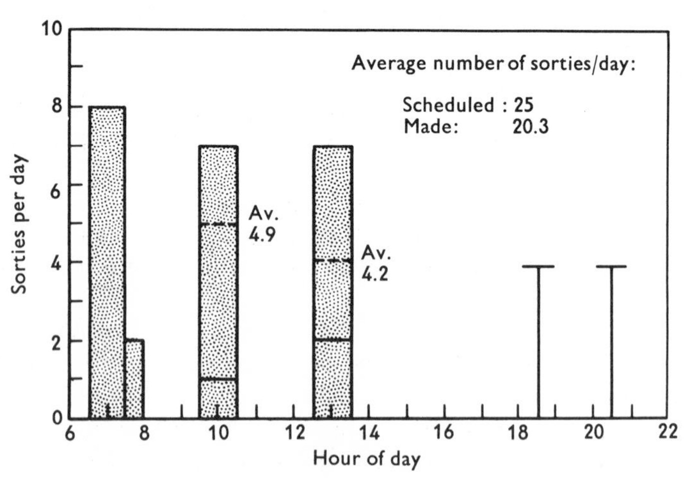

Fig. 13.2 – Simulated daily flying program (typical manning)

easily use the computer model to determine the upper capability limit possible with the type of aircraft and the extent of the flying day. If we do so – instructing the model to fly all available aircraft beginning at 0630 and continuing until 1845, using manpower as needed – we will find that nearly 30 sorties a day are possible. Thus, the previous failure to reach the 25-sortie goal is not due to the type of aircraft being used or a lack of manpower, but rather to the proposed flying schedule. We can skip some of the steps involved in arriving at a reasonable schedule and look at one (Fig. 13.4) which allows the goal to be reached, with only some reassignment of existing manpower to different workshifts.

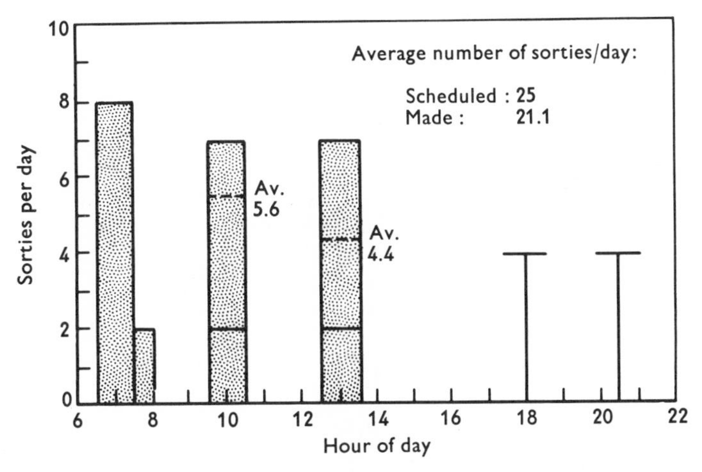

Fig. 13.3 – Simulated daily flying program (manpower as needed)

While this is a fairly simple and straightforward example of a proposed change in operational plan, the point to be emphasized is that examination of weapon system capability as well as support requirements can be made early in the development cycle with much greater confidence than is possible without models of this type. SAMSOM was used early in 1964, for instance, during the first phase of a study of V/STOL aircraft in the 1970's. By examining cost-effectiveness trade-offs early in a program, not only are specific operational and logistics planning factors made available, but weak links in the design – problems that might have a particularly adverse impact on operations or support requirements – are identified, so that corrective action may be initiated, literally years before the difficulties themselves might come to light otherwise.

Clearly, a wide variety of logistics support questions can be answered by a model such as SAMSOM. Of course, if we were to attempt to find the appropriate number of men for each shift for each shop, we would obviously have to go through a large number of iterations, which would be cumbersome and, from an economic viewpoint, needlessly expensive. In such cases, it makes sense to develop other cost-effectiveness models which treat such specific problems, and then check out or validate the results using the more comprehensive model.

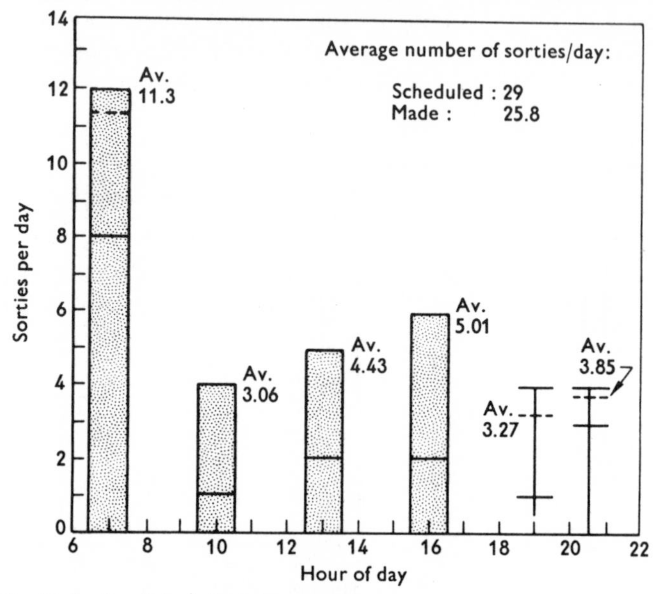

Fig. 13.4 – Possible daily flying program – typical manning reallocated

ESTIMATING REQUIREMENTS FOR MANPOWER AND EQUIPMENT

One important problem is that of determining economic quantities of maintenance resources, both base manpower and maintenance ground equipment (MGE). Only manpower will be discussed here, but MGE is handled in like manner. Figure 13.5 shows the important considerations in establishing such resource requirements, and all are explicitly treated in our methodology. Considering the dramatic example of the effect of flying schedule on operational capability that we have just seen, probably no one would object to the presence of any of the items on the list. It might be noted in passing that most of these considerations are not specifically included in present Air Force manning requirement determinations. Therefore, this is an area where marked improvement is possible.

A part of the problem lies in data availability under the current Air Force maintenance management system[3] – a fact that will come as no

[3] This system is described in AFM 66-1, *Depot, Field, and Organizational Maintenance Management*, Headquarters, Department of the Air Force, 15 June 1966.

Frequency of Occurrence of Demand
Time That the Resource Is Occupied/Occurrence
Flying Schedule
Randomness of the Demand Pattern
Workshift Policy
Cost-effectiveness Trade-off

Fig. 13.5 – Important considerations in establishing resource requirements

surprise to many readers, since shortcomings in reliability and maintainability data have been discussed frequently in the past, and reliability and maintainability are key factors in establishing resource requirements. An inexpensive method of treating these considerations in AFM 66-1 was developed by RAND and was under test for nearly three years at Oxnard Air Force Base, and for a lesser time at Williams Air Force Base. It was also implemented by the Tactical Air Command at MacDill Air Force Base in 1964 in order that proper manning, equipping, and other operational and material decisions could be made on the F-4C. In addition, it has been used in a number of operational field tests in the last three years. The key to RAND's solution is time-oriented data, illustrated by Fig. 13.6. We added clock times to the several types of information already being collected, identified delays in returning aircraft to readiness, and recorded maintenance team sizes. This permits relating maintenance demands to flight

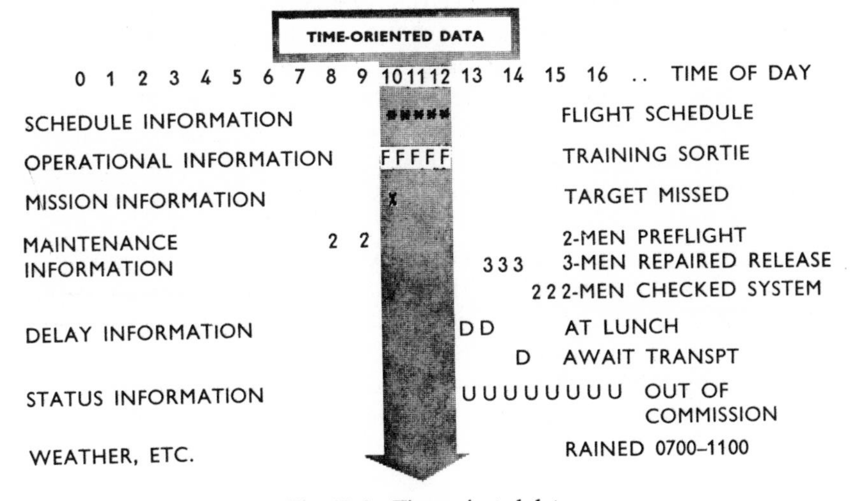

Fig. 13.6 – Time-oriented data

schedules, determining elapsed times to perform maintenance, and so on.

Given time-oriented data, computer routines have been developed to display what work was actually performed by a particular shop, at what time the requirement was known, and when the work was performed. Since one day's operation is not adequate for planning purposes, an appropriate sample can be compiled, like that shown in Fig. 13.7, which happens to be for a 24-day sample. "Adjusted" means that the work is shown as if accomplished without delays due to shortages of manpower. It should be emphasized that a computer display like this can be generated either from flight operations (this is a random sample from Oxnard Air Force Base), or from a simulated run of SAMSOM.

If we look across the bottom of Fig. 13.7 at the number of times that no one is needed, we see, on the one hand, that between 0400 and 0700 hours no one was needed 21 or 22 days out of the 24 days in the sample; on the other hand, in only 6 days out of 24 was no one needed between 1500 and 1600. Looking at the 1700–1800 time period in detail, we find that on one occasion 9 men were needed, yet 11 times out of 24 (46 per cent of the time) there was no work at all. This, of course, merely emphasizes the earlier statement regarding the impact of flying schedules and randomness of demand patterns on workload, and should be convincing evidence of the futility of attempting to use averages, or utilization rates, or validation teams in arriving at good manning or equipment requirements.

To establish that level of manpower which is neither excessively costly, nor likely to leave expensive aircraft standing idle awaiting manpower, is an appropriate task of a cost-effectiveness analysis. Given an approach such as this, which is oriented toward time and the probability of having a workload, the principle of measuring the costs and savings can be applied. For instance, if we provide 8 instead of 9 men on the night shift, we save one man at, say, $600 per month, at the cost of losing one ready aircraft hour per month, worth perhaps $100 in "lost" investment and operating costs. Another man can be saved at a cost of 4 aircraft hours per month, and so on.

Each shop's manpower can be computed in this fashion, and the expected loss in capability recorded. The resource quantities thus tentatively decided upon can be reintroduced in the SAMSOM model, previously run with "as needed" quantities to establish maximum capabilities under the desired operational program and the loss in effectiveness noted. If the results are not satisfactory, further adjustments may be made.

Conclusions

Since models such as these can be used in all stages of the weapon system's

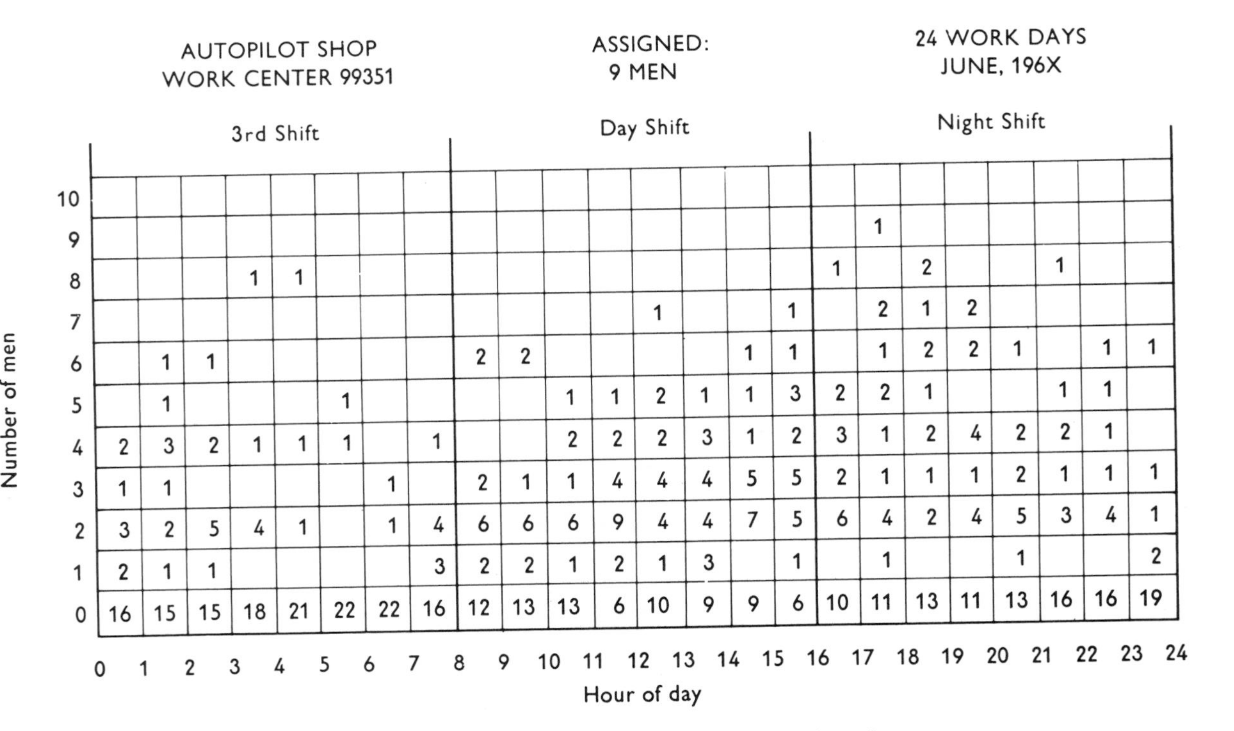

Fig. 13.7 – Manpower utilization analysis (adjusted)

life cycle in order to find ways of reducing expensive system downtime and to minimize required quantities of scarce and expensive resources, we believe they will find wide use as soon as their potential is fully recognized and adequate data are available. Development work is continuing at RAND to broaden their potential application and simplify their use.

Chapter 14

GAMING

M. G. WEINER

War gaming is widely used in many military analyses. This Chapter discusses some of the different types of war games and some of the differences and similarities between them. A limited-war gaming technique is described in terms of the steps involved in preparation, play, and analysis. Several examples of analyses using this technique are included.

Introduction

Many people seem to think of war gaming as a never-never land, something like Southern California – widely misunderstood, but at the same time frequently visited. This Chapter intends to say nothing about California. It will attempt to describe briefly the background of war gaming and identify its different types. To do so, it will use as illustrations some of the gaming that has been going on at RAND over the years, concentrating on some of our limited war or tactical war gaming. Several examples of this work will be presented to highlight certain of the ways in which gaming can be useful as a research tool.

What war gaming is not

Because a good deal of the misunderstanding of war gaming lies in overestimating its value, it is important at the outset that its limitations be considered. War games are not, at least intentionally, attempts to hypnotize people into believing results that the war gamers obtain either by hand calculations or on the computer. In most cases, the games are done by people who are using one of the available tools of analysis, which, like all tools, has advantages and limitations. As such, there is little or no conscious charlatanism in war gaming.

Similarly, games do not constitute an adequate substitute for test or experiment. The field exercise and the laboratory still have to be used as active parts of any research program on weapons, on vehicles, on equipment, on doctrine, or on policy. And, above all, gaming is not a technique for predicting the future. War gamers can hypothesize about some future environment, weapons, or force capabilities, and can explore the impact of these on military operations as they would be conducted in the war game. But that is all they can do.

What war gaming is

If war gaming is not any of these things, what is it?

First, war gaming is a fair amount of work. It is not something that can be undertaken quickly and easily. It is also relatively expensive. On the other hand, there are some jobs that cannot be done without attempting to represent both our own capabilities and the enemy's capabilities, showing how they interact in a military confrontation. So the primary uses of gaming are as an organizing device to pull things together, as a training and indoctrination technique, and, finally, as an analytic tool by which different concepts, different pieces of hardware, or different military plans can be investigated in a two-sided confrontation.

Types of war games

War games take many different forms. They can be field exercises, which may involve literally tens of thousands of real troops in mock combat. They can be map exercises, in which people work around sand tables or on maps and conduct combat operations in a symbolic fashion. It is also possible to put certain aspects of games on machines and to carry out some of the formal and routine aspects by computers. They can also be mathematical games of a highly precise form which arrive at determined solutions. And, finally, they can be parlor games, like the paper and pencil game, BATTLESHIP. Obviously, with this wide variety of games, from the parlor games involving two participants to the field exercise involving thousands of troops and hundreds of tons of equipment, what is meant when one uses the term "war gaming" should be specified rather clearly.

Varying requirements

It is particularly important to remember this point because different war games have quite different requirements. They differ in the number of personnel that are involved, in the facilities that are used, in the equipment that supports them, in using or not using computers, and in the kinds of data that are accumulated and handled in the war gaming situation. What determines the type of war game that will be conducted? Obviously, it is the purpose or objective of the game. Not all war games have the same purpose. But all war games have or should have a purpose. They are ordinarily not undertaken for the sheer joy of playing friendly versus enemy forces. There is usually an identified and specific research or training question connected with any of the games played.

Essential characteristics

Thus, one of the characteristics which all games should have is a purpose or an objective. There are a few others. In games there are usually "sides," frequently two, but sometimes more, and the basic characteristic of these sides is that they have some conflicting objectives. Both sides are trying to attain these objectives, and they frequently have different resources which they can use to try to achieve them. Moreover, all games have rules. Whether it be the parlor game that children play with very simple rules, or a mock war played as an elaborate field exercise with impressive manuals and rule books to cover as many of the events or contingencies that will occur in the course of the game, the rules are critical. Otherwise there is chaos – as any mother or field exercise director can testify. One of the problems in war gaming is that the latter two requirements – that there be rules and sides with conflicting objectives – are often subject to misinterpretation. The validity of a war game does not ordinarily lie in the accuracy with which the mathematical computations or the arithmetic is done, but rather in the extent to which the sides can be faithfully represented and the rules be designed to bear some relationship to real operations.

Even with these few simple requirements, games can be very elaborate or very simple. As an illustration of some of the variety in types, we can look at just a few of the games that have been conducted at The RAND Corporation.

Some war games at RAND

These games have covered a fair spectrum, from some elaborate machine games to some very specific map exercises. We have engaged in machine gaming of strategic war in our so-called Strategic Operations Model, which, as N. C. Dalkey has pointed out,[1] became the basic model adopted in the U.S. Air Force for its strategic analyses. We have also had other formal games – that is, games in which all the rules were explicitly stated – but they included human players. Some of them – such as our STRAW, or Strategic Air War games – have been used to examine force requirements and different strategies. Others – such as SAFE, or Strategy-and-Force-Evaluation game – looked at some of the implications of budget planning.[2] Still others have included COW, or the Cold War game, in which certain military aspects were subordinated to some of the political and economic aspects of military operations. And finally, we have had considerable experience in conducting map exercises for limited war, of which the earlier SIERRA and REDWOOD studies and the present Tactical Operations

---

[1] See Chapter 12, p. 248.
[2] See Chapter 15 for a detailed description of SAFE.

Group game, MAGIC (for Manual Assisted Gaming of Integrated Combat), are examples.

In what follows, we can skip over the rather large formal games that RAND has played, and concentrate on the map exercises and the limited war games. These are of great interest because they involve a large number of political, economic, and logistics factors that influence military operations and we see no way of eliminating the human player from them. In short, these games, while interesting in their own right, help to bring out our ultimate dependence, in systems analysis, on human judgment and decisions.

Special conditions for limited war game play

In the course of our limited war or tactical gaming we have used a number of techniques to facilitate the play of the game and to examine a variety of different situations. The techniques include the following:

1. The classic *two-sided play*, in which a Red and Blue side are identified, each with different resources, and each conducting military operations which are evaluated in order either to identify important factors that contribute to the outcome of the confrontation or to provide data for some other analytic problem.

2. The *seminar technique*, in which an experienced staff of war gamers plays through the war in seminar fashion, exposing all the information, including information about plans and actions, to all the participants, as in a game of showdown poker. This is a technique for rapid play of a game.

3. The *open-play technique*, in which there are no intelligence restrictions on either side. The sides are aware of the planning, the positioning, and the capabilities of all forces. The purpose of the game usually is to get a general assessment of the course of the operation.

4. The *closed-play technique*, in which uncertainty and intelligence are included. This play usually simulates the "fog of war," with the participants knowing only that information that is passed to them by a Control Team, and with their operations constrained by lack of complete information on the enemy's forces, location, capabilities, intentions, and plans.

5. The *branch-point analysis technique*, in which a game is carried to a certain point and then, because several major courses of action could occur at that point, divided into essentially two different branches. Sometimes each of the branches is played out concurrently, but more often separately. Using branch-point analysis it is often possible to make a "tree" of games, using this tree to look at a fairly complex number of variables in the particular game situation – different plans, different kinds of weapons, and so on. In branch-point analysis the initial conditions or assumptions of the game are not changed.

6. The *series and variation technique*, in which the initial conditions are changed and a second game played under these different circumstances, or different assumptions. Such characteristics as the geographical area, the time period, and the sides involved in the conflict are not changed, however. In this manner, much of the preparation that went into the gaming can be used over again.

These techniques, which vary from rather elaborate detailed two-sided game play to the very aggregated, rapid-play seminar technique, provide a spectrum of gaming tools which can be adapted to the particular research problem, time limits, and personnel levels available for carrying out the gaming.

PHASES OF WAR GAMING

These techniques refer, however, only to the playing of the game. Gaming is more than just "play." It is a three-phased activity. The three phases are preparation, play, and analysis.

The following sections describe briefly the major activities in preparing, playing, and analyzing limited-war games at RAND. These phases are, however, typical of all gaming.

Preparation Phase

The first phase is preparation. Preparation must start with the definition of the purpose or objective of the game. Without a definition of purpose it becomes extremely difficult to control all of the possible data required, all the possible events that might be considered, all the possible moves that might be made, and all the possible analyses that might be done.

Once the purpose is established, it usually dictates a geographical locale and time period which are adequate to provide answers to the research questions. Although it is possible to use an entirely fictitious locale and build up a hypothetical world, it is generally a very time-consuming task if one wishes to assure consistency in all of the different aspects of the fictional geography, weather, location of forces, targets, lines of communications, and so on, as well as the fictional political, economic and logistics characteristics. Increasing emphasis seems to be falling, therefore, on using real-world locales and situations rather than specially constructed, make-believe, ones.

The time period selected for most RAND games is two to four years beyond the present, because games which are too immediately tied to the present may well end up producing results whose utility is already out of date by the time the game has been set up, played, and analyzed.

On the other hand, projecting into time periods beyond five or ten years leads to a great many questionable assumptions about the political situation, about force postures, and about weapon systems that will be in existence, and about the capabilities and performance of these weapon systems. In view of these uncertainties, it is advantageous to try to confine the time period to something between two to four years in the future. It should be noted, however, that there is a class of games, at least in theory, for which one would deliberately select a time period of five, ten, or more years in the future. In this type, which might be called "research and development games," the long-term uncertainties have to be faced.

Once the locale and time period for a game have been selected, there is a variety of inputs which must be developed. Almost any war gaming publication that discusses the methodology of gaming provides a detailed description of these inputs. Briefly, there are four major types, each of which has to be prepared in whatever degree of detail is appropriate to the research objective of the game. The first is the military inputs, including the objectives of the military forces, their sizes and locations, the military and support facilities available, and so on.

The second is the political inputs, including the national or international objectives of the participants and any restraints or constraints on military operations, such as the bases or ports that can be used, the areas that can be overflown or otherwise transited, the weapons that can be introduced, the targets that can be attacked, and so on. Frequently, it is necessary to consider other countries besides the original combatants which may enter the conflict, and to develop the appropriate military inputs for these countries.

The third is the economic inputs, including the capabilities of the countries involved to support the military operations, and any economic vulnerability of the countries in question.

The fourth is the technical inputs, including the performance capabilities of the weapons and the weapon systems and the technical characteristics of any equipment used by the military forces.

Having developed the inputs, the next major part of preparation is to develop the rules of the game. These rules include not only any political restrictions on the forces, weapons, or geographical areas that are included in the game, but also all the "planning factors," performance data, attrition factors, and other rules for assessing the outcome of the military confrontation. In many cases, a game will involve individual engagements that occur under specific conditions of geography, weather, force position, objective of the forces, and so on, and appropriate rules, tailored to the particular game situation, have to be developed.

Once all this preparatory work has been done – the objective of the game is defined, the data are collected, and the rules are developed (whether for use by the human participants or by computing machinery) – it is possible to initiate game play. In the following we describe only the play of games which involve human participants, that is, games where the actions or rules are incomplete, and human decisions are necessary.

Play Phase

Typical play for a two-sided game requires a Control team, which is the ultimate authority for all moves made in the game; a team, usually called the Blue team, representing one or more friendly forces; and a team, usually called the Red team, representing one or more unfriendly forces. Both teams have military forces, objectives, and different courses of action open to them. These operate in a game setting frequently called a scenario,[3] which provides the political-military environment for the conflict, along with the events or conditions that led to the confrontation.

The game play is started by an intelligence briefing to both sides. The intelligence briefing describes the general and special situations, as well as any "precipitating event," that initiate the conflict. Each of the teams prepares its plans separately. The plans usually take the form of the familiar five-part Estimate of the Situation, which includes an analysis of the courses of action available to both sides and the decision on which course of action each side elects.

Having selected the course of action, the plans are passed to the Control team, which evaluates and assesses the outcome. In the course of this, the Control team is often in back-and-forth communication with the sides in order to make clear any possible points of confusion, to replay any events

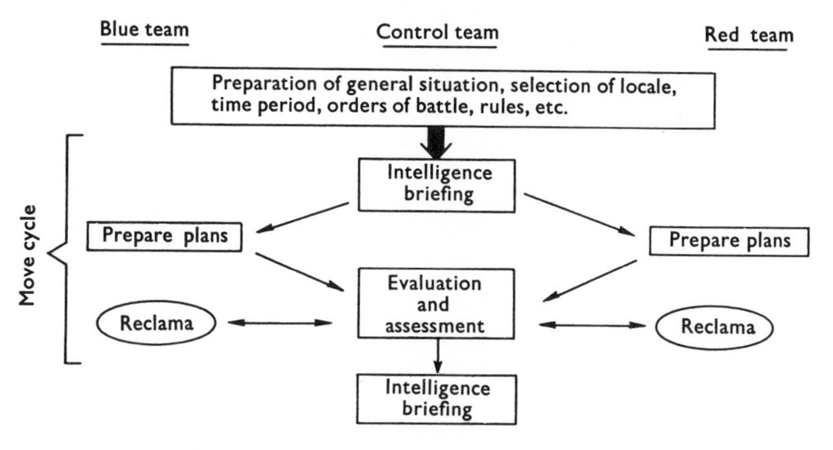

Fig. 14.1 – Play phase (two-sided games)

[3] Scenarios are discussed in Chapter 16.

that have to be changed, and so on. When any sources of confusion, error, or argument have been reconciled and the outcome of the engagement has been determined, the move is terminated by writing up the appropriate documentation. The sequence of events from the initial intelligence briefing through the preparation and assessment of the plan constitutes a "move cycle" (see Fig. 14.1 p.278). Following the completion of one move cycle, a second one is initiated by a second intelligence briefing, which carries the war into the next stage. The game continues through a series of move cycles until some defined termination point is reached.

Analysis Phase

When play ends, the analysis phase begins. Since the type of analysis that will be done is determined by the objective or purpose for which the game is being played, it is not possible to describe specific analyses without describing specific games. It is possible, however, to describe some of the typical kinds of analyses that are done. These are the outcome-oriented analyses, the special analyses, and the follow-on work.

Outcome-oriented Analyses. The first and most common type of outcome-oriented analysis is a narrative of what took place in the course of the war. It is the history of the war situation. It can be regarded as a "synthetic history" and is useful for drawing certain kinds of conclusions about events that might have happened in the real world.

A second type of outcome-oriented analysis involves looking at various alternatives in posture, in weapons, and in force employment that have become apparent during the course of the game, and comparing their effect on the outcome of the game.

A third type of outcome-oriented analysis includes a review of the game in which all the important decision points are selected and an effort is made to identify ones critical to the particular outcome of the game. This is a form of sensitivity testing that frequently leads to a branch-point analysis, in which a portion of the game is replayed as if some other decision had been made at a particular point.

Special Analyses. All of the outcome-oriented analyses generally use the entire game situation. In addition, there are some special analyses that look at only one particular aspect of game play. One of these is force-effectiveness analysis. For an analysis of an air operation, this may involve examining all the data on the sorties that were flown against all the different classes of targets to determine the kinds and numbers of weapons delivered and what parts of the force contributed most effectively to the conflict. Sometimes, as a result, ideas about different force requirements arise and the game data may be analyzed to identify the impact of changing force requirements on the game situation. This is the force-requirements analysis.

A second type of special analysis deals with performance capabilities. Frequently, performance capabilities, such as range, payload, and loiter time, may be examined to see whether there are indications of how changes in these performance capabilities (for example, added payload capabilities) might have affected the outcome of the game.

Similarly, a third type of special analysis considers the tactics or operational concepts that were used in the game. As a result, different tactics or concepts may be introduced in order to examine their possible effects.

However, the outcome-oriented analyses and the special analyses have, in a sense, simply defined some problems. They have not solved them.

And this is one of the major sources of misunderstanding about gaming, even among war gamers. Games *prove* nothing about the real world. They only prove things about the game world, a difficulty from which no logical analysis is free. To relate the conclusions drawn from the game world to the real world additional work is necessary. This is the follow-on work needed to make gaming useful.

Follow-on Work. One frequent form of follow-on work is a replay of the game, using different concepts, forces, or equipment, in order to derive further evidence for any one of the special analyses. Another is to specify some operational or field tests that might be conducted to reduce uncertainty as to the validity of the game results and to make them more reliable. There are frequently technical studies that can be defined by the gaming. Such studies translate the implications derived from the game world into the hardware of the real world. Finally, there are other types of analyses, such as other types of systems analyses, operations analyses, and special side studies, that can be conducted to supplement the conclusions drawn from the gaming. Thus, even when the game and the initial analysis are finished, a good deal of follow-on work is frequently necessary to turn any of the findings into useful material as far as military operations or capabilities are concerned.

Some examples

Up to this point, we have kept to a rather general description of the types of war games, some of the techniques used in limited-war gaming, and gaming's three major phases – preparation, play, and analysis. Perhaps now several examples of the use of war gaming are in order. The first example illustrates the use of tactical gaming to determine the comparative utility of different weapons; the second illustrates the use of gaming to examine force requirements; and the third illustrates the use of gaming to compare two different concepts of military assistance. We will describe the first two briefly and develop the third in somewhat greater detail. All three are based on actual research.

A Tactical Game for Comparing Different Weapons

The purpose of this game was to determine the comparative utility of several different types of aircraft-delivered weapons for use in interdiction. These included non-nuclear, biological, chemical, and nuclear weapons. The specific situation involved an enemy invasion of a friendly country through a relatively mountainous area. The enemy, Red, prepared a plan of action that indicated the invasion routes he would use, the size and disposition of the forces he would commit, the routes and depots he would employ to support his combat forces, and the timing of his operation. Blue prepared a counterplan indicating the forces that he would commit, the interdiction targets he would attack, and the weapons he would use. Additional data available to both sides indicated the detailed geography, the weather conditions, and a variety of other factors.

In the game, Blue used different types of aircraft-delivered weapons to interdict the Red movement, and comparisons were made of the number of weapons required, the number of sorties flown, the cost of the operations, and the comparative effectiveness of the weapons as influenced by weather, target conditions, and the ability of the Red forces to continue their military operation.

In this type of analysis, the game provides a setting or situation in which many of the important factors influencing a military operation are incorporated. It goes beyond a "target analysis" since it allows both sides to have different options as to how they will use their forces or react to the operations of the other side.

A Tactical Game for Examining Force Requirements

The second example concerns a game that was played in order to examine force requirements. For this problem a novel type of military unit – different in weapons, organizational structure, and combat capabilities from any existing military unit – was hypothesized. The game examined some of the conditions under which this unit might be employed in combat and particularly what effect the use of this new unit might have on the existing military forces. That is to say, the game paid as much attention to the impact of this new force on existing forces as to its impact on the enemy's forces.

The game investigated some of the missions that this unit might undertake, the support requirements for deploying it into the combat theater, the bases from which it might operate, the unit's effectiveness in combat, and, specifically, how its use might require different numbers and types of aircraft sorties in attack and air resupply missions, or otherwise affect the operations of the existing forces.

This game provided both qualitative and quantitative indications of some of the major effects of introducing this new military unit. It indicated some of the counteractions the enemy might take, and provided a context for evaluating how, and to what extent, the new unit might generate requirements on existing units.

A Tactical Game for Comparing Programs of Military Assistance

The third example involves a series of games that was intended to compare two concepts or programs of military assistance. The games were part of a larger evaluation that incorporated economic and political factors in the comparison. For the military evaluation, the gaming examined the capability of two countries to operate in three levels of combat under each of the two programs. That is, the gaming involved 12 different games; each of two concepts was compared in three levels of conflict for two countries. The method used in the study had five separate steps:

1. Alternative programs for spending the same hypothetical four-year military air budget were drawn up. The amount of the budget was based roughly on experience in the particular underdeveloped countries for which case studies were conducted. The programs were designed to be of equal cost, but they were significantly different in their content. One program, which was called the "A" program, generally stressed fairly large forces, armed and trained conventionally. It followed rather closely the lines of recent military aid programs and force structures in the major underdeveloped recipient countries. The other program, which was called the "B" program, advocated smaller, more lightly armed forces, with the dollar savings resulting from these reductions used hypothetically for expanding internal security forces, increasing ground and air mobility, providing additional ground and airfield installations intended to facilitate effective intervention by free-world forces if this should be necessary, and, finally, expanding the technical training of military manpower.

In effect, under the "A" and "B" programs, the same four-year dollar budget was hypothetically expended in different ways for initial equipment (that is, force improvement); for four-year replacement, operating, and spare-parts costs (that is, force maintenance); for military construction; and for military training in United States technical service schools. Standard cost factors were used for equipment, maintenance, and training costs, and generous estimates were made for the construction costs of roads, airfields, and other infrastructures in the countries under study when accurate information was not available.

2. The second step consisted of formulating a range of threats, covering differing levels of violence: a major insurrection, an invasion by a minor neighbor with only marginal support from one of the large adjacent Communist powers, and, finally, a larger invasion with overt participation by one of the latter powers. The threats were sketched out in game scenarios that gave the game players a set of initial conditions to start from, as well as a plausible sequence of hypothetical events through which these conditions might have evolved.

The scenarios projected events several years into the future in order to allow time for the hypothetical "A" and "B" programs to be carried out. Although effort was devoted to making these projections sufficiently realistic to motivate the play, detailed realism was not the primary consideration in the design and choice of scenarios. The scenarios were kept at a fairly macroscopic level, and details, to provide a semblance of added realism, were excluded if they were not judged to be essential to the games' purpose. The primary consideration in formulating the scenarios was their relevance in helping to span the differing levels of violence needed to test the military performance of the contrasting aid programs.

3. Next, the research group, consisting of two teams of senior retired military officers, and a Control team conducted the game operations, using the military resources available to them to try to achieve objectives specified in the game scenarios, which were then played *seriatim*. Because the free-world, or Blue, team was assumed to have expended military aid dollars in differing ways in the pre-game period, Blue's order-of-battle and logistic support resources were markedly different under the two programs, and these differences were made known to the Red team. In formulating strategy and carrying out operations, the Blue team used, in sequence, the two different force-and-facilities packages represented by the "A" and "B" programs, while the enemy team used his "best" strategy against each of the Blue alternatives.

4. In the fourth step, the military performance of the alternative packages was evaluated. This was done primarily in terms of three measures: the area occupied in a stipulated time period (or the time required to occupy or defend a stipulated area), the number of casualties, and the materiel and property damage of the combatants. Occasionally, the military performance was also evaluated in terms of the bargaining position of each side when the game hostilities ended, and the relative probability that a particular contingency (for example, an insurrection) would have broken

out at all, depending on whether "A" or "B" had been implemented in the pre-game years. The evaluation technique used standard planning factors and simple quantitative models where they were applicable (as, for example, in assessing air-to-air combat, the effects of interdiction attacks, and the movement of ground forces), but it also relied on discussion and experienced judgment where they were not.

In conducting and evaluating the game, play was divided into segments or phases, usually based on convenient blocks of time or space. Each phase was played under both of the program assumptions before either of them was evaluated. This eliminated the feedback that would have distorted the results if one program had been played and evaluated before the other.

It is worth noting that the evaluation was less concerned with the *absolute* outcomes (that is, who "won" or "lost," and by how much?) than with *comparative* outcomes (that is, how did program "A" perform in comparison to program "B"?)

5. Finally, independent of the war games, a separate evaluation of the economic and political side-effects of the two different, but equal-cost programs, "A" and "B," was conducted. The purpose of the economic evaluation was to provide a quantitative indication of differences between the two programs in their effects on economic development in the countries studied. The purpose of the more general political assessment was to get at least a qualitative indication of how the alternative programs would be likely to be received by key groups and individuals comprising the leadership of these countries.

A LESSON FROM PRINCE LADISLAUS

These three examples of the use of war gaming are illustrative of the variety of problems to which gaming may be applied. The first example involved the use of gaming in a rather limited manner, that is, in weapons comparisons for a specific situation. The second example was somewhat broader and involved examination of force requirements. The third was very general and involved the use of gaming as an integral part of a much larger systems analysis, which incorporated political and economic as well as military evaluation.

There are many other examples that might be cited. All of them involve representing a conflict situation and using that representation to throw light on military questions. But, above all, they are representations. The ultimate test of the validity of any war gaming conclusion lies in the real world, not in the play of the game. And in the real military world, as Prince Ladislaus teaches us (Fig. 14.2), no one can accomplish the impossible.

Fig. 14.2 – A princely failure

Chapter 15

THE ANALYSIS OF FORCE POLICY AND POSTURE INTERACTIONS

ROGER LEVIEN

This Chapter briefly examines the nature of the interactions between national defense policy and force posture as it affects strategic forces. It considers the difficulties associated with the analysis of those interactions and establishes some requirements that a technique for such analyses must satisfy. It then presents the SAFE (Strategy-and-Force-Evaluation) game as a particular tool of analysis that meets the requirements established. The use of SAFE as one aid in the study of alternative future defense strategies is discussed and the limitations and benefits of national policy and force procurement games as tools of analysis are presented.

This Chapter has three objectives. The first is to discuss the interaction between policy and posture as it affects the strategic forces; the second is to discuss the possible role of analysis in the examination of strategic force posture and strategic defense policy; and the third is to discuss a particular tool of analysis – the SAFE (Strategy-and-Force-Evaluation) game – as it may be used in studying this interaction.

INTERACTION OF POLICY AND POSTURE

What is the nature of the interaction between policy and posture? If "strategic defense policy" means our national goals with respect to our strategic forces and "posture" means the exact composition of the strategic forces, then the interaction appears to be that policy *guides* posture, and posture *constrains* policy.

There are some familiar examples of policy guiding posture. A deterrence policy requires us to have adequate second-strike forces. It does not uniquely define what would be adequate or what are the conditions for a second strike, but it does require that we be able to retaliate after an initial blow. Similarly, a policy that anticipates a careful conduct of strategic conflict so as to reduce damage and increase our advantage – in short, a warfighting policy – requires us to have controllable and survivable forces. So policy acts as a guide to posture.

Conversely, posture constrains policy. Now, this is obvious in the case of nations without large nuclear forces – their policies are limited regarding strategic conflict. But even for the large nations, posture can constrain policy. An example or two might make this clear. If our posture were not likely to contain an antiballistic missile (ABM) system, then, for that reason alone, it is highly unlikely that we could ever develop a policy that depended on the threat of general war to coerce the Soviets into acting against their own interests. Similarly, complete strategic superiority in the usual sense is impossible without an ABM system. Or, to take another example, if our future posture did not include a civil defense program, it is possible that our freedom to choose a policy that attempted to limit damage through wartime targeting restrictions would be constrained by the knowledge that fallout from attacks on military targets would inflict great population damage regardless of the restrictions.

Other Determinants of Posture

This emphasis on the interaction of policy and posture is by no means intended to suggest that the two are rigidly linked, one to one. They are certainly not. Many other factors enter into determining posture, some of them under our own control, some under our opponent's control, and some whose control we share.

In the first group, we would have to include – besides policy – the budget that is available to us. We also cannot ignore the fact that we have a present posture from which must evolve the future posture we desire. Too frequently that initial condition is ignored during analysis. And we cannot forget the role that skillful strategy and tactics might play in developing force postures. It is not only analysis, but also that indefinable quality that makes the difference in chess that counts here as well.

In the second group are those determinants of posture that are under our opponent's control. Thus, his policy, the size and structure of his budget, his present posture, and his future capability will all have an obvious effect on the character of our forces.

And in the third group there is the influence of technology, a factor whose control we share with our opponent and with the world scientific community and which can affect the feasibility of a policy or a posture quite significantly. In this group there are also the influences of nature and of third nations, the effects of which can be significant. Obviously, neither is under the complete control of any major power.

So, stated very simply, the problem of posture planning for strategic forces is to plan and procure forces that will enable policy to be carried out, subject to the constraints that are exercised by budget level, by technology, by the opponent, and by present posture.

ROLE OF ANALYSIS

Now, what role can analysis play in helping to solve this problem? What help can an analyst be? Our first observation about the interaction between policy and posture provides a clue to the answer. It suggests that the analyst has two roles to play. First, insofar as posture guides policy, the analyst can help the decisionmaker in determining what forces are required to support a pure deterrence policy or a warfighting policy or a damage-limiting policy; or he can provide estimates of how much it will cost to carry out any of these, or other, possible policies. Second, he can help the policymaker to understand the constraints under which he operates. He can, for instance, try to indicate to the decisionmaker whether or not, with current and foreseeable technology and a moderate budget, strategic superiority can be achieved, or whether deterrence can be maintained. Or he can attempt to clarify the intensity of the threats and counterthreats under which a policy must be carried out by describing how much damage will be suffered in a first or second strike. To put it briefly, the analyst can either explore the implications of a given policy for a posture or explore the policy limitations that result from a given posture.

What is Needed?

In playing these roles, the analyst needs techniques – analytic techniques – that will raise his answers above the level of intuition. But the problem, as we have already seen, is a difficult one. There are many factors influencing the posture that will evolve, and in choosing his technique the analyst has to make sure that he does not underestimate, or eliminate, or define away any that are important. Before we look at particular techniques, therefore, it might be worthwhile to list some of the characteristics that a useful technique for exploring the interaction between policy and posture should have.

First of all, it should cover a span of years – a ten-year or five-year range. A technique that simply examines a single instant at some time in the future ignores the dynamic character of the interaction between policy and posture; it ignores the fact that in planning for an optimal force in 1972 we may inadvertently procure weak forces in 1970 and in 1974. Thus, we must not only attain a good force posture, but, while we are attaining it, we must maintain our policy and support it.

Secondly, since we have had, and hope to have, experience only with the peacetime aspects of strategic force posture, an analysis should really not be concerned strictly with the wartime aspects, not simply with the wartime uses of force, but also with its peacetime deterrent value, with its influence on the transition to war and on the ending of war.

Thirdly, an analysis that is concerned with force posture should consider *all* the forces affecting an aspect of policy. For instance, if we are talking about strategic defense policy, all the strategic forces interact. Factoring out for analysis just the bombers, or just the bombers and the missiles, or just the offensive forces ignores some of the strong interactions that occur between bombers and missiles, or between offense and defense.

Next, analysis should recognize that the situation is two-sided and competitive; we cannot ignore the fact that our opponent is constantly observing *our* force posture as it evolves and is continually trying to adjust his response to *our* actions. We must not make the mistake of imagining that our opponent's forces remain fixed while ours vary.

The method of analysis must also admit of the opportunity for skillful strategy and tactics; it has to allow human judgment to apply in force posture evolution, for analysis simply cannot answer all the questions that must be answered in evolving a force posture.

In addition, it should be capable of handling the difficult problems of research and development, not only those concerning what forces we should buy, but also those concerning the kinds of R&D strategy we should follow and the preparations we should make for future procurement.

It must also account for the many sources of uncertainty, not only in R&D, but in costs, in our opponent's actions, in the amount and kind of intelligence we are going to receive, and so on.

And, finally, it should involve an awareness of the fact that we are going to interact with our opponent and he with us through the medium of intelligence, and, therefore, that the value of some of the forces we may procure will depend on how their eventual use is affected by their vulnerability to enemy intelligence or our ability to benefit from our own intelligence.

In short, we need an analytic technique that reflects the complexity, the uncertainty, the competition, and the dynamics of the problem. This is a tall order, of course.

Traditional Tools of Analysis

How well do the traditional tools of the systems analyst fill that order? Consider, for example, three of them: computational modeling and analysis, war gaming, and scenario writing. In the first of these – that is, in an analysis that constructs a mathematical or computer model of the situation to be studied – an attempt is made to represent all the relevant aspects of the problem in the model. There are probabilities of kill, sortie rates, reliability estimates, and so on. The model is then manipulated so as to obtain measures of effectiveness and of cost. Quite frequently, the opponent is assumed to have fixed forces, and the analysts then choose various postures for their own forces to see how effective each is against the unresponsive opponent.

The second tool, war gaming, involves the simulation of the wartime conduct of opposing forces. War gaming is two-sided and employs a great deal of human judgment.

The scenario, a description of a possible future history, frequently aids the analyst. It may take the form, for example, of a two- or five-year history of opposing force posture developments and of the associated developments in international relations. Such histories may then be studied to try to identify how the forces helped or hindered policy.

Computational modeling and analysis, war gaming, and scenario writing represent three points along what really is a spectrum of techniques available to the analyst.[1] At one end, precise computation is emphasized and intuition is played down; at the other end, computation is de-emphasized and intuition is fully exploited. In the middle, war gaming depends upon both formal models and human intuition. Good analysis, of course, will frequently employ several of these techniques simultaneously. However, for the moment, let us examine them individually to see how each stacks up against the requirements we listed earlier.[2]

Can the Traditional Methods Meet the Requirements?

Figure 15.1 summarizes a personal assessment of the degree to which these techniques satisfy each of the demands placed on them by the problem of relating force posture and policy. This assessment assumes neither the best nor the worst, neither the unusual nor the unique application of each technique, but rather only the level one might expect them to reach in the typical case. And although, in principle, the techniques may often be able to overcome the limitations they are credited with, the time, cost, and manpower limitations of the real world usually prohibit them from doing so in practice.

Let us consider each requirement in turn.

> Computational modeling usually does not take into account a span of years, but is rather concerned with the design of forces for a particular date in the future – say, 1970. War gaming, too, is usually concerned with a short time span – the length of a particular war or some battles in it. Scenarios do, however, extend across a period of years.

[1] I am indebted to Richard Nelson of The RAND Corporation for this observation.
[2] For additional discussion of these methods, see especially Chapters 10, 11, 12, 14, and 16.

Requirement	Computational modeling and analysis	War gaming	Scenario writing
Considers a span of years	No	No	Yes
Includes both peacetime and wartime	No	No	Yes
Considers all the factors affecting policy	No	Yes	Yes
Is truly two-sided, competitive	No	Yes	No
Provides for skillful strategy and tactics	No	Yes	Yes
Takes account of R&D	No	No	No
Takes account of uncertainty	Yes	Yes	No
Takes account of intelligence data	No	Yes	Yes

Fig. 15.1.– Can the traditional methods do the job?

Computational modeling usually considers only what happens during a war, as does war gaming. Scenario writing includes peace.

Computational modeling usually takes a segment of the forces – the missiles or the bombers, or the missiles and the bombers – but ignores the defenses, for example. War gaming includes all of the forces, of course, and so can scenario writing.

Computational modeling is usually not two-sided and competitive. War gaming is. Scenario writing may be, but usually is not, at least not in the sense that an attempt is made to optimize all of an opponent's responses. We would not usually find two people, each taking a side, constructing a scenario; there is typically only one scenarist, who tries to imagine what each side will do.

There is little room for the introduction of skillful strategy and tactics in a computational modeling study. There is, of course, in war gaming, and there can be in scenario writing.

Computational modeling does not usually take into account the varying possibilities of success in an R&D venture; war gaming certainly does not; scenario writing usually does not.

There is, of course, some account taken of uncertainty in computational modeling and in war gaming. Scenario writing usually tries to make a single projection of the future, although there may be branch points to allow the study of several alternatives in selected cases. But, as a general rule, scenario writing does not explicitly account for real-world uncertainties.

In computational modeling, analysts customarily ignore the role that intelligence will play in the war. While they usually do make use of intelligence data in the analysis, this seldom extends to the intelligence aspects of force structures. War gaming, however, does examine the use of intelligence data, as does scenario writing.

We can conclude, therefore, that none of these methods of analysis meets all the demands that the list of requirements places on it. Thus, if the validity of the requirements is granted, we must conclude that traditional methods alone cannot do the job.

The SAFE game

But if traditional methods cannot do the job, what can be suggested in their place? We have had some experience at RAND in the last few years with a technique that is related to scenario writing and war gaming and that employs the results of computational modeling studies, yet is more than just a combination of those techniques. In one sense, it is an extension of war gaming to include the peacetime aspects – the procurement and R&D aspects – of the interaction between force posture and policy. This technique is called the SAFE game – the Strategy-and-Force-Evaluation game.[3] It provides a technique that allows all the requirements on our list to be met – that is, it is a tool that will enable analysts to carry out useful analysis of policy and posture interactions.

Thus, first of all, the SAFE game covers a ten-year span. It is a game that is concerned with the development of weapons, force posture, and strategy over ten years. It is divided into five two-year periods. It includes both peacetime and wartime aspects. It includes in the forces to be procured and used all the strategic forces – the bombers, missiles, defenses, payloads (that is, the various warheads and missiles that may be carried by aircraft and missiles), and some others (command and control, civil defense, and related aspects). It is a two-sided game. There is a Blue team representing the United States; there is a Red team representing the Soviet Union; and, of course, there is a Control team that represents the

[3] The SAFE game is the latest of a series of strategic planning games that have been developed at The RAND Corporation. The version described here was developed under the leadership of Olaf Helmer and Thomas Brown.

environment – all the other nations, technology, the high-level national policymakers, diplomats, and so on. The planning cycle for the game requires the teams to operate and deploy their forces, to procure and retrofit those forces, to set up the factories and technology to produce those forces, and to engage in R&D on new forces. (In the game, as in the real world, R&D includes a wide variety of options. The many R&D choices available to the teams represented, at the time the game was designed in 1962, those that we felt were reasonable for the next time period. Moreover, in SAFE the success of R&D is uncertain; there is no guarantee that when an R&D program is undertaken it will be successful.) Uncertainty also enters into the intelligence reports the teams receive and into the effectiveness estimates of their own forces that the teams are given. The players never know the exact effectiveness of any of their weapons. They have estimates, but these change when evaluations are made. Uncertainty appears as well in the system costs and in budget estimates. The costs of the items change from time period to time period, representing the increasing certainty about costs as weapons progress from R&D to operation. Budgets vary, and a team is never quite certain how much budget it is going to get, although there is not too much variation. And then there are intelligence data exchanged between the teams on such things as R&D, actual developments, procurement, and budgets.

The Menu of Options in SAFE

Figure 15.2 presents an example of the set of options, the choices, that are available to one team in this game. Among the choices are weapon systems that may in some cases already be in the force – say, the Atlas and Titan. In some cases, they may be in procurement – not yet in the force but coming in – for example, the improved Minuteman. Then there are other forces that require R&D if the teams hope to procure them – say, the long-endurance multipurpose aircraft. And then there are some forces that may be retrofitted. So, facing each team when it starts this game is a veritable military feast; it can choose from among all of these items those that its budget and its policy suggest it should have.

Figure 15.3 presents a detailed view of two typical items that might appear on the menu: the menu descriptions of a long-endurance aircraft and of a hardened missile.[4] When a team must decide which of the menu items to procure, to deploy, to set up production for, or to do R&D on, it has available this kind of information.

To the left of each description in Fig. 15.3 is indicated the effectiveness

[4] The entries are representative, not real, to avoid the need for security classification.

BOMBERS	MISSILES	DEFENSE	PAYLOADS	OTHERS
B–47	Atlas & Titan	Fighter	Armed recce option	Offense C & C
B–52	Minuteman	Advanced fighter	Subsonic decoys	Ultra-hard offense C & C
Tanker	Minuteman-options	BOMARC	Supersonic decoys	SAGE
B–58	Titan II	Fighter shelters	Hound-dog	Distributed C & C
B–70	Titan II options	Fighter dispersal	Advanced ASM	Warning satellite
Mach 2 heavy	Heavy payload ICBM	Nike-Hercules	ALBM	Satellite decoys
Long-endurance aircraft (LEA)	Mobile ICBM	Hawk	Accountable warheads	Phased-array warning
LEA-parasite	Advanced cruise missile	Nike-Zeus	MIRV	Low-altitude radar
Low-altitude aircraft	Glide missile	Nike X	Penetration aids	Blast shelters
Nuclear-powered aircraft	Polaris	Advanced AICBM I	Guided warhead	Fallout shelters
Bomber shelters	MRBM	Advanced AICBM II	Counter-ABM satellite	Civil defense preparation
Hard service facility		ABM satellite	Antiaircraft missiles	
Bomber dispersal		Satellite interceptor	Antisubmarine missile	

Fig. 15.2 – Blue menu of options in SAFE (circa 1962)

Item	R & D	Capacity	Procurement	Operation	Retrofit
Long-endurance aircraft Unit: 45; 15 on airborne-alert Reliability: III; Speed: .3 Range: 13,500; Endurance: 3 days Usable offensively or defensively	Score: 12 R_1: 200 R+4: 300	140	750	380	
Missile (hard) Unit: 20 Hardness: 200 psi Firing time: 5 min ($2\frac{1}{2}$ soft)	Score: 3 (Mod B) 10 (Mod C) R_1: 500 R+4: 600	20	Mod A, B: 90 Mod C: 120	30	10

	CEP	Yield	Reliability
Mod A	1	1	II
Mod B	0.5	2	III
Mod C	0.3	4	III

Fig. 15.3 – Two Blue menu items

of the weapon as described for this game. Included here are such things as the unit size. There are 45 long-endurance aircraft in each unit that one buys; there are 20 hardened missiles. Of the 45 aircraft, 15 are on airborne alert; their speed is Mach 0.3; their endurance is three days; their range is 13,500 miles. The aircraft may be used either offensively or defensively, depending on the payload that is procured for it. The missile has a hardness of about 200 psi; it has a firing time of five minutes, of which two and a half minutes are soft; that is, it may be destroyed much more easily during those two and a half minutes than it can be while it is in its hardened launcher. There are three modifications of the missile available – Mod A, Mod B, and Mod C – and they are characterized in CEP and yield as shown by the small table in the lower left-hand corner of the Figure. The reliability figure, given in roman numerals, identifies a reliability class. Class III has a moderate reliability, between 65 and 75 per cent in the first period of its operation. This improves in subsequent periods, however. A reliability figure of IV is generally better, between 75 and 85 per cent, and a reliability figure of V is the best, 85 per cent.

These data define the effectiveness part of the evaluation that each team must make. The players must ask, Do we need a weapon system that has those properties? And they must balance those properties against their costs. For the aircraft for which R&D has not been completed yet, the team must spend research and development money before it can consider doing anything else.

A few words about how R&D is modeled in SAFE will explain the entries in the R&D column in Fig. 15.3. Each researchable weapon system is assigned an R&D score, a number between 1 and 33. The score assigned to the aircraft is 12, and, in the first period, R&D on the plane will cost the team $200 million. If it elects to spend that $200 million, the Control team will draw a random number between 0 and 9. The team's objective is to exceed the R&D score as quickly as possible, that is, to have drawn a series of random numbers whose sum is 12 or greater in as few periods as possible. When that happens, R&D is successful. In the case illustrated in Fig. 15.3, success is impossible in one period; the team must pay R&D for two periods. That is, it will take at least four years for R&D to be successful in this development project. But in the second period that the team pays R&D money, which will then be $300 million because of the greater complexity of the R&D project at that stage, they get a bonus score of 4. So if they have scored 8 or 9 in their first period, they automatically get a successful R&D development program in the second period. In any event, they have to score a sum of 8 or greater in two periods in order to achieve successful R&D in the minimum time for the aircraft. That is a fairly probable situation, and the score was chosen so it would be; it represents an estimate of how probable it is that a long-endurance aircraft could be developed successfully. For the missile, the Mod A is available immediately, but if the team wants to get a Mod B or a Mod C with the improved CEPs and yields, it must spend some R&D money.

After successful R&D, the next step is to procure capacity for production. For the aircraft, Fig. 15.3 indicates that the capacity charge of $140 million will procure the capacity to produce one unit each period. When enough capacity exists, the team can undertake procurement, which will cost $750 million for one unit, and operation, which will cost $380 million for one unit per period. Now notice that there are blanks left below the costs of procurement and operation. These costs change with time. When the team undertakes the R&D project, $750 million and $380 million are the estimated costs. In the next period, however, there will be new costs in the blanks, representing what the team has learned as a result of its R&D project. The costs may be greater or they may be less, depending on the outcome of a random process that is, however, weighted toward the high side, a sad but apparently realistic simulation of actual events. Subsequently, additional costs will be entered as the team approaches procurement. The cost it finally pays will be the one that is located in the lowest boxes. So in the development of the aircraft, there is a great deal of uncertainty, which, we feel, gives a reasonable model of the sort of uncertainty that the decisionmaker faces in real life. For the missile, similar costs apply.

Routine of Play

The routine of play for the SAFE game has three phases: the beginning of the play; five periods, all of which are roughly the same except that the first period is somewhat simpler than the four that follow; and the post-game analysis. At the beginning of play the Red and Blue teams each receive from the Control team policy guidelines defining the national policy that they are supposed to support, and budget estimates indicating the level of budget that they can expect to receive to support this policy during the coming ten-year period. Keeping these in mind and looking at the menu, they will decide on the kind of forces that they feel are necessary. Then they will start play in the first period.

They are given an initial posture that represents the current status of their forces. Now, they can begin by phasing out those forces that they feel they no longer want. For those that remain, they must pay the operational costs, subtracting that expenditure from their budget allotment. The remainder of the budget they then allocate among R&D, capacity expansion, procurement, and retrofit of forces. They have to consider all of these activities in allocating their budget. They cannot limit their attention simply to what they think are the best R&D choices, or to what they would like to buy; they must plan ahead for the subsequent time periods as well as act for the present one. They then deploy their old and new forces and receive from the Control team a description of some general war outbreak circumstance, such as "Red initiates in a time of tension." With such a circumstance in mind, they write a strike plan for the forces they have at the end of that period, basing their estimate of the opposing forces on intelligence data received through Control. The strike plan describes the timing, the route, and the target of each of their offensive weapons, and indicates as well what declarations are to be made during the strike. They do not see the strike plan again until the end of the game. And they do not know what the outcome of the interaction between that strike plan and that of their opponent would be. They just use the strike plan to suggest to themselves the capabilities that they would like to have, and they use that information in the second through the fifth periods to guide their R&D, capacity, and procurement choices.

The second through fifth periods differ from the first period only in that the teams receive some news of successes or failures in R&D ventures they had undertaken earlier, learn the modified costs of systems in development, and get intelligence reports on what their opponents are doing. They then go back and repeat the steps carried out in the first period. When the "ten years" have elapsed the game is over.

At the end of the game, three things can be done. The first is an immediate review, at which the teams receive accounts of wars fought according to their strike plans. In the five periods, they will have written five strike plans, two or three of which will have been taken and analyzed by the Control team during the course of play. Only at the end of the game does the Control team indicate to the Red and Blue teams what the estimated outcomes of those wars were. The second possibility for post-game analysis is an immediate critique of the game, a search for the obvious successes, the evident failures, and the direct lessons of the play. The third involves using the play as a starting point for further analysis, developing interesting situations or systems in greater detail.

This, then, is a rough picture of the SAFE tool and how it may be used.

A Set of SAFE Plays

In order to explore the role of SAFE in analyzing the interaction between policy and force posture, we might now examine a round of plays that were carried out with the SAFE game in 1962 at RAND (under the leadership of F. S. Hoffman). The game policy and budgets are shown in Fig. 15.4. We were concerned with what U.S. strategic defense policies would

	Red policy	Peacetime deterrence *and*			
		Defense	Force parity	Counterforce	Force Superiority
Blue policy	Red \$ / Blue \$	Moderate growth	Low	Moderate growth	Rapid growth
Peacetime deterrence & wartime deterrence & counterforce	Constant	F			
	Moderate growth	A		B	
Peacetime deterrence & counterforce	Rapid growth	E			
Peacetime deterrence & wartime deterrence	Low		D		C

Fig. 15.4 – Six SAFE plays

be feasible in the coming decade – in particular, in 1970. The project leader made up several sets of policy guidelines for the Soviet Union and for the United States, and some budgets for the Soviet Union and for the United States, and chose certain combinations of these to investigate. The purpose of the investigation was to determine which of these U.S. policies were feasible.

In all cases, both the Soviet Union and the United States had peacetime deterrence as part of their policies. In addition the Soviets in one case had a policy that comprised peacetime deterrence and an emphasis on defense. That policy was an attempt to model what appeared to us to be their posture – a high emphasis on defense, much higher than that in the United States. This policy was supported by a moderately growing budget.

Another Soviet posture that was analyzed was one in which they had a policy that combined peacetime deterrence and the objective of maintaining force parity with the United States. A third case had a policy that comprised peacetime deterrence and an attempt to develop a good counterforce capability against the United States, supported by a moderately growing budget. And, finally, there was a case in which the Soviets had as their policy objectives peacetime deterrence and an attempt to obtain superiority, supported by a rapidly growing budget that might allow them to do so.

Against these there were Blue policies that had various combinations of peacetime deterrence, of wartime deterrence (that is, limiting damage to oneself by threatening to inflict similar damage on the opponent), and of counterforce (that is, trying to limit damage to oneself by destroying the opponent's forces). The most extensive policy included, in addition to peacetime deterrence, both wartime deterrence and counterforce. Another one had just peacetime deterrence and counterforce. And a third one had just peacetime and wartime deterrence. There were appropriate budgets for each of these three policies. Six cases were explored in game plays. The combinations are represented in Fig. 15.4 by the letters A, B, C, D, E, and F.

Types of Analysis of SAFE Plays

Having played these games, what is it that an analyst can do with the output? Six games were played; force postures were generated; the teams each thought that they had followed their policies moderately well. What is it that we can learn from this kind of experience?

There are three types of analysis that can be carried out. The first is an analysis of the resulting games, game by game, looking at each and trying to answer such questions as: Were the objectives of the team achieved? Did they effectively support their policy? What were the best postures for achieving them if they did so? What forces did they find were useful for carrying out a wartime deterrence policy, for instance? What were their failures and their successes? There are lessons in the answers to each of these questions, but perhaps the most important result is that situations are identified for further analysis. The SAFE game is inaccurate in many respects and an analyst would not want to draw conclusions simply from the results of one or even a set of plays. However, situations arise that look so interesting that subsequent analysis by other techniques – by scenario writing, by war gaming, or by computational modeling – seems warranted. And the game-by-game analysis can pick out these situations for further analysis.

The second type of analysis is a comparison of the games in pairs or triples. And as Fig. 15.4 suggests, certain of the pairs in the six games we played were set up particularly for that purpose. For instance, in the pair A and F the Blue team faced a Red team whose policy was chosen so as to represent what was believed to be actual Soviet policy at that time. The Blue team's objective was to obtain a moderate degree of U.S. superiority. The pair was chosen so as to test the sensitivity of that aim to Blue's budget level. In one case, the budget was kept fairly constant at current levels; in the other, it grew moderately. And so these two games could be examined to see whether or not the policy objectives were better achieved in game A than in game F.

To take another example of the same sort of analysis, we might check to see what the sensitivity of policy is to an opponent's policy. The U.S. policy remained constant in cases A and B, but in case A the Soviets were still emphasizing defense; in B the Soviets had switched to a counterforce objective. How difficult does that make U.S. achievement of the policy of superiority? Similarly, in C and D the United States has taken a relaxed view of the cold war. Its policy comprises just peacetime deterrence and prevention of damage during wartime by counterthreat rather than by counterforce, so the required missile forces are low. There is a lower budget. In game D the United States is facing a Soviet Union that is also interested in reducing tension; it is interested only in parity, and has a low budget itself. In game C, however, it is actually interested in superiority and a rapid military growth, but it is maintaining secrecy about these policy objectives. And so games C and D can be examined to see what happens. As it turned out, some very interesting things happened in those two plays. In game C the Soviets were able to achieve a superior force and they did it rapidly, principally because, in this example, the United States did not do all the R&D and capacity procurement that it should have done and was, therefore, caught napping when Red's objectives finally became evident.

Still another instance of this type of analysis might involve an examination of the relative feasibility of alternative policies. Thus, against a policy that was thought to represent the actual Soviet policy, three different U.S. policies were tested to see how feasible they were for the present time. Games A, E, and F were the result.

The third type of analysis that can be performed on a set of SAFE games is an examination of the entire group to see what general lessons it provides. For example, the plays can be studied to determine which of the many R&D possibilities were favored by the teams and for what reasons. Do these reasons transfer to the real world? Do the plays suggest that certain weapon systems are especially attractive R&D choices? The plays can also be examined to see what lessons for strategy they may teach. The six SAFE plays that we conducted strongly suggested to the players and the analysts the benefits of maintaining flexibility through extensive R&D. Again, the plays can be studied for what they reveal about specific force structure questions. What was the role of defense or of bombers in the set of plays? What were the players' attitudes toward defense or toward bombers? What did they feel was the relation between such systems and their particular policy statements?[5]

Limitations of SAFE as an Analytic Tool

Despite this glowing description, SAFE has a number of limitations as an analytic tool; by now many of them are probably apparent. Some of the more important ones can be grouped together into three general classes.

First of all, SAFE is *an unrealistic model* of the real world. This is not to say that it is completely unrealistic. A serious effort was made to get as accurate a simulation as would be practical, but no matter how serious our effort, we could not avoid having to use imprecise data. The effectiveness of most weapon and support systems is just impossible to determine with high confidence, even if very extensive efforts are made. The models are, of course, simplified, as the R&D model illustrated. We are unable to predict much of what will occur ten years in the future, and this shows up most obviously in the menu of options. The R&D choices are those that we anticipated in 1962, but in 1970 there will be others that are not modeled in the SAFE game, even though we were considering the decision situation in 1970. And an important limitation is that we cannot model motivations. The people who were playing the part of the Soviet Union decisionmakers were not dialectical materialists or Russian nationalists. Since they lacked that view of the world and it is impossible to simulate it, they tended to think and to act like RAND analysts, which we know the Soviets do not. Still another limitation arises from our inability to model the rigidities and irrationalities of the real world. Service tradition and politics, unchanging doctrinal precepts, and internal politics strongly affect military force postures, but are largely absent from SAFE.

Secondly, SAFE is *a restricted model* of the real world. It excludes the limited war forces, even though, as we noted earlier, it is important to include all the forces that enter into a policy judgment. And while it may be a good idea to factor out the strategic forces for analysis in many cases, we know now that there is a strong interaction between limited war forces and strategic forces that we must take into account in many other circumstances. SAFE is also restricted in that it does not include other countries,

[5] An attempt to describe and generalize the defensive lessons learned in the plays is presented in R. E. Levien's *An Appreciation of the Value of Continental Defense*, The RAND Corporation, RM-3987-PR, March 1964.

especially those in NATO, even though they play a large role in our strategic policies. That exclusion is probably one of the most important weaknesses of the model. And there is an exclusion of diplomacy and small wars, both of which influence, and are influenced by, strategic posture.

Finally, SAFE *cannot provide unambiguous answers.* The outcomes of SAFE plays require further extensive analysis of the type we have mentioned – and even more than that – before they can be used as guides to policy. And there is the ever-present danger of techniques for gaining artificial experience, a very important danger: The decisionmakers and the analysts may learn the wrong lessons from the games. We always have to be careful that what we are learning, what the models show, is not an artifact of the model completely divorced from the real world.

Benefits of the SAFE Game

As opposed to these limitations, what are the benefits provided by the SAFE game? Perhaps the most important one is that it is *a tool for providing consistent force posture histories.* It is fairly easy to do back-of-the-envelope calculations employing parts of forces; it is fairly easy to consider any one of the particular factors of policy, or budget, or responsive opponent, or uncertainty. But to bring them all together and ensure that alternative forces are developed under their total, interacting influence is extremely difficult. The SAFE game provides a tool that facilitates the development of consistent force postures. Each team has specific policy objectives. It has a specific budget to spend in supporting those objectives. It faces and must take into account the actions of a responsive opponent. The time factors are vitally important, for the forces evolve over ten years and do not appear full-grown in a single moment at some time in the future. Uncertainty is ever present. The team never knows exactly what the R&D successes will be, what the costs of a system in development will be, how effective its weapons will be, and so on. It must work with its initial posture choice, developing it into the posture it wants without becoming weak along the way. And there is a place for strategy. The teams can be clever and cautious or naive and foolhardy; what they are will influence the outcome of the play.

Another benefit of SAFE is that it is *a tool that allows us to consider broad force trade-offs.* The teams rarely think only in terms of which is the best bomber or the best missile. They think also of bombers versus missiles, of more missiles versus better defense of those in existence, of antimissile defense versus hardening of the missiles themselves. The many interactions that are brought out influence the entire strategic force posture, and it is difficult to get computational modeling or analysis that can do that in a reasonable amount of time. In studying the results of SAFE games we can examine such trade-offs as antibomber versus antimissile defense, bomber versus missile offense, active versus passive defense, R&D expenditures versus procurement.

Finally, another important benefit of SAFE is that it offers *a valuable background for decision.* It provides artificial experience for decision-makers. It provides a framework for detailed analysis. It suggests important areas for further study. And it is a vehicle for obtaining human judgment. All of these characteristics are valuable when the SAFE game is imbedded in a larger analytic environment – when there are other types of analysis being carried out at the same time. We have found at RAND that as part of a large project SAFE can be an extremely valuable tool.

Chapter 16

SCENARIOS IN SYSTEMS ANALYSIS

SEYOM BROWN

This Chapter points out that the function, form, and content of scenarios are determined by the specific research task at hand. Different levels of analysis have differing requirements for detail and for scenario credibility. But the construction and use of scenarios with political content is increasingly regarded as a crucial aspect of systems analysis at most levels of analysis in the DOD, and especially in the analysis of major force posture alternatives.

INTRODUCTION

In the hearings on the Defense Budget for fiscal 1965, the Secretary of Defense was quizzed on his decisionmaking techniques by Congressman Melvin Laird. The Representative from Wisconsin was of the opinion that OSD was relying more and more on cost-effectiveness studies, and he wanted to know if his opinion was correct.

"No; I think not," answered Secretary McNamara. "We are relying more and more upon sophisticated analyses of potential political-military conflicts and an appraisal of the advantage to the United States of alternative force sizes in relation to those contingencies, and the various applications of those forces in those contingencies. The cost-effectiveness study as it would be narrowly defined comes into importance only in choosing between alternative means of satisfying an established force requirement."[1] The force requirements, Mr. McNamara told the Congressional committee, were derived from analysis of potential contingency war plans for a variety of such political-military situations.

We need not get into a discussion here of whether the Secretary was describing an ideal or real process. But we can agree that the weight he attributed to the analysis of potential political-military situations is warranted.

The top military planners with authority to make budget decisions affecting over-all force posture, and to allocate roles and missions with

---

[1] Subcommittee on Department of Defense Appropriations, Committee on Appropriations, House of Representatives, *Department of Defense Appropriations for 1965*, February 17, 1964, pp. 304-305.

the services; the Commander-in-Chief and his subordinate commanders doing contingency planning; the systems analyst who provides inputs to these decisions – all of these people *require*, at some stage in their work, pictures, imaginings, fictions, if you will, of the circumstances under which the military systems they are concerned with will have to operate.

What will be the tasks these systems will be called upon to perform? Under what physical conditions? Under what political constraints? When we ask these questions, we are asking for a scenario.

Definition of scenario

There are many notions floating around of what a scenario is or ought to be. More often than not these notions, or attempted definitions, are the product of the specialist's acquaintance with those things which are called scenarios in his special field of work, and exclude those things which other specialists choose to label scenarios in their own fields.

Often with great conviction the champions of various definitions try to convince others that their particular notion is the correct one; that the other animals that may be presented to them bearing the label "scenario" are really something else in disguise – possibly "contexts," "situations," "plans," "assumptions," "parameter values," but "certainly nothing you ought to be calling a scenario."

It appears that the phrase, "Now *this* is a scenario . . .," can be found to be variously applied to

An outline of a sequence of hypothetical events;

A record of the actions and counteractions taken by parties to a conflict;

A plan of actions to be taken during a projected exercise or maneuver;

The estimate of the situation by Commander "X" at time "Y" in a war or war game;

A specific set of parametric values selected for a given run of the computer.

How, then, can we proceed to talk about "scenarios" without excluding someone's notion of what scenarios really are?

If we dwell for a moment on these various notions, we do find a common thread. They all refer to descriptions of the *conditions under which* the systems they are analyzing, designing, or operating are assumed to be performing. The system may be a weapon, it may be a component of a weapon, it may be a vast complex of weapons and support facilities (such as NORAD), it may be an organization (such as the National Military Command and Control System), it may even be the entire national security establishment. Whatever the scope and properties of the specific system, a scenario – in systems analysis – can be defined as a statement of assumptions about the operating environment of the particular system we are analyzing.

Of course, this definition is so widely cast that the net may drag up a wide assortment of fish. One way of making the discussion more specific might be to *classify the fish*. The only purpose in insisting that we start from a broad notion first is to guard against getting ourselves entangled at the outset in a narrow definitional net that lets some of the more interesting catches swim away.

Importance of discussing scenarios systematically

Before we turn to the main theme of this chapter – the political content of scenarios – we might underscore the importance of treating the scenario aspect of systems analysis *systematically*. After all, it is from our anticipations of the environments in which our systems are to operate – the state-of-the-world, the conflict situations, and the tasks these systems are expected to accomplish – that many of our criteria for *evaluating* the *performance* of a given system emerge. Thus, having a casual attitude toward scenarios is often tantamount to having a casual attitude toward the selection of our evaluative criteria. If we accept the proposition that our analyses can be no better than the criteria we employ, then we must accept the corollary proposition that, where criteria are dependent upon scenario assumptions, our analyses can be no better than our scenarios.

A complete systems analysis probably requires many such scenarios – some stipulating typical tasks, typical conditions, and typical constraints for the system under investigation, and some stipulating unique, and even extreme, situations.[2]

The problem of finding or constructing the proper scenarios, whether one wants either typical or extreme situations, is a tremendously difficult analytical task in itself. All we can attempt to do in this Chapter is to scratch the surface of this problem. A systematic approach to any actual scenario to be used in some specified project would require many working sessions. Unfortunately, this aspect of our work is too often handled in a hurried or slipshod manner. There are notable exceptions, however – one of the most outstanding being the scenario work done a few years ago for a RAND study on limited war in Iran. The scenario work was carefully and conscientiously handled; those who participated in this effort have some small reward in the knowledge that the reports from this study are being used

[2] See Chapter 4.

today as a basis for deriving scenarios for some current tactical air studies elsewhere in the defense analysis community.

For present purposes, it might be best simply to characterize the essential requirements of scenario form and content at different levels of systems analysis, so as to provide a backdrop and framework for any detailed evaluation we may want to make of specific scenarios.

FORM AND CONTENT DETERMINED BY RESEARCH TASK

What should a scenario look like? It is a misconception to assume that the scenario is what is left after everything that can be quantified has been quantified – that the scenario is the words, and the Analysis, with a capital "A," is the numbers.

In some cases, the scenario may need to be presented in computer language; in others, the most useful scenario may resemble an historical essay, rich in detail, with the purpose of conveying not only the tangible features of a situation, but its tone and mood.

Frequently a scenario is solicited in "raw" form, from which the systems analyst then abstracts inputs for his analytical model. And sometimes, when there is very close collaboration between the systems analyst and scenarist, or when they happen to be one and the same person, the original scenario itself may be a highly formalized statement. But, ultimately, what determines the final form of a scenario in any of these cases is the form of the analytical model into which the scenario is fed.

For example, a user of the Strategic Operations Model, which N. C. Dalkey considers in Chapter 12, may want to compute the results of a missile exchange under differing assumptions concerning the shelter protection of the U.S. population and the amount of strategic and tactical warning received of an impending Soviet strike. Now, he needs a "scenario" only to provide some plausible rationale for assigning certain numbers to these parameters. Thus, one such scenario is likely to postulate a high state of tension and U.S. involvement in an on-going low-level war before the strategic nuclear war begins. But when converted into terms usable by his model, this scenario will be expressed in *numbers* designating, first of all, the degree to which people in cities have taken shelter and, second, the state of readiness of our missile forces – further refined into generation rates.

But say the model is a manual war game, with a wide range of options – tactical, strategic, possibly diplomatic – open to the players. In form and content, the scenario, even as it is fed into the model, will probably be complex; it will be written on pieces of $8\frac{1}{2}''$ by $11''$ paper, mostly in English words, and combine the attributes of a Basic National Security Policy Statement drawn up by the Policy Planning Council of the State

Department, Order of Battle documents of all sides, and the New York *Times*' "News of the Week in Review." Obviously, such a "book" will be so far removed from the sparse form of the scenario used in the Strategic Operations Model that many would regard it as unhelpful to talk of them in the same breath. But this purely semantic or definitional debate is really unimportant to our work.

What is important is the basic point that the research *questions* determine the design of the testing *equipment*. If they are questions of grand-strategic effectiveness, as measured by, say, U.S. ability to impose terms on a hostile enemy, the situations to which the analyst would want to subject the system must be more widely and richly drawn than in analyses in which the research questions have to do with the costs and effectiveness of alternative logistics networks, where costs are measured by dollars spent in manufacturing, installation, and maintenance, and effectiveness is measured by the volume and rate of flow of specified supplies under a range of physical conditions.

The character of a scenario – its language, numbers of words, or other symbols; the kinds of detail it presents – cannot be settled in principle, but only with reference to the specific research task at hand. The content itself – the bits of information that are required or irrelevant in the scenario – is usually determined by the capability of the analytic model to handle various kinds of information.

Levels of analysis

After we have cleared away the semantic underbrush, we are compelled to admit that there is no established doctrine for creating the scenarios used in systems analysis. One cannot set up a list of *do's* and *don't's* for scenario content, form, detail, style – even credibility – without reference to the specific systems analysis task at hand. Indeed, we might go one step farther and argue that the relevance of a scenario is part and parcel of the relevance of the analytic model to the research objective. Perhaps we can see this point more clearly if we attempt to relate the problem of scenario construction to the four levels of analysis suggested by E. S. Quade in Chapter 1.

Quade related the work we are doing and the analytical tools we are developing to four levels of decisionmaking in the field of national security:

1. *Management of Operations* – which is essentially no different from management science: an attempt to increase the efficiency of some particular man-machine system, where "efficiency" is something as straightforward as maximizing profits.

2. *Choice of Tactical Alternatives* – which is sometimes called "conflict design in the small," and where the objective of the operation is usually clear, and some reasonably satisfactory measure of effectiveness already is assumed to exist.
3. *Systems Engineering, Systems Design, and Systems Research* – where the problem is to find better ways, or the best way, of implementing a system requirement. The operations to be performed are given, are already specified; the analysis assumes these operations are important, it accepts the requirement, but the inquiry *may* question established criteria of costs and effectiveness, or indeed may be *explicitly* charged with the responsibility for coming up with better measures for evaluating the performance of alternative systems.
4. *Determination of Major Policy Alternatives* – which is sometimes called "conflict design in the large": the analysis of alternative means for implementing basic strategies, or the analysis of the impact of force posture choices and strategic alternatives upon the nation's ability to achieve its foreign policy objectives.

Analyses of military systems, on this continuum from management science to grand strategy, have need of vastly different analytical models, and consequently are likely to call for scenarios which differ just as widely in their appearance and substance. Now, most of the scenarios that political analysts are interested in are those which call for some political content. Usually, this means that they are unlikely to get involved in work at the lower end of the continuum. Thus, much of the work of RAND'S Social Science Department is at Level 4 – "conflict design in the large" – but this work often involves the construction or evaluation of scenarios to be used in cost-effectiveness studies of alternative means of accomplishing some stated military mission. Somewhere in these studies, be they systems design studies or studies related to the choice of tactical alternatives, there is a need to stipulate the assumptions, the conflict conditions, under which the stated mission is to be accomplished.

For example, suppose the analyst's purpose is to compare the effectiveness of various systems within essentially the same force structure in conflicts of the same scale, fought under essentially the same rules of engagement; and suppose further that effectiveness is measured by some clear criteria, such as the number of casualties inflicted and the number of enemy weapons destroyed. The scenario may then be no more than a relatively straightforward statement of the general military situation, the ground rules that are assumed to exist, the disposition of enemy forces, the natural constraints (like weather and terrain), and, finally, the sequence of decisions up to and including the one to resort to a given targeting option. But already we have a requirement for filling in the blanks with items that say or imply something about the political state of the world and political constraints that U.S. policy may impose on military operations.

Now, for such analytical tasks, in which the constraints are given, or subject to wide variation for analytical purposes, it is often not necessary for the scenario *itself* to state the political assumptions behind the parametric values selected.

But let us change our criterion of effectiveness to one, not of measurable attrition, but to the effect upon enemy decisions to cease and desist from his attack and return to the *status quo ante*: now we are clearly on Level 4, the determination of major policy alternatives, and in addition to the parameters already called for, the scenario will need to say quite a bit about what makes the enemy tick – why he attacked in the first place, and how high a value he places on the success of his venture as compared with other competing values. Our assumptions concerning enemy motivations, his willingness to run risks, his tolerance for deprivations of different sorts – whether stated explicitly in the opening scenario or allowed to unfold in the form of improvisations by war gamers – will determine the *output* of our model, and may also affect the *design* of the model, particularly the choice of which outputs to display and on what evaluative scales.

Thus, as we ascend the levels-of-analysis continuum, the political content of the scenario assumes more and more weight in the over-all analysis.

Sometimes the military planner may find it useful to leap-frog formal analytical models, and do a verbal scenario *first*, because, intuitively, he perceives that the particular scenario is immediately relevant to important planning problems – that is, even before any formal model has been constructed that is able to handle the scenario. It seems probable that many of our crisis scenarios – for Berlin or Cuba, hypothetical or real – are of this sort. The scenario itself becomes, so to speak, the model. Military forces, concepts, and policies are judged in terms of their usefulness in these crises. The criteria for usefulness are often vague and subjective, and sometimes remain unarticulated; but nonetheless they may become the operating criteria at the OSD-Presidential level. Program Change Proposals may or may not win acceptance on the basis of these implicit judgments of the worth of various systems and strategies in crisis situations. But once these criteria have been found useful, the systems analyst then needs to develop models which are relevant to the scenarios implicit in the minds of the decisionmakers. To do so, he has need of the services of scenario writers, who tell *him* what is relevant, rather than the other way around. And the scenario writer is only as good as his ability to perceive the world as the top decisionmakers perceive it.

There is yet a grander task for scenario writers. The trouble is that there are too few fools willing to rush in where most angels, having been burned, fear to tread. Essentially, this task involves viewing the scenario as a device for altering not only the systems analyst's model, but such criteria as those of costs and effectiveness held by the top decisionmakers. Where such an approach is meaningful, the scenario writer (or scenario-writing group) operates autonomously, not directly coupled to any existing systems analysis project. He conceives it to be his purpose to serve as a kind of advance sentinel, to be able to alert the decisionmaker, the military planner, the systems analyst, to state-of-the-world changes and specific situations which may require the application of military force in ways that are not now planned for, or which may alter prevailing expectations of what planned-for applications of military force can accomplish. In a word, the scenarist at his grandest (and most insufferable) is an iconoclast, a model breaker, a questioner of assumptions, and – in rare instances – a fashioner of new criteria.

The problem of political credibility

The question of what kind and what amount of information should go into a scenario leads us head on into the devilish question of political credibility. If the systems analyst and the scenario writer have already largely agreed upon criteria of relevance, kinds of detail, and form of presentation, it will be easier to break up the question of credibility into manageable parts.

The strictest standards of credibility, of course, ought to apply to those aspects of a scenario which are deemed relevant to the research purpose at hand. Let us first take inquiries at the level of choosing tactical alternatives. Suppose that in a field exercise a major purpose is to compare the effectiveness of certain fighter aircraft and surface-to-air defensive systems against enemy fighters. Given this objective, the fact that Blue and Red are assumed to have equally high motivations to take risks in order to achieve air superiority over the ground battle zone need not, in this instance, be subjected to a political credibility test, even though in the real world Red may be assumed by many reputable analysts to be very unlikely to engage Blue in a tactical air duel over Blue territory without first attacking the SAM sites. The gap between the behavior of Red in the real world and in the exercise, even if it did exist, might not be a defect of this particular scenario. But the economic and technological credibility of Red or any plausible opponent having systems with the performance characteristics of those used in the exercise definitely *would* be important to determine, particularly if the exercise results were expected to provide inputs to cost-effectiveness studies directly related to forthcoming DOD decisions on alternative tactical air deployments to Europe.

Or take another analytical task, at the level of systems research and design, where the research purpose is to evaluate the physical vulnerability of our missile systems under varied but extreme conditions. The feeling that these conditions are highly unlikely is not always germane; all they need be is not inconceivable. Some of our important investigations do need to take into account the worst cases, however incredible they may seem to the political analyst. No one really regards an all-out Soviet strategic attack against our U.S.-based weapons as at all likely. We assume the Soviets are deterred. Yet few would deny the need for hypothetically subjecting our strategic forces to such attacks, under various assumptions concerning the strategic balance, Soviet firing doctrine, weather conditions, the state of alert of our forces, and the amount of tactical warning received.

For some of these narrower research tasks – those designed to evaluate the physical performance of systems – we are saying nothing very helpful if we render a general verdict for or against the usefulness of a scenario simply on grounds of credibility, plausibility, or some other vaguely phrased standard. In some of these studies we are really only concerned with the technical feasibility and physical realism of the postulated events. But, again, at the level of analysis described earlier as "conflict design in the large," the credibility of the political assumptions in a scenario are not at all an *incidental* aspect of the analysis.

We are talking here of military planning and force posture decisions which have a major impact on foreign policy – military decisions which can have a significant effect on deterrence of various kinds of aggression, and on damage limitation, escalation, and war termination, if deterrence fails. The relevance of any proposed new system or system alteration to these functions, as J. R. Schlesinger points out later,[3] is becoming more and more evident across the entire spectrum of force planning.

Force posture decisions are increasingly looked at for their effects on present and future political-military situations; they are assessed for their ability to affect enemy cost, risk, and benefit calculations significantly.

The kinds of political situations that are postulated, as a context for assessing the function of our military tools, must be realistic when these questions are asked, since it is precisely the *political functions* of these tools that are then being evaluated.

When it is said that the postulated political situations must be *realistic*, this does not mean that every particular event or military engagement contemplated must be a highly likely occurrence. That would be to confuse the

---

[3] See Chapter 20.

prediction of specific events with realistic forecasting of alternative political-military contexts. The scenario writer who is servicing a systems analysis task concerned with the political functions of force posture choices needs to be concerned with the *consistency* of his postulated military situations with the political-military context he assumes; he needs to assume political-military contexts *which are expected to materialize* (this is a form of prediction – and we cannot get away from it); and he needs to be able to point out the *connections between political-military conditions in his picture of the general context, and the specific sequence of actions contained in his scenario.*

Now, obviously, this means that there is a tremendous range of not implausible, not unrealistic events which can be ginned up by an imaginative and skillful scenario writer in order to provide hypothetical conflict environments for observing the behavior of military systems. But it does not mean that we have no way of determining how much *importance* to attach to certain anticipated conflict situations as opposed to others, particularly when we are assessing the political-military worth of various force posture choices. At this level of analysis, where we are evaluating major policy alternatives, the considerations determining the attention and weight we should give to certain cases include more than our estimates of the likelihood of their occurrence. At least two other major considerations, each involving a subset of complex judgments, must enter into our attribution of importance to specific scenarios: (1) estimates of consequences to be suffered if the situation did in fact occur; (2) notions concerning the factors that affect the probability of the occurrence of the contemplated situation.

Applying these considerations to the assumedly very-low probability event of a full-scale Soviet surprise attack on the continental United States, we can see that our willingness to hypothesize descriptions of such an extreme situation results from something other than a morbid pre-occupation with apocalyptic fantasies. Rather it is the result of the terrible consequences we foresee as accompanying the situation, and our impression that the Soviets' estimates of the price *they* would have to pay for bringing about such an event is, to a significant degree, dependent upon what we can convincingly claim to be able to do to them in case they try.

Where systems analysis involves *strategic* operations, these remarks on the usefulness of scenarios in describing unlikely situations will probably be considered uncontroversial. But when we get into the question of the worth of certain scenarios for low-likelihood limited-war situations, there is often considerable disagreement. Apart from parochial service-oriented considerations, the judgment in military planning circles of the worth of analyses based upon such limited-war scenarios is again the product of real differences in estimating the consequences if the postulated events did in fact occur, and of differing notions about how our being prepared to fight effectively in such conflicts might affect the likelihood of their occurrence.

Most likely it is because of such considerations that political analysts get few calls these days for scenarios describing a massive Soviet march into Western Europe. Some of those who consider such scenarios to be of very little use for intelligent military planning assume the event to have so little likelihood that they are even prepared to label it "inconceivable." Moreover, they sometimes assume that if the event ever did occur, it would be futile to expect its consequences to be contained by our capabilities for fighting in Western Europe; rather, it would be our capabilities for expanding the conflict – our power to escalate, if you will – that would be relied on to limit the consequences of the massive aggression, presumably by coercing the Soviets to call it off. On this view, the extent of our military capabilities in the theater is less relevant to the probability of such an event than many nonmilitary factors. Insofar as we need a hedge against the failure of these nonmilitary factors to keep the event from occurring, there is a rather wide consensus among those belonging to this school of thought that our threat to escalate such a conflict is sufficient to keep the Soviets from trying.

These two extreme situations – one of all-out surprise strategic strike, the other of a massive land aggression in Europe – are relevant here because they illustrate how the usefulness of a scenario, by which we mean the market for it, is only partially affected by judgments concerning the likelihood of the events it describes. In the strategic war case, other considerations keep the market good. In the big European war case, these other considerations – consequences of a failure of deterrence, and marginal increments to pre-war deterrence – are assumed by many analysts to be weak; but this judgment is certainly not beyond controversy.

The really controversial scenarios are to be found among those that describe events which are assumed to have relatively low likelihood and consequences that may or may not be critical. Major controversies arise once the likelihood and consequences of these events are assumed, by at least some influential members in the military planning community, to be actually quite *sensitive* to contemplated changes in military posture. We noted before that it is difficult to find very many influential military planners today who would put a massive Soviet ground assault on Western Europe in this category. But as we slide down the scale-of-aggression continuum, we get more and more disagreement.

A limited conflict of this character might be one growing out of a West German military intervention in an East German revolution, which brings on a series of military clashes between West German and Soviet troops across the zonal boundary, the start of a Soviet military advance toward Schleswig-Holstein, and the blocking of Western access to the city of Berlin.

The political credibility of any such scenario will surely be hotly debated, particularly if it is well done – if, that is, it provides an internally self-consistent picture of a not unlikely political and military global context, in which the particular conflict situation is consistent with the general context. But even more important, from the standpoint of relevance to systems analysis, there will be some who will argue that the consequences of such an event *would* be vitally shaped by the characteristics of the military systems available for application in the particular situation, and that the range of consequences sensitive to feasible alterations in military systems or changes in force posture is of such significance that top national decision-makers *ought* to be interested. Some will even go so far as to argue that the likelihood of such a large military conflict growing out of an East German revolt will be importantly *reduced* if, by taking the scenario seriously as one of the hypothetical environments for our systems analyses, we are able to come up with system, force posture, and strategy recommendations that will improve our ability to control the political and military variables in such situations.

In fact, the existence of heated controversy surrounding the political credibility of a scenario to be used in systems analysis is often less a cause for its rejection as an analytical tool than it is an indication that the scenario describes situations within that region of significant contingencies about which we still do not know what to do. And it is precisely such contingencies that the systems analyst, being an applied scientist, ought to be looking for as testing environments.

Furthermore, when today it is still highly debatable among respected analysts whether a scenario is credible or incredible, and when few can agree in advance on what aspects of the scenario are likely to be significantly sensitive to contemplated changes in military systems, the results of military systems analysis – tentative as they must be when the criteria themselves are controversial – can be a significant input to decisions of high national policy. Translated into practical terms, this is like saying that if the Air Force comes up with a really "good" concept for a long-range reconnaissance-strike bomber, it may yet make it.

Scenarios and specialists

If the scenario is not a trivial part of our systems analysis – if our results are sensitive to scenario assumptions – we will find it very difficult to avoid using controversial scenarios. But, returning to the theme that opened this

Chapter, this counsels for taking whatever scenarios we do use very seriously, rather than casually. The scenario is a part of our data base – and, if it is a critical part, it requires the same laborious back-up work required by any critical segment of our data base. For special aspects of the scenario, political no less than technical, it is important to get all the specialized assistance available. To the extent that German nationalism, for example, is a factor in the expansion or limitation of the geographical confines of a limited conflict, we should go to the experts on Germany. Needless to say, the same holds for scenario assumptions concerning the objectives and behavioral style of our principal enemies. The experts on these controversial matters are likely to disagree. But they will know why they disagree and will be able to defend truly controversial aspects of a scenario against unqualified charges of political incredibility.

But after all is said and done, *Caveat emptor*: Let the buyer beware of the scenarios he uses in his systems analysis. They are sure to come back to haunt him at the day of reckoning.

Chapter 17

U.S. SPACE POLICY: AN EXAMPLE OF POLITICAL ANALYSIS

ALTON FRYE

Despite the development of useful quantitative techniques for systems analysis, the most critical and pervasive issues for military planning, those centering on uncertainties of human behavior, defy quantitative definition and analysis. Virtually every important decision of politics and strategy demands an assessment of (1) the basic political values being sought and (2) the probable intentions and behavior of other nations. The analyst dealing with these matters is thrown immediately into a qualitative process of weighing, evaluating, and judging a vast array of nonquantifiable factors. Illustrative of the difficulties and deficiencies, the quality and utility, of such political analysis are efforts to devise a satisfactory U.S. policy prescribing the military role in space.

Some analytical distinctions

Expressing his faith in the possibilities of natural science, Albert Einstein once commented, "God may be subtle, but he isn't plain mean."[1] By that aphorism Einstein meant that the scientist can pursue his investigations with confidence that nature will behave with reasonable consistency. One need not fear the malevolent intervention of some supernatural being, for God will not change the natural laws governing the physical universe merely to frustrate the inquiring scientist. The relative stability of the physical environment facilitates systematic, quantitative research.

But the analyst who is concerned with strategy and politics is not dealing exclusively with problems of natural science. The most important feature of the politico-strategic environment is its adversary quality: this environment is basically the creation of men, the product of their competition to acquire and protect the values, material and spiritual, which they cherish. Human decisionmakers may consciously alter their behavior patterns in order to confound efforts to understand the course of events and to forecast likely trends. The inherent idiosyncrasies of human behavior are

---

[1] "Der Herr Gott ist raffiniert, aber boshaft ist Er nicht." The translation follows that of Norbert Wiener, *The Human Use of Human Beings*, Doubleday, Garden City, New York, 1956, p. 188.

compounded in politics, for in the contest for power, whether between individuals or nations, there may be a premium on deception. The uncertainties concerning human behavior pose the most severe obstacles to effective contingency planning. However systematic and sophisticated our treatment of the quantifiable components of a given problem, the most critical and pervasive questions bearing on strategic and political decisions defy quantitative definition and analysis.

Merely to pose some of the current problems confronting U.S. strategy demonstrates this point. What will be the effects on the cohesiveness of NATO of U.S. insistence on establishing a firebreak between conventional and nuclear war? How can the United States maintain a credible strategic deterrent against Soviet aggression in Europe? How will an opponent respond if the United States employs tactical nuclear weapons against him in a particular military context? Such issues abound. And the working hypotheses which serve as answers to them have critical implications for many aspects of national security policy and planning. Since definitive solutions to these questions are seldom forthcoming, the task of analysis becomes that of perpetual re-examination and refinement of the tentative guidelines on which policy is necessarily framed. Research on these matters, like that on more quantifiable problems, seeks to aid the decision-maker to develop more informed subjective probabilities with which to guide his action.

Political-strategic analysis: a perpetual process

Two central tasks face the political analyst who seeks to contribute to military planning. To begin with, he must attempt to identify, clarify, and help apply relevant national values to decisions regarding specific goals and objectives. But, especially in a pluralistic society, those national values are themselves subject to debate. Often they offer little guidance beyond such vague and ambiguous concepts as that of the "national interest" or "national security."[2] Just what constitutes the national interest is not self-evident. Lacking fully operational definitions for such concepts, the analyst is thrown immediately into a qualitative process of weighing, appraising, judging individual plans and policies against criteria which both permit and encourage many different interpretations.

Is the national interest served by deploying an antiballistic missile system? Or would such an action contribute to heightened international ten-

[2] A classic exposition bearing on this topic is Arnold Wolfers, " 'National Security' as an Ambiguous Symbol," *Political Science Quarterly*, Vol. 67, No. 4, December 1952, pp. 481–502.

sions, to an increased probability that war would actually occur, and hence to a net reduction in American security? Current discussion of these questions finds similar values and symbols being invoked by both opponents and proponents of ABM.

Furthermore, our own values, goals, and objectives are not determined in isolation. They are greatly influenced by the prevailing political and strategic environment, in particular by our expectations of the behavior of other nations. Thus, the first function of the political analyst is critically dependent on his second principal task, to assess the probable intentions and behavior of other states in the international arena. This requires that the analyst gain some notion of the scheme of values governing the decisions of these other political entities. Much of the analyst's work thus deals with the "intentions" side of the "intentions-capabilities" dichotomy familiar to all members of the defense community. In spite of the obstacles to achieving reliable insights into a potential enemy's intentions, it would obviously be irrational and unwise to ignore them in our planning.

In shaping our attitudes and policies toward others, the key role played by our projection of other nations' intentions is easily appreciated. For example, although Great Britain has always had significant *capabilities* with which to affect U.S. security, our confidence in her friendly disposition toward us and our awareness of our countries' mutual interests have kept us, for a century and a half, from worrying unduly about her probable behavior. As a matter of fact, during much of the nineteenth century Britain's control of the Atlantic provided the United States with an era of free security. Relying on British benevolence, this country felt no necessity to invest heavily in warships; instead we were able to pursue our commercial interests at home and abroad with extremely modest expenses for military forces to defend our shores and to protect our merchant marine.

A contrasting but comparable relationship which places little burden on the U.S. military establishment is illustrated by Albania's current animosity toward this country. Although we assume that Albania's *intentions* toward us are hostile, we are not particularly alarmed by her unfriendly attitude because she has such limited capabilities to act upon it. Although one must be wary of growing capabilities on the part of a hostile power, and of changing intentions on the part of a capable country, the immediate problems for political and strategic planners are posed by the combination of hostile intentions and major capabilities to threaten U.S. interests. It is this combination that has made Soviet Russia the most significant potential enemy of our time. Yet neither Russian intentions nor Russian capabilities are static and, if U.S. policy is to be effective, it is imperative to keep abreast of both.

TYPICAL CHALLENGE: MILITARY ROLE IN SPACE

A typically difficult challenge to policy analysis, in its effort to deal with the inseparable relationship between Soviet intentions and capabilities, has been presented by the need to devise suitable U.S. policy for the development and use of space technology. In terms of the origins of the national space program and of its sources of sustained support, U.S. activities in outer space reflect our apprehension that national security may be adversely affected by developments in space technology. To meet the danger of a breakthrough that might jeopardize peace and security, the United States has mounted an extensive effort to advance our general competence in this field. At the same time, we have attempted to obtain some idea of probable Soviet intentions in this new arena and to take them into account in determining our own objectives, especially with regard to establishing an appropriate role for military operations in space. The problematic nature of the latter issue and the various approaches to designing a balanced policy on the subject reveal something of the difficulties and deficiencies, the quality and utility, of this type of analysis.

The basic charter of the American space program, the National Aeronautics and Space Act of 1958, typifies the kind of broad guidelines within which analysts and decisionmakers must operate. That Act prescribes as U.S. policy that "activities in space should be devoted to peaceful purposes for the benefit of all mankind," but it provides no comprehensive definition of what qualifies as "peaceful." Considerations of national security loomed large in Congressional passage of the legislation, and the Act assigned to the Department of Defense specific responsibility for space activities "peculiar to or primarily associated with the development of weapons systems, military operations, or the defense of the United States" (including necessary research and development).

However, the emphasis on "peaceful purposes" generated considerable ambivalence concerning the proper scope and character of the military space program. It even contributed to the unfortunate tendency in some quarters to contrast peaceful activities with military activities. Only in recent years has this artificial dichotomy been explicitly rejected; Presidents Kennedy and Johnson have made clear that peaceful activities include those military efforts required to help us keep the peace. The recurrent confusion over this matter suggests some of the difficulties of interpreting even those values and goals which Congress has formulated in legislation.[3]

To conclude that a military space program is desirable does not resolve the complex problems of what kind and what size of program should be

[3] See Alton Frye, "Our Gamble in Space: The Military Danger," *The Atlantic*, Vol. 202, No. 2, August 1963, pp. 46–50.

undertaken. The techniques and criteria for evaluating these issues are generally vague and crude. But there are several devices, some of them simple to identify but laborious to apply, which may permit an analyst to get a moorage in an ocean of uncertainty. Some of these have already been reviewed in previous chapters.

A fundamental question, and one to which the national space effort itself has been addressed, is, What are the technical possibilities for operations in space? Knowledge of the technological context in which policy must function provides at least a partial perspective on the kinds of policy problems that may arise in the real world. Few political analysts are worrying about the security implications of progress in perpetual motion machines. Similarly, if space boosters could never orbit satellites larger and more useful than the Vanguard payload, there would be little need for intensive planning to meet potential space-based threats or to exploit opportunities in space to enhance existing military capabilities.

In fact, of course, the possible military applications of space technology are more numerous and significant than Vanguard would have implied. A feel for these technological opportunities is obviously prerequisite to sound policy. Given a reasonable familiarity with the probable technical trends, the policy analyst can then try to identify those that are likely to have political and military ramifications and to estimate their importance. At the same time the analyst can begin to suggest any political factors that might indicate a preference for one technological avenue or goal over others.

A case in point is the military communications satellite program. While it might be technically feasible to devise secure communications links as part of the civilian system being developed by the Communications Satellite Corporation, there were sound reasons for pursuing an independent military effort. Direct military participation in the Comsat system would have been an impediment to negotiating global arrangements for the civilian undertaking, since many countries, particularly nonaligned states, would have had reservations about cooperating in an enterprise of immediate military significance to one of the great powers. This political consideration, together with substantial technical advantages in terms of coverage, reliability, and security, argued strongly for the separate military communications satellite network that is now contemplated. The influences of politics and technology are reciprocal; each stimulates and inhibits the other in complex and subtle ways.

U.S. Policy of Restraint

The political constraints on technology have rarely been so evident as they are in the space program. In deciding which areas of space technology the United States should exploit for military purposes, political and strategic considerations should properly be primary, although conflicting values may compete for dominance in our decisionmaking. For example, one can conceive of space weapons to which the United States might be attracted for purposes of deterrence; satellites deployed in random orbits might provide highly invulnerable systems that could only be used effectively for retaliation and would not raise fears that the United States was preparing a first strike. Moreover, by diverting counterforce fire away from the continental United States, deployment of weapons to outer space might serve the U.S. goal of damage-limitation, although associated facilities on earth would still be inviting targets.

Although analysts have recognized these possibilities, the controlling factors in U.S. policy have been to avoid provocative innovations in the strategic forces and to prevent a new spiral in the arms competition.

While some space systems, notably those promising unique or superior capabilities for observation, communication, and navigation, have been deemed acceptable and valuable from the standpoint of stable deterrence, the United States has chosen to exercise restraint in deployment of weapons to space. Several factors have encouraged this country to adopt such a course and to attempt to elicit similar restraint on the part of the Soviet Union.

Enjoying preponderance in present types of strategic weapons, the United States has not had a strong incentive to shift the arms race to a new environment and to novel technologies. Furthermore, although either the United States or the Soviet Union could place thermonuclear weapons in orbit, there have appeared to be no decisive military advantages which would make deployment of bombardment satellites a rational strategy. America's strategic superiority has also provided a margin of confidence that the country could afford the risks that might be involved in a policy of self-denial. This situation contrasts with a number of previous junctures in the contemporary arms race. For example, to most analysts and to the responsible decisionmakers the competition in space technology differs markedly from the race for the thermonuclear weapon. The policy debate of 1949-1950 produced a consensus that the United States could not forego development of the H-bomb in hopes that the Soviets would do likewise.

Although the space age has witnessed more than one display of Soviet rocket-rattling and attempted nuclear blackmail, there have been indications that the Russian government might be genuinely interested in moderating the cold war. The enunciation of the peaceful coexistence line and the public abandonment of the Marxist-Leninist doctrine of the inevitability of war were impressive departures in Soviet political posture. To a number of policy planners, these radical deviations from previous Communist positions seemed worth testing, and an attempt to induce mutual restraint in deployment of space weapons appeared one comparatively safe method of doing so.

Analysts have, of course, recognized that the low value assigned to space weapons by the United States might differ drastically from Soviet estimates. Space policy planners have had to be constantly alert to the different strategic criteria and doctrines of the Soviet Union, giving special attention to prospects that might enable the Soviets to overcome their current military inferiority. One must always appraise the utility of space systems not only in terms of a retaliatory second strike that might be consonant with U.S. strategy, but also in terms of a possible first strike against U.S. targets, or in terms of a bold campaign of nuclear blackmail rather than a strategy of stable deterrence.

Appraising Soviet Intentions in Space

Thus a cardinal question for U.S. policy planners has been, Will the Soviets reciprocate our restraint? Or, even if the Soviet Union saw no overwhelming strategic advantage in space weapons, would it seek to exploit such capabilities for psycho-political effects, augmenting its reputation as an invincible modern power?[4] To obtain a rational and realistic opinion on these questions, one must consider a large body of heterogeneous data, including statements of various types by Soviet leaders, Soviet actions in political forums concerned with outer space, technical initiatives, scientific discoveries, and so forth.

To gain some insight into Soviet intentions in space, the United States has long monitored Russian propaganda, official statements, and other relevant communications. From such material analysts seek to infer probable Soviet intentions. Obviously, public expressions of Soviet leaders and other political information may be deliberately deceptive. One must attempt to discern relationships between symbolic behavior – what the Soviets say – and nonsymbolic behavior – what they actually do. Political information must be measured against data on trends in Soviet capabilities to arrive at an improved estimate of their probable intentions. Even if one discovers that political signals are in fact misleading, it is useful to know in what

[4] Should it appear likely that the Soviets will not reciprocate, a new set of issues would arise for the United States, forcing re-examination of the policy of restraint. The disclosure in November 1967 that the U.S.S.R. is developing a Fractional Orbital Bombardment System (FOBS) emphasizes this problem. But FOBS would presumably be based on earth, not in space; it is probably best construed as a special type of extended-range missile capable of striking from any direction. While the Soviets may have hoped to circumvent U.S. warning systems by such a weapon, Secretary McNamara has announced a new, over-the-horizon radar that will provide some capability to detect FOBS launchings.

respects the Soviets may wish to deceive us, for that knowledge may provide clues to their real purposes.[5]

One must, however, resist the temptation to assume that deception, even if identified as such, is a valid measure of hostile intentions. It may merely reflect mistrust, suspicion, or fear. Perhaps you recall the story of the Russian who stopped to admire an American automobile parked near the U.S. embassy in Moscow. "What a beautiful Russian car!" he said; "What glorious Russian chrome!" Another passerby, overhearing him, remarked, "You fool, don't you know that's an *American* car?" To which the first man replied, "Yes, I know that, but I don't know *you*." In our search for explanations of Soviet behavior, we should remember that the deceptive practices in which Moscow engages may be a form of political camouflage, protective coloration designed to hide a weakness rather than to disguise an aggressive move.

To evaluate the credibility and significance of Soviet political communications, it is necessary to compare a great many statements by different spokesmen intended for various audiences. Are communications to the domestic Soviet audience consistent with those meant for international consumption? Is there general agreement on space policy in the statements of different Soviet authorities? Does public information on these matters correspond with any covert information we may obtain concerning in-house Soviet debates on such topics? Even in a police state with rigid control of its communications, a study of the over-all context of political discussion can throw light on the probable motivations and goals of decisionmakers.

Previous experience in the analysis of Soviet political and strategic pronouncements has conditioned U.S. analysts to expect the ambiguity and exaggeration common in Russian discussion of their accomplishments in space. Outright lies, however, have been rare. Soviet disclosure of activities in science and technology often seems to follow a pattern. Moscow tends to say nothing about its efforts in space or elsewhere until significant progress has been made and a substantial accomplishment offers an opportunity for political exploitation. Together with the characteristic emphasis of Soviet technologists on intensive laboratory work, scale model testing, and component development – activities which help to delay the visibility of a program until it is well advanced – this kind of disclosure policy contributes to the shock effect frequently achieved by Soviet technical feats.

Once having announced the existence of a particular technology or

[5] A notable study relevant to these issues is Alexander L. George's *Propaganda Analysis: A Study of Inferences Made from Nazi Propaganda in World War II*, Row, Peterson and Company, Evanston, Ill., 1959, especially pp. 3–90, 125–138.

program, the Soviets often convey vague, exaggerated, and ambiguous claims concerning it. Ambiguity is an invaluable asset to Soviet leaders, for it permits them to extract maximum credit from a given development without an irrevocable commitment to further action and without loss of face should additional progress in that area prove halting.

In this fashion the Soviet Union sought to conceal its strategic weakness and to bolster its bargaining position in world affairs by its overblown early boasts of long-range missile capabilities. Such claims have declined in recent years, as evidence of the slow pace of Soviet ICBM deployment has undermined their credibility. Similar misleading exaggerations marked initial Soviet statements on their antiballistic missile program. Premier Khrushchev announced that his forces could intercept a "fly-speck" in space. President Kennedy promptly punctured that balloon by noting that there was a great difference between swatting a single fly and defending against a whole swarm of them, attacking all at once, and employing the most modern penetration aids. In subsequent months, Soviet statements on the ABM grew more modest, paralleling Marshal Malinovsky's cautious claim that the problem of destroying a missile in flight had been solved *in principle*. Despite indications that the Soviet Union was actually engaged in some kind of ABM deployment, its military leaders revealed significant doubts about the system's potential effectiveness.

Soviet commentators and officials have also maintained a pattern of ambiguous allusions to possible military implications of Russian activities in space, while keeping details of the program under strict security wraps. Without claiming a bombardment satellite capability, Premier Khrushchev, Cosmonaut Titov, and others declared that the rockets which launched the Vostoks could orbit "other payloads" for other purposes. In early 1963 the late Marshal Biryuzov offered a more explicit remark on this subject when he announced that the Soviet Union could launch rockets from satellites at any time in their orbits and in any direction. Following the military parade in Red Square on November 7, 1965, Soviet commentators specifically alluded to the orbital bombardment capability of some of the rockets on display. These hints of offensive space capabilities were supplemented by vague threats to take action against U.S. observation satellites, presumably by some kind of antisatellite weapons.

These menacing overtones in Russia's discussion of her space program, together with her apparent capability actually to deploy a Vostok-class bombardment force and to jury-rig an antisatellite system, led most analysts to the conviction that the United States required at least a limited antisatellite capability. It was hoped that a system that could interrupt any Soviet attempt to deploy bombardment satellites and that could take reprisals for any attacks on American payloads would deter hostile action by the Soviets in space. In late 1964 the President and Secretary of Defense announced U.S. development of certain ground-based weapons for the antisatellite mission.

As the *détente* began in 1963, those following Soviet discourse on space detected a noticeable shift in public pronouncements related to military activities. The earlier crescendo of implied threats to exploit space weapons gave way to greater Soviet insistence on international cooperation in outer space and on restraint in deployment of orbital weapons. The altered tone of Soviet discussion of these matters gained credibility by new departures in the cold war. Not only did the nuclear test ban treaty prohibit nuclear explosions in space, but the Soviets even joined in the unanimous U.N. resolution of October 17, 1963, expressing the members' intentions not to orbit weapons of mass destruction. The substance of that resolution and other agreed principles regarding activities in outer space were incorporated in a treaty unanimously endorsed by the General Assembly in 1966, and ratified by the United States Senate on April 25, 1967.

However, the treaty makes no provision for verification, and contemplates inspection only of facilities on celestial bodies – and then only on the condition that the nations operating such facilities receive prior notification of all planned inspection visits. No arrangements exist for pre-launch inspection of all payloads or for inspection of satellites after they are in space. Debate continues among American planners as to what contingency preparations the United States should take to guard against violation of this international commitment. The ambiguity of the Soviet space capability persists. Unlike ICBM or ABM systems, space weapons might be deployed without the United States discovering them. Although the United States might know that satellites were in orbit, we could not be sure whether or not they contained weapons.

In spite of these difficulties the United States has been encouraged to maintain her policy of restraint by the general improvement of the cold war climate. The Soviets have proved especially accommodating of late in the field of international space policy. They have backed off from their adamant opposition to observation satellites and have even published the fact that the Soviet Union has a program of reconnaissance from space. Although the United States has refrained from public comments on the changing Soviet position on observation until it is clear how far Moscow has moved on this point, disclosure of Soviet efforts in this field is in some respects welcome, for they reinforce the long-standing U.S. argument that such operations are legitimate.

These variable trends in the public face of Soviet space policy are obviously not a firm and final guide to appropriate U.S. policy, and many uncertainties remain. But it is inevitable that some of our decisions have to be taken before all the data are in. In dynamic relationships between self-governing units, each must try to anticipate and respond to choices and acts of the other even before they are decided upon.

There is some evidence that the Soviets also are having a difficult time reaching decisions on the military role in space. An internal debate comparable to our own may be under way. The Soviets may have adopted a more conciliatory tone regarding efforts in space because they are now less optimistic than formerly that space technology is a promising avenue for strategic innovations. They may also see the extraordinary pace of the U.S. program as likely to deprive the Soviet Union of possible political or military advantages in space. But no responsible analyst assumes that any one of these hypotheses is *the* explanation for the apparent shift in Soviet attitudes and behavior in this area. It remains possible that Russia's less bellicose behavior on these matters will prove transient. The present display of sweet and reasonable amiability may obscure more malign trends actually taking place in the Soviet space program.

Outstanding Issues for U.S. Space Policy

For this reason, among others, many issues remain for those engaged in devising a satisfactory U.S. space effort. In order to anticipate possible breakthroughs by other powers which might alter the current expectation that orbital weapons will not be deployed, what type and degree of exploratory development should the United States perform? Now that we have made a start on an initial antisatellite capability, what additional R&D is needed and how large a force should we procure? What relative priorities should we assign for nuclear and non-nuclear kill mechanisms for such a system? What are the implications of the nuclear test ban treaty of 1963 for the usefulness of a nuclear warhead?[6] What would be an appropriate U.S. response to an ambiguous Soviet deployment which might – or might not – involve bombardment satellites? These and many other complicated questions demand serious study and sober judgment – and all require the most informed estimate we can attain of probable Soviet intentions, behavior, initiatives, and reactions.

---

[6] The treaty prohibits not only nuclear tests but "any other nuclear explosion" in the atmosphere, underwater, and in outer space. It is difficult to see how a nuclear weapon could *legally* be used for the antisatellite mission, unless the party contemplating such use invoked the right to withdraw from the agreement on the ground that some event jeopardized its "supreme interests." In the latter case three months' notice is required, an impractical delay if one is faced with an urgent need for a prompt antisatellite response.

It is clear that the test ban has greatly increased the *political* importance of a non-nuclear mechanism, particularly for use against any ambiguous or uncertain threats that would not justify abrogation of the treaty.

For this purpose we require multiple analyses by many individuals on a continuing basis. Such assessments are inevitably imperfect, but for that very reason they are all the more essential.

Rational decisionmaking in a technical-political environment

This brief review of some of the issues involved in planning U.S. space policy and of some of the responses to them suggests a few of the handles one may hope to get on problems of this kind. To the extent that one is dealing with problems across the interface of technology and politics, it helps enormously to begin with a firm grasp of projected technical trends. Although one might prefer to specify political goals and then inquire of the technologist how to attain them, more often than not the process is inverted; technology presents itself to policy and asks what good it is. In any given case, the analyst and the decisionmaker are obliged to discover relevant and hopefully reliable criteria and to ferret out as much information as possible that will help in applying them to the issues at hand. The political values and guidelines that are employed are usually extremely broad and rarely provide a conclusive standard for action on complex problems. The values themselves may not be compatible, and one may have to make trade-offs and sacrifices among them. Deterrence may not always comport with damage-limitation, and avoidance of new spirals in the arms race may be difficult to manage if one is intent on hedging against a competitor's treachery.

However frustrating and difficult may be the attempt to perceive and project the probable intentions and behavior of a potential adversary, prudence demands that the effort be made. Only with some appreciation of the other side's motives and some notion of the underlying values which influence its decisionmakers can we hope to induce behavior on their part that is favorable to us. Thus, at virtually every juncture of the military-political network of decisions, from questions of force structure to targeting doctrine, from problems of preconflict deterrence through those of intrawar escalation and war termination, the necessity for intensive study and evaluation of a host of nonquantitative factors is unavoidable.

Obviously the policymaker must use this kind of qualitative research and analysis with the greatest caution. Since all the important variables may not even have been identified, much less analyzed, there is always the danger that the entire study may be 180 degrees askew. In many cases, several explanations or projections may be plausible, and it hardly needs saying that policy should not be tied irrevocably to any one of them.

In this world of subjective probabilities, of ill-defined and personally ascribed confidence levels, the investment of individual egos and reputations in support of positions which cannot be scientifically verified may tend to distort the analyical context, to polarize discussion along artificial lines, and to pervert the entire process into sterile controversy. Both analyst and user must constantly guard against the inclination to form unjustifiably rigid convictions and to develop a false confidence that any particular model or solution accurately conforms to the real problems of politics and strategy.

The Greek philosopher, Agathon, offered the proper admonition to all policy analysts and decisionmakers when he observed, "It is probable that the improbable will happen."[7] With that maxim in mind, perhaps we can maintain humility in studying problems of international behavior, and flexibility in implementing our current political and strategic hypotheses.

---

[7] Abraham Kaplan, *The Conduct of Inquiry*, Chandler, San Francisco, 1964, pp. 403–405.

Chapter 18

WHEN QUANTITATIVE MODELS ARE INADEQUATE

E. S. QUADE

This Chapter surveys the problem of handling those aspects of analysis for which quantitative models are inapplicable. To this end, it discusses ways of identifying experts and of utilizing their expertness – individually or in concert. Operational gaming and scenarios are considered briefly; the Delphi technique is discussed in detail, with examples.

INTRODUCTION

In national security affairs, no matter what problem the analyst investigates, there will always be aspects for which purely quantitative techniques are clearly inapplicable. Often this is of little consequence; the aspects that can be treated quantitatively dominate and the advice given to the decision-maker can be based on purely analytic models. Many problems of logistics, military operations, and even weapon selection are of this type. There is, however, a large class of problems for which analytic models that can realistically take into account the dominant organizational, political, and social factors have not yet been developed or even conceived. Problems related to general purpose forces, military assistance, revolutionary war, and arms control are likely to be this latter type.

This Chapter discusses various ways to tackle those aspects of analysis that are ill-suited to a purely quantitative approach. In scope, these approaches range from an appeal to individual judgment and intuition to fairly elaborate schemes for bringing to bear in a systematic fashion the opinions of many people. Most of the schemes treated here were mentioned earlier in the Chapter on models, and two of them, war gaming and scenario writing, were the subject of individual Chapters.[1] We will add some points to these discussions, and then consider one of the more promising methods – the Delphi technique – in detail.

THE BASIC ROLE OF JUDGMENT

As emphasized earlier both in this book and elsewhere, systems analysis

[1] See Chapters 10, 14, 15, and 16.

is based on the systematic and efficient use of expertise. As Charles Hitch has written:

> Systems analyses should be looked upon not as the antithesis of judgment but as a framework which permits the judgment of experts in numerous subfields to be utilized – to yield results which transcend any individual judgment. This is its aim and opportunity.[2]

The standard framework for systematizing this judgment – some form of mathematical model – has become the standard because of its remarkable success in cases ranging from astronomy to economics. But for the analyst to provide sound advice, it is not essential that the framework or model be expressed in quantitative terms. What is essential is his reliance on judgment. As Olaf Helmer puts it:

> While model-building is an extremely systematic expedient to promote the understanding and control of our environment, reliance on the use of expert judgment, though often unsystematic, is more than an expedient: it is an absolute necessity. Expert opinion must be called on whenever it becomes necessary to choose among several alternative courses of action in the absence of an accepted body of theoretical knowledge that would clearly single out one course as the preferred alternative. [This can happen if there is either] a factual uncertainty as to the real consequences of the proposed courses of action, or, even if the consequences are relatively predictable, there [is] a moral uncertainty as to which of the consequent states of the world would be preferable. The latter kind of doubt often arises even when there is a clear-cut basic ethical code, because the multiple moral implications of a complex change in the environment may not be directly assessable in terms of the basic code.[3]

IDENTIFYING EXPERTS

We know that we need expert judgment, but do we know who qualifies as an expert, or even what sort of expert is needed? Is there any method of determining expertness on the basis of past performance? One of the most obvious ways is to use a scale of "reliability": An individual's *degree of reliability* would be the relative frequency of instances in which he ascribed a greater personal probability to the alternative that eventually was correct than to the other choices open to him. On this basis, the more often he proved himself correct, the greater his authority as an expert. This sort of measure has its use, but, as Helmer remarks, it

> . . . must yet be taken with a grain of salt, for there are circumstances where even a layman's degree of reliability, as defined above, can be very close to 1. For instance, in a region of very constant weather, a layman can prognosticate the weather quite

---

[2] C. J. Hitch, "Analysis for Air Force Decisions," in E. S. Quade (ed.), *Analysis for Military Decisions*, Rand McNally & Co., Chicago, 1964, p. 23.
[3] Olaf Helmer, *Social Technology*, Basic Books, Inc., New York, 1966, p. 11.

successfully by always predicting the same weather for the next day as for the current one. Similarly, a quack who hands out bread pills and reassures his patients of recovery "in due time" may prove right more often than not and yet have no legitimate claim to being classified as a medical expert. Thus what matters is not so much an expert's absolute degree of reliability but his relative degree of reliability, that is, his reliability as compared to that of the average person. But even this may not be enough. In the case of the medical diagnostician . . . the layman may have no information that might give him a clue as to which of diseases A and B is the more probable, while anyone with a certain amount of rudimentary medical knowledge may know that disease A generally occurs much more frequently than disease B; yet his prediction of A rather than B on this basis alone would not qualify him as a reliable diagnostician. Thus a more subtle assessment of the qualifications of an expert may require his comparison with the average person having some degree of general background knowledge in his field of specialization. One method of scoring experts somewhat more subtly than just by their reliability is in terms of their "accuracy": the *degree of accuracy* of an expert's predictions is the correlation between his personal probability p and his correctness in the class of those hypotheses to which he ascribed the probability p. Thus of a highly accurate predictor we expect that of those hypotheses to which he ascribes, say, a probability of 70%, approximately 70% will eventually turn out to be confirmed. Accuracy in this sense, by the way, does not guarantee reliability, but accuracy in addition to reliability may be sufficient to distinguish the real expert from the specious one.[4]

This distinction is helpful as far as it goes. But both reliability and accuracy as Helmer defines them are based on the assumption that one can identify a class of similar issues or questions for which an expert is good or not so good. In fields not so well-defined as weather or medicine, this may be hard to do. Where does one turn for an expert to estimate the accuracy of a missile whose development is still being contemplated?

There are, of course, several signs of an expert's qualifications other than his specific pronouncements, such as years of professional experience, number of publications, academic rank, and so forth. The value of such objective indexes is not above doubt, but usually they are all we have. In some cases, however, there are reasons to believe that better measures exist; for example, a fairly recent experiment suggested that it might be meaningful to select an expert to participate in a specific project on the basis of his own appraisal of his competence to carry out that assignment.[5] Whether or not there is anything to this hypothesis needs to be determined by further investigation.

---

[4] O. Helmer and Nicholas Rescher, "On the Epistemology of the Inexact Sciences," *Management Science*, Vol. 6, No. 1, October 1959, p. 40. Over the past few years Dr. Helmer has devoted considerable attention to a search for better ways to make use of expert judgment. The procedures described in this Chapter are based on the work of Helmer and his collaborators, and much of the material not directly quoted is taken with minor changes from his papers. Relevant references in addition to those cited directly in this Chapter are included in the Bibliography.

[5] See B. Brown and O. Helmer, *Improving the Reliability of Estimates Obtained from a Consensus of Experts*, The RAND Corporation, P-2986, September 1964.

UTILIZING INDIVIDUAL EXPERTS

Once we have identified the experts whose judgment we wish to bring to bear on a problem, what ways do we have of drawing on them for advice, assuming no analytic model can be devised?

For a broad question, like determining an efficient allocation of resources to U.S. military and nonmilitary programs in space, the range of expertise required cannot be provided by a single individual. Nevertheless, an individual expert can be very helpful both as a source of information in a larger study and as an advisor to a decisionmaker. One method we have for obtaining his contribution is through an explicit personal analysis, such as the political analysis presented earlier by Alton Frye.[6] Even though it is the work of a single expert, it can bring together many opinions and provide an extremely useful input to a systems analysis.

Another device for using one man's expertise is to ask him to prepare a scenario. This is often appropriate when political, economic, or social questions are being considered. While scenario writing may not, strictly speaking, be a form of model building, it is certainly a closely related activity. Starting with some present state of the world, the expert would attempt to show, step by step, how the future might evolve in plausible fashion. Here the purpose would not necessarily be to predict, but merely to demonstrate the possibility of a certain future state of affairs by exhibiting a reasonable chain of events leading to it.

One new technique that offers a way to at least partially model organizational behavior at the national level (with all its inefficiencies) is the "multidimensional" scenario.[7] This type of scenario attempts to depict recurrent national decisionmaking processes as being crudely analogous to the decisionmaking processes observed in executive committees. The important participants are identified as institutions whose underlying characteristics and motivations can lead them – when the nation faces important domestic and foreign problems – to group themselves into temporary factions that recommend specific actions. These national institutions are the Foreign Ministry, displaying a recurrent tendency toward conservatism in foreign affairs; the internal administrative bureaucracy, shifting its position on international problems in accord with possible domestic consequences; the military structure, demanding freedom from political control; the technical and industrial managerial group, seeking commitment of national resources to industrial growth and production; and the political control structure, viewing other national insti-

[6] See Chapter 17.

[7] William M. Jones, *Fractional Debates and National Commitments: The Multidimensional Scenario*, The RAND Corporation, RM-5259-ISA, March 1967.

tutions as technical service agencies whose recommendations are likely to be based on improper and inadequate ideological considerations and knowledge of the domestic political situation.

By manipulating these institutions, over time, in each of the affected nations, a generalized scenario is produced. Its characteristics are that decisions are portrayed as being the result of compromises between temporary and shifting factional alliances among institutions, with each institution advocating those actions that promise to enhance its own power position.

Scenarios – whether conventional or multidimensional – can be extremely useful. By providing an insight into possible futures, they can make us aware of the potential consequences of particular policies. They can help the analysts working on other aspects of a problem to discern the important relationships among the elements of the situation and to eliminate irrelevancies. Thus they can lead to formal models. In addition, the discovery that no plausible scenario leading to some predicted situation can be prepared may be evidence that the situation is unlikely to happen.

In working with individual experts, it is important to insist that each one make the logic behind his opinions or judgments explicit. For only when the reasoning is explicit can someone else, whose information and perspective may be different, use the work of the first to modify his own opinion. Often it is the analyst himself who must serve as the bridge between experts, going from one to another to get them to explain the limits each has set to the problem, and to find counterarguments or better schemes. This effort sometimes even leads the analyst to propose naive or outrageous schemes, for experts often tend to cling to the conventional wisdom of their specialties.

Utilizing groups of experts

These are some ways of obtaining help from individual experts in cases where analytic models are not feasible. How, in the same context, might groups of experts be used?

Experiments have shown that the best use of a number of experts is not the traditional method of having the issues presented to them and debated in open round-table discussion until a consensus emerges or until they arrive at an agreed-upon group position. Committees, as noted earlier, often fail to make their assumptions and reasoning explicit. Sometimes the opinions of dissenters are not even recorded. What is needed is a way to avoid the psychological drawbacks of a round-table discussion – such as the "bandwagon" effect and the unwillingness to abandon publicly expressed opinions – and thus to provide a setting in which the pros and cons of an issue can be examined systematically and dispassionately.

Unfortunately, experiments have not yet shown the optimal way to do this, particularly where different disciplines need to be brought to bear on a problem:

> When dealing with a multi-faceted problem with the aid of a variety of experts of different backgrounds, perhaps the most important requirement in the interest of an efficient use of these experts is to provide an effective means of communication among them. Since each of the participating experts is likely to have his own specialized terminology, a conceptual alignment and a real agreement as to the identity of the problem may not be easy to achieve and it becomes almost imperative to construct a common frame of reference in order to promote a unified collaborative effort.[8]

As emphasized earlier, the ideal way (in our view) to enforce a common usage and understanding of concepts among experts is to present the problem to them in terms of an analytic model, or, even better, to have them participate in formulating the model as well. When this cannot be done readily (but not excluding all cases when it can!), other schemes can substitute with surprising effectiveness. These range from simple organizing devices to elaborate pseudoexperiments.

The Contextual Map

A simple device for facilitating cooperation, first described in the literature of operations research in terms of an application to an anthropological experiment conducted in Peru in 1955, is the contextual map:

> As an interdisciplinary team of planners faced with the complexities of a large interacting cultural system and its own problems of internal communication, we needed a method for systematically utilizing the special talents and experiences of the planners despite the frustrations of having to establish a common vocabulary, an agreed-upon ideology, a set of reasonable goals, a common context for symbols, and ways of translating ideas into actions. Our solution was to design and make up a "map room," whose walls contained a large matrix with time (in years) on the ordinate and the "variables" the group was interested in along the abscissa. This matrix was the "contextual map."[9]

Such a contextual map ("matrix" might have been a more descriptive term to use than "map") was used to display goals, predictions, and achievements, thus furnishing each member of the team at all times with an up-to-date exhibit of the project's status – its accomplishments and its remaining tasks.

Operational Gaming

The contextual map has been useful in a number of studies, but often

[8] Helmer, *Social Technology*, p. 17.

[9] J. L. Kennedy, *A Display Technique for Planning*, The RAND Corporation, P-965, October 1956.

something stronger is needed. For this purpose vicarious experimentation in the form of simulation by operational gaming is a powerful technique.

> Past experience with simulation models suggests that they can be highly instrumental in motivating the participating research personnel to communicate effectively with one another, to learn more about the subject matter by viewing it through the eyes of persons with backgrounds and skills different from their own, and thereby, above all, to acquire an integrated overview of the problem area. This catalytic effect of a simulation model is associated, not only with the employment of the completed model, but equally with the process of constructing it. (In fact, the two activities usually go hand in hand. The application of the model almost invariably suggests amendments, so that it is not uncommon to have an alternation of construction and simulation.)
>
> The heuristic effect of collaborating on the construction and use of a simulation model is particularly powerful when the simulation takes the form of an operational game where the participants act out the roles of decision- and policy-making entities (individuals or corporate institutions). By being exposed within a simulated environment to a conflict situation involving an intelligent opposition, the "player" is compelled, no matter how narrow his specialty, to consider many aspects of the scene that might not normally weigh heavily in his mind when he works in isolation . . .
>
> We note, in passing, that a player's assignment in an operational game may be either optimization or simulation. In the first case, he is to attempt, within the constraints of the game rules, to maximize a personal score (his "payoff" in game-theoretical terminology). This tends to put the verisimilitude of the game model, which after all is intended to be only an abstraction of the real world, to a severe test and to suggest amendments in the underlying assumptions. The second mode in which a player may function, namely as a simulant, is more likely to utilize his expertise properly; for in this role, he is required to contribute constructively to the developing scenario by feeding in such simulated decisions which, in his estimate, would most faithfully reflect the decisions that his actual counterpart would make in the corresponding real situation.[10]

The use of gaming as a research tool was discussed earlier by M. Weiner and R. Levien. In the examples they gave, the players were used to simulate the activity of two military opponents. Operational gaming, however, may be used to tackle other types of problems. To illustrate, here is how manual operational gaming might be used to study policies for aiding the economy of an underdeveloped nation.

This approach might involve the following steps.[11] First, the game model would have to be constructed so that it simulates the reactions of a certain aggregate of the economy to changes in the environment – say, the introduction of foreign aid, a change in tax policy, or a long period of adverse weather. To do this would require decisions as to which elements of the economy will be simulated by specialized experts, and what degree of

---

[10] Helmer, *Social Technology*, pp. 17–19.

[11] The description given here is a modification of one presented in an earlier paper: O. Helmer and E. S. Quade, "An Approach to the Study of a Developing Economy by Operational Gaming," in *Recherche Opérationnelle et Problems du Tiers Monde*, Dunod, Paris, 1964.

industrial and governmental aggregation will be employed. In addition, there would need to be government players, who could introduce new fiscal or monetary policies and regulations affecting taxes, subsidies, tariffs, price ceilings, and so on. These policies, in turn, could affect social and political innovations that have only indirect economic implications (social security, education, appeals to patriotism, universal military service, and so on). In addition, there would need to be a control team to interpret the rules governing the players' options and constraints and to evaluate the actions taken within these rules.

Next, having chosen the roles of the "players," it would be necessary to specify exactly what options are available at given stages of the game. For instance, the players representing the goods-producing sectors of the economy might be allowed, within stated constraints, to shut down, expand, or modernize manufacturing facilities, to change the raw material and labor inputs, to vary the prices of their products, and so on. In acting out their roles, these experts or specialists would be expected not so much to play a competitive game against one another as to use their intuition as experts to simulate as best they could the attitudes and consequent decisions of their possible real-life counterparts.[12] The game rules would have to describe the consequences arising from the options available to the players.

In playing the completed game:

> Early runs of the game would mainly serve to refine the underlying model and to permit the players to gain some insight into the sensitivity of the behavior of the economy to various assumptions and manipulations. Later, after a sufficient amount of synthetic experience has been accumulated and the model has been made to perform in a manner intuitively acceptable to the participating experts, the game may be used to explore the gross consequences of various alternative economic policies.
>
> Needless to say, there is no guarantee that economic projections obtained in this manner would be highly reliable. But a great deal would already be gained if the reliability is not quite so low as that of less systematically obtained forecasts. Moreover, if the trend of events in the real world should disconfirm part of the developmental pattern predicted by the simulation, it should be easy to trace the disagreement to its source and amend the model accordingly for future use.
>
> The intuitive judgment appealed to at several stages of the approach described above would undoubtedly be aided considerably by various analytical techniques. So-called no-change projections would establish reasonable bounds for projections based on postulated changes. Specific measures having empirically well-established consequences (e.g., the rise in soil productivity resulting from the application of fertilizer) could be taken into account systematically in estimating economic progress. Also, by now, there may be enough historical data on distinct methods of furnishing economic aid to underdeveloped countries by the United States and other

[12] I do not wish to imply that this is always an easy role to play. An expert in the role, say, of a Minister of Agriculture in the underdeveloped nation may find it necessary to act both as an expert and as a typical or an actual Minister of Agriculture.

economically advanced nations to discriminate in a statistically significant manner between their relative efficacies (the definition of "efficacy" admittedly offering some conceptual difficulties that may reintroduce a judgmental element at a higher level).[13]

No matter how carefully they are designed, games fail to achieve realism in many respects. For instance, the typical military game[14] represents the decisionmaking bureaucracy as an organization with a well-defined, consistent set of objectives, reflecting a clear interpretation of intelligence, coherent policy, and the ability to eliminate ineffective alternatives rapidly. These assumptions lead to plans or postures that are far more efficient in their use of resources than are found in real life. In other words, the decisionmaking process modeled in the ordinary game is too rational to reflect the many limitations on the decisionmaking process in real-world bureaucracies.[15]

Operational gaming, like other systems analysis techniques, undoubtedly is most fruitful when applied with a clear objective in mind to well-structured problems about which there are abundant data. But other, less time-consuming techniques are usually available when these conditions hold. Its major utility lies elsewhere, as an educational device, providing both ideas and insights, useful for the generation and preliminary comparison of alternative policies.[16]

In the analysis of major questions of public policy, it may be well worth the sacrifice of precision in handling some of the elements that can be readily quantified to gain other benefits. Among these would be some indication – though perhaps with inadequate emphasis – of the relevant political, economic, social, and psychological factors that might otherwise be overlooked or considered unimportant. Another benefit would be to provide the analysts with a greater opportunity to take into account "feedback" of the type that might lead one to want to modify the model in accordance with changes in beliefs about the real world it strives to simulate. One objective of such an "unsophisticated" simulation as a game is to get some clue as to how to model the situation in the first place.

The formal structure of an operational game automatically subjects any notion or theory to detailed critical review: the players are forced to take

---

[13] Helmer, *Social Technology*, pp. 30–31.

[14] The typical scenario also; see Jones, *Fractional Debates and National Commitments: The Multidimensional Scenario.*

[15] A. W. Marshall, *Problems of Estimating Military Power*, The RAND Corporation, P-3417, August 1966.

[16] For a recent survey of the uses of gaming in the military, see Lt. Col. A. W. Banister, "The Case for Cold War Gaming in the Military Services," *Air University Review*, Vol. XVIII, No. 5, July–August 1967, pp. 49-52.

active roles, to take specific and concrete actions in situations where a man sitting in his office or participating in a discussion might fail to consider the full range of possibilities or to carry the argument beyond the opening steps. It is easy to be vague in talking about theory or doctrine, but a game shares with the analytically formulated computer model the quality of concreteness – there can be no vague moves in a well-formulated and well-run game. Moreover, controversial parts of the model which are likely to be buried and forgotten in a computer program remain visible.

In short, the technique of gaming, as a way of bringing experts together, can do much to facilitate a systems study. Admittedly, the predictive quality of such an exercise is very clearly a function of the quality of intuitive insight provided by the experts involved. On the other hand, by allowing for the introduction of judgment at every step, the game provides an opportunity to take into account those intangible factors often considered completely beyond the pale of analysis. This is true both of the player, who can let his decisions be influenced by his appraisal of the human effects of the simulated environment, and of the expert on the control team. For example, the success or failure of a plan may depend upon assumptions about cooperation from the population or flexibility in the command structure. For an analytic formulation or a computer simulation, decisions about these things must be made in advance; in a game they can be made *seriatim*, as the need arises.

A great disadvantage of a simulation using human participants is the time required to carry it out. A computerized simulation can run through hundreds of thousands of cases in far less time, once it has been programmed. But the gaming process can be speeded up by introducing a computer for routine and well-understood phases; whether this would be economical or not would depend, of course, on the scale and nature of the exercise.

One additional point. Even where the more conventional analytic techniques, such as computer simulation, can provide correct guidance, they may still be unpersuasive. In an area like military planning, any solution to a problem that has been exclusively formulated by "outsiders," using what is essentially a "black box," may not be readily accepted as a solution. In contrast, an important aspect of an unsophisticated simulation by gaming – and one that has not been much exploited – is that the decision-maker or his representatives can actually participate.

The Delphi Technique

Gaming seems most appropriate when the experts represent different disciplines or professional interests. For a situation in which the experts are all of the same speciality, a somewhat different approach is called for. One possibility is to use the so-called Delphi technique, a process that might logically be called "cybernetic arbitration" – "cybernetic" because the process of deliberation is steered, through feedback, by a control group.[17]

The Delphi technique attempts to improve the panel or committee approach in arriving at a forecast or estimate by subjecting the views of individual experts to each other's criticism in ways that avoid face-to-face confrontation and provide anonymity of opinions and of arguments advanced in defense of these opinions. In one version, direct debate is replaced by the interchange of information and opinion through a carefully designed sequence of questionnaires. The participants are asked not only to give their opinions but the reasons for these opinions, and, at each successive interrogation, they are given new and refined information, in the form of opinion feedback, which is derived by a computed consensus from the earlier parts of the program. The process continues until further progress toward a consensus appears to be negligible. The conflicting views are then documented.

To clarify the principles of the technique let us consider two examples – one that illustrates the procedure that would be followed in seeking an answer to a fairly narrow question, the second, the procedure when a much broader question is tackled.

Example: Choosing a Number by Delphi. Consider the common situation of having to arrive at an answer to the question of how large a particular number N should be. (For example, N might be the estimated cost of a measure, or a value representing its over-all benefit.) We would then proceed as follows: First, we would ask each expert independently to give an estimate of N, and then arrange the responses in order of magnitude, and determine the quartiles, Q_1, M, Q_3, so that the four intervals formed on the N-line by these three points each contained one quarter of the estimates. If we had eleven participants, the N-line might look like this:

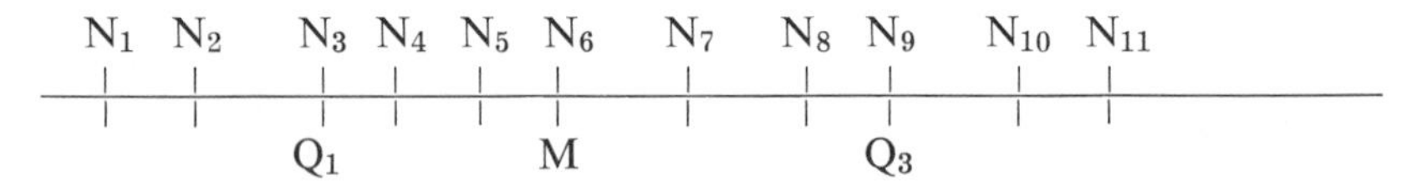

Second, we would communicate the values of Q_1, M, Q_3 to each respondent, ask him to reconsider his previous estimate, and, if his estimate (old

[17] The originators of this method for using a group of experts called the approach "Delphi," after the Greek oracle, for they first thought of it as a scheme for better forecasting. (The first experiment using this approach – about 1948 – was an attempt to use the daily racing forms from several different handicappers to obtain better "win" predictions.) As we shall see in the following pages, however, the uses of the Delphi technique are by no means limited to forecasting.

or revised) lies outside the interquartile range (Q_1, Q_3), to state briefly the reason why, in his opinion, the answer should be lower (or higher) than the 75-per cent majority opinion expressed in the first round. Third, we would communicate the results of this second round (which as a rule will be less dispersed than the first) to the respondents in summary form, including the new quartiles and median. In addition, we would document the reasons that the experts gave in Round 2 for raising or lowering the values. (As collated and edited, these reasons would, of course, preserve the anonymity of the respondents.) We would then ask the experts to consider the new estimates and the arguments offered for them, giving them the weight they think they deserve, and, in light of this new information, to revise their previous estimates. Again, if the revised estimates fell outside the second round's interquartile range, we would ask the respondent to state briefly why he found unconvincing the argument that might have drawn his estimate toward the median. Finally, in a fourth round, we would submit both the quartiles of the third distribution of responses and the counterarguments elicited in Round 3 to the respondents, and encourage them to make one last revision of their estimates. The median of these Round 4 responses could then be taken as representing the group position as to what N should be.

Example: Policy Advice from Delphi. The Delphi technique can also be applied to broad policy problems. For example, let us consider how it might be used to uncover and evaluate measures that might help to speed recovery of a nation after a thermonuclear war.

There are a number of reasons why an approach to this problem via the development of a mathematical model or a computer simulation might not be the most desirable way to proceed. If we had in mind six or eight fairly well-defined and promising alternative postwar measures, we might consider adding a "recovery" model to one of the many models that have been constructed to compute the damage caused by a nuclear attack. Assuming this could be done, the alternatives could then be compared in the traditional way used for comparing alternative force structures, employing a range of different war initiation scenarios and undertaking sensitivity analyses of the uncertain parameters.

But the concept of "recovery" is not very well defined. Very few of the many measures that might aid the survival of a nation or an area after a thermonuclear attack have been studied extensively. The emphasis so far has fallen primarily on measures such as shelters and active defense, which seek to reduce the *immediate* effects of the attack, rather than on measures to speed recovery after the initial effects of an attack have been experienced. Almost everyone has ideas about recovery measures of this type that might be helpful, but seldom any well-developed notion of their relative effectiveness and cost. Thus there is a need to survey these ideas – to create an atmosphere in which they may be brought forth, subjected to critical review, modified and ordered according to various criteria with respect to their possible effectiveness, acceptability, and costs, including social costs. The Delphi technique is well-suited to this task.

In addition to the presence of so many ill-defined alternatives, and the difficulties with the notion of recovery, there are a number of other reasons why an approach to the problem that puts emphasis on informed judgment is desirable. The decisionmakers who would use the study would clearly be in the best position to judge the acceptability of measures that might either require radical changes in the prewar way of life or imply such changes for the postwar period – for example, how far to violate the rights of privacy or favor one sector of the economy or country over another if nuclear war were to come. But their decisions would necessarily be based on many lowly but important relationships that require the intuition and judgment of specialists. Determining objectives – what we want to accomplish in the way of recovery and how we might distinguish one type of postwar world from another – must also be the responsibility of the decisionmaker. But how to attain these objectives would require contributions from many disciplines.

The alternative provided by the Delphi technique is to give up for the moment any attempt to *compute* the state of the postwar environment at various times after hostilities have ceased and instead to try simply to rank alternative prewar policies on the basis of the qualities that promise, in the judgment of specialists, to contribute the most to postwar recovery. This procedure cannot demonstrate beyond all reasonable doubt that a particular course of action is best. At most, it can assess some of the implications of choosing certain alternatives over others. But the systematic searching out and partial ordering of promising steps could be extremely valuable.

We should be under no illusion that for this problem a Delphi procedure would be the easiest thing in the world to carry out. In order to persuade the proper people to authorize or to participate in such a study, the following points would have to be brought to their attention. One, the effort would not be intended as a substitute for other research. Two, if nothing else, it would highlight areas needing detailed study and, in general, stimulate further work. Three, ideas provided in the course of the study – because of their possible half-baked character – would be kept anonymous unless attribution was specifically authorized. And four, the entire effort, in terms of manpower, could be kept quite minor, even though as much as ten months might be needed to complete the study, since getting responses to questionnaires is just slow business.[18]

Since the kind of survey being proposed is not a statistical survey of the Gallup type, but an attempt to generate ideas and to use the respondents to trace out the interrelationships among these ideas and the consequences of their adoption, it is immaterial whether the respondents form a representative sample of the initially known points of view. What matters is that the viewpoints of persons with all major relevant backgrounds have a chance of being voiced.

Assuming that our study would involve a range of experts both within and outside the organization conducting it, the respondents might be organized into several "units," so that the administrative task of running the experiment could be kept simple. Each unit might consist of a central committee of three plus a panel of six to twelve respondents. The committee chairman would be the person responsible for organizing his unit's activity, for maintaining liaison with the project director, and for transmitting the responses of his unit. One or more units might be located within the organization carrying out the study and the other units at some of the various places where there is a concentration of respondents. Alternatively, the respondents might be dealt with directly or split into functional groups or disciplines such as ecology, economic growth, and so on.

The inquiry itself could be broken down into four to six successive rounds, each based on a suitably formulated questionnaire. Only round one would necessarily involve all respondents.

The first questionnaire would contain, in addition to the questions themselves, a brief background statement explaining the purpose of the study. It would include a statement that responses will be handled anonymously, except that approval for the use of names may eventually be asked in case certain suggestions are deemed worthy of being recommended for further action. Only the members of the steering committee would initially be cognizant of the authorship of ideas. In the statement suggestions would be included about keeping the proposals in practical operational terms and avoiding generalities. The respondents would be urged to include

[18] Incidentally, there exists an Act of Congress (5 U.S.C. Sec. 139, c-e [1942]) that forbids a government agency to conduct or to sponsor a study in which identically worded questionnaires are circulated to more than nine respondents without prior permission of the Bureau of the Budget. Since the intent of the Act is to keep businessmen from being bothered with a continuous stream of government forms – not to hamper scientific investigation – users of the Delphi technique whose support comes from government funds should not have difficulty obtaining such permission. Of course, one could confine the respondents (except for at most nine outsiders) to the research organization (this includes consultants) or the sponsoring agency.

all suggestions that they think should be examined, even though they might be dubious about advocating them.

The following sample questionnaire incorporates a number of these suggestions. Since it is addressed more to the readers of this book than to potential respondents, considerable reworking would be required before it could actually be used.

Questionnaire 1

This questionnaire is being submitted to you in an effort to elicit fresh ideas on what steps should be taken to reduce the problem of postattack recovery after a thermonuclear exchange. We are not looking for measures that reduce the number of weapons impacting (ABM, for example) or measures that reduce their efficiency (such as shelters). Primarily we are looking for ways to help restore agriculture and manufacturing and the structure of society and government. An earlier study has suggested that the measures we are seeking to identify and weigh fall into three classes: *preventive*, which would aim at reducing the damage to our resources, such as food stocks and water and power sources; *emergency*, which would attempt to deal with the distribution and management of supplies to sustain the population after the war; and *long run*, which would deal with recovery proper. Regardless of your feelings about the probability of nuclear war and the futility of such actions – in themselves or in contrast to the results we might obtain if we contributed equal resources to deterrence – ask yourself what measures should be considered.

This effort is being conducted very much in the spirit of a brainstorming session, except that it sets out to collect ideas in written form rather than through the give-and-take of open debate. At this stage, therefore, it would be entirely in order for you to submit ideas even if you yourself consider them half-baked, or if you merely regard them as worthy of further exploration without wishing to endorse them, or if they would only gain full meaning within an adequately elaborated context. Remember that this survey is in no way intended as a substitute for other research; indeed, its chief virtue might be to highlight areas needing detailed study and, in general, to stimulate further work.

Question A. If you were a close advisor to the President, what actions would you advise him to consider taking (including recommendation of legislation to Congress) that might speed recovery after a thermonuclear attack?

The following considerations – the list is by no means complete – seem relevant to this question. You may wish to delete or modify some items or add others. They are offered only to spark thought, and are listed randomly to avoid prejudging the order of importance or the feasibility of any measures.

1. Since the control of infectious diseases could be a serious problem in the disrupted postattack environment, should current public health policies be reviewed for possible changes that would improve their effectiveness in a postattack situation? What policies? What changes?

2. A number of studies indicate that fires, both urban and wildland, as well as their sequelae of floods, erosion, and additional fire hazards, could be serious long-term problems in the postattack environment. Is there a need to review current fire prevention and control practices for possible changes and innovations that could improve our postattack capabilities to cope with these problems? What changes might be made? It has been suggested, for example, that we might undertake controlled burning prewar and also create appropriate firebreaks to prevent wildland fires from encroaching on contiguous urban areas or to keep urban fires from spreading to the countryside. We might also consider some steps to provide for re-seeding burned areas postattack to reduce erosion and flooding.

3. How serious a problem would it be to find feasible alternative postattack land uses that would be keyed to postattack requirements for food and other agricultural products? For example, what other crops could be grown on land too heavily contaminated with fallout to grow food, or what food crops could be grown on land not heavily contaminated but now used to grow non-food products?

4. What priorities should be observed in restoration of facilities postattack?

5. Should differential protection be provided for different segments of the population?

6. Is organizational damage likely to be a serious problem in the postattack environment?

Question B. What research should be undertaken by the scientific and technical community that might either lead to or accelerate the discovery of measures that would help speed postwar recovery?

Again, here are a few possibilities that you may wish to consider in your response.

1. Develop models. It might, for example, be important to build a flexible modular fallout model, or a model of the ignition and spread of urban fire and its impact on population in the fire area, including the protection afforded by available shelters against heat and carbon monoxide poisoning. A model of wildland fire that would relate ignition and spread to plant cover, season of year, weather, geographical region, and the nature of the nuclear attack might also be useful, as would models of a disrupted economy, since current models all seem to assume an organized society.

2. Perform further research. Research in atmospheric physics, for example, might give us a way to estimate the effects of nuclear exchanges on weather and climate. Similarly, research might be undertaken on ecological disturbance or on the long-term genetic effects of radiation on man. (Both of these problems have already been studied in some detail, but much ignorance remains.)

3. Develop technologies for food storage and synthesis.

4. Develop contingency plans for priorities in resource allocation by age, by sector of the economy, or by some other standard.

Once the responses to this first questionnaire had been received, the next, and hardest, step would be for the steering committee to sort and collate them, clarifying their meaning through checks with the respondents if necessary, eliminating obviously nonoperational suggestions, doing some minor editing and, hopefully, generating useful additions to the list.

The list of proposals thus produced might then be submitted either directly to the original respondents or, as an intermediate step to obtain further refinement, to the "unit" committees. The result of this review might be the elimination of, say, two-thirds of the proposals as being less promising. The remainder would then be annotated by the steering committee with brief arguments pro and con; they might also be ranked by merit according to some consensus formula.

Because the wording of every questionnaire but the first depends on the outcome of preceding rounds, we can at best indicate only the *form* the remaining questionnaires might take. The second might look something like this:

Questionnaire 2

The tabulation given below contains a list of tentative proposals to speed postwar recovery. We would like you to give us your judgment of each item in terms of its desirability, its feasibility, and its potential importance (assuming feasibility).

For each item, check one box under Columns A, B, and C. In making this evaluation, consider the *intrinsic* rather than *relative* merits of the proposal.

	A Desirability					B Feasibility					C Importance			
No. Proposal	Desirable	Mildly Desirable	Doubtful	Mildly Undesirable	Undesirable	Definitely Feasible	Possibly Feasible	Doubtful	Possibly Infeasible	Definitely Infeasible	Very Important	Important	Slightly Important	Unimportant
1 Establish contingency plans for priorities in allocating resources														
2 Modify current public health policies to increase the possibility of controlling infectious diseases after nuclear attack														
m. . . .														

This questionnaire would, of course, be accompanied by written arguments, pro and con, for each proposal listed.

If the results of this appraisal indicate that an item ranks no higher than "doubtful" in any category, it would be eliminated from further consideration.

For the remaining items, some of which would obviously be controversial in one or more respects, more exacting standards of acceptability would need to be set. The next questionnaire would explore the reasons for any divergence of opinions; it might take this form:

Questionnaire 3

The following items out of the list previously submitted to you have been eliminated for the reasons checked:

Item	Description	Reason for Elimination		
		Undesirable	Infeasible	Unimportant
1	.. .. ..	X		
3	.. .. ..		X	
4	.. .. ..	X		X
–				
–				
–				

The following items have been accepted as being desirable, feasible, and important.

Item	Description
11	.. .. ..
17	.. .. ..
—	
—	
—	

The remaining items are controversial in one or more respects. In those cases where a check mark is circled, your previously expressed opinion was at variance with the opinions of several of the other respondents. For each, please indicate very briefly why you hold this particular opinion. (For example, if, in Item 6, a check mark in the Desirability column is circled, please explain why you gave Item 6 the desirability rating you did in response to Questionnaire 2.) Alternatively, if on reconsideration you do not feel strongly enough about your previously expressed opinion to defend it, please indicate this by stating a revised rating.

Item	Description	Controversial as to			Reason for Previous Rating or Revised Rating
		Desirability	Feasibility	Importance	
2	.. .. ..				
5	.. .. ..				
6	.. .. ..				
–					
–					
–					

If the replies to this questionnaire continue to move toward a consensus on some of the proposals, or if for some reason the apparently irreconcilable differences of opinion seem inadequately documented, one or more additional questionnaires may be worthwhile. In form, these would resemble Questionnaire 3.

What might the final result tell us that we did not already know or could not obtain from less unconventional types of analysis? The answer can be very brief. Many aspects of the postattack recovery problem cannot be handled by standard cost-effectiveness techniques. For example, how can one assess the effect on the arms race of a prewar measure such as the storage of materials for the recovery period? Our example suggests that the Delphi technique offers, at the very least, a way to approach such questions.

The principal drawback of the technique as we have illustrated it here (and as it has been used in several actual experiments) is that it is cumbersome: several weeks may elapse before questionnaires are returned or an interviewer can poll the panel. Moreover, the amount of material each respondent must process for each round may be considerable, and, because of the lapse of time, he may have difficulty reproducing his earlier reasoning. And those who are running the experiment have their own difficulties in digesting and collating what becomes an increasingly formidable amount of material.

Hopefully these problems will be overcome. Eventually it should be possible to have each expert equipped with a console through which he can feed his responses into a computer that would process them, possibly adding relevant information automatically drawn from an existing data bank, and then would feed the results back to each respondent. But computer editing of the type required may be a long time coming.

Although such a computer system does not yet exist, the search is on. The U.S. Navy has tried out a simplified version that preserves the anonymity of the estimates but not of the reasons. The procedure is to have the experts meet in one room, each equipped with a device permitting him to select one of a set of numbers (say, from 0 to 10) by pushing an appropriate button. The set of responses then appears in scrambled order on a screen visible to all participants. They engage in a free debate, which produces reasons for raising or lowering the estimates, as well as a critique of the reasons. This is then followed by another anonymous vote. At The RAND Corporation, Delphi experiments are presently being conducted that use a number of personal electric typewriter consoles connected through an on-line time-sharing computer system; inputs and outputs are in natural English.

Much remains to be learned about the use of expertise and about procedures like Delphi in particular. For example, although there are some experimental indications that the Delphi procedure can be made to produce better estimates by introducing weighted opinions, we do not know if it is possible and easy to measure the relative trustworthiness of the various experts objectively. Experiments with the technique suggest that if the experts themselves are asked to rate their own expertise on each question, the estimates they give are fairly reliable. On the basis of this information, it may be possible to obtain better results by using as the group consensus, not the median of all responses to a question, but the median of only those responses that came from some fraction of the respondents who had declared themselves relatively most expert with regard to that question.[19]

We would also like to know how much of the convergence that usually takes place is induced by the process itself rather than by elimination of the basic causes of disagreement. Placing the onus of justifying extreme responses on the respondents clearly tends to have the effect of causing those without strong convictions to move their estimates closer to the median, but those who originally felt they had a good argument for a "deviationist" opinion may tend to give up their original estimate too easily; this may result in increasing the bandwagon effect instead of reducing it as intended.

Delphi is still in the experimental stage. It has not yet been tested at RAND in any major systems study. Elsewhere, among other applications, it has been used to study educational innovation,[20] to survey technological developments of interest to a commercial organization,[21] and to provide short-range forecasts of business indices.[22] Except for the last, however, the value of such exercises is hard to assess. Erich Jantsch, for one, seems to feel that Delphi may realize its greatest successes in "setting up goals on high levels: social goals, national goals, corporate goals, major military goals, etc."[23]

Further experimental work is planned. This includes using the Delphi questionnaire and feedback technique in conjunction with in-depth interviews, structured conferences, and operational gaming. Imperfect as it is today, the Delphi process or some future modification appears to be one of the most promising approaches under development for the investigation of problems with a high social and political content. Because it can be used to allocate resources rationally and to force explicit thinking about the measurement of benefits, it offers a hope of introducing cost-effectiveness thinking into these problems.

Summary

Roger Levien earlier pointed out that mathematical modeling, computer simulation, operational gaming, and scenario writing represent points

---

[19] See Brown and Helmer, *Improving the Reliability of Estimates Obtained from a Consensus of Experts.*

[20] Marvin Adelson (ed.), "Planning Education for the Future," *American Behavioral Scientist*, Vol. 10, No. 7, March 1967, pp. 1–31.

[21] H. Q. North, *A Probe of TRW's Future: The Next 20 Years*, TRW Systems, Redondo Beach, California, July 1966. See also, "Setting a Timetable," *Business Week*, Issue 1969, May 27, 1967, pp. 52–61.

[22] Robert H. Campbell, *A Methodological Study of Expert Utilization in Business Forecasting*, unpublished doctoral dissertation, University of California, Los Angeles, 1966.

[23] Erich Jantsch, *Technological Forecasting in Perspective*, Organization for Economic Cooperation and Development, Paris, October 1966, p. 146.

along what really is a spectrum of techniques available to the analyst.[24] He did not mention Delphi and explicit personal analyses like Alton Frye's political analysis in Chapter 17; the first enters somewhere before scenario writing and the last somewhere afterwards. At one end, as Levien puts it, precise computation is emphasized, and judgment and intuition, while present, are played down; at the other end, computation is de-emphasized and intuition and judgment are fully exploited. In between, as we have seen in this Chapter, operational gaming and the Delphi approach attempt to achieve the direct application of expert judgment and intuition through a formal structure. Depending on the complexity, the uncertainty, the competitive aspects, and the dynamics of the problem, an appropriate technique or combination of techniques can be applied.

Intuition and judgment permeate all analysis – not only as to which hypothesis is better than another, or which approach is likely to be more fruitful, or what facts are relevant, but also in fulfilling the role of the model when quantitative models are not adequate. As questions get broader, intuition and judgment must supplement quantitative analysis to an increasing extent. To make such judgment and intuition more effective, a greater use of systematic techniques for the direct involvement of experts – in particular, techniques like the Delphi process and its extensions – seems inevitable.

---

[24] See Chapter 15, p. 283.

Chapter 19

PITFALLS AND LIMITATIONS

E. S. QUADE

This Chapter attempts to alert the reader to some of the causes of error and sources of misunderstanding in systems analysis. It illustrates errors made in past analyses and indicates ways of avoiding similar errors in the future. Special attention is given to the pitfalls and inherent limitations in the use of models to aid policy decisions.

ORDINARY ERRORS AND EXPERT BLUNDERS

The fact that systems analysis depends so strongly on judgment and intuition, that it still lacks a complete theoretical foundation, implies that it is pointless for any analyst to expect success merely by following a set of definite rules. The problems the systems analyst is asked to tackle are particularly frustrating. Usually, they are urgent, complicated, and ill-defined – and sometimes those that are finally investigated are far different from those originally posed. The environment, the circumstances under which these problems are analyzed, may compound the difficulties, as witness the decisionmaker who is in such haste for an answer (even though there may be no real need for haste) that he actually makes it impossible for the analyst to conduct – or complete – the study properly. The analyst himself, after a period of fruitful work, may have become so entrenched in his organization or so saturated with a decisionmaker's ideas that he has lost his objectivity. Indeed, the very aim of the analysis, which is ordinarily to suggest a course of action to someone else, tends to introduce all the difficulties and contradictions associated with value concepts, human behavior, and the communication of ideas. These conditions, whether inherent in analysis or external to it, are common to most work that is not, strictly speaking, scientific. For systems analysis, they combine to create a situation full of pitfalls.

One purpose of this Chapter is to alert analysts and the sponsors, evaluators, and users of analysis to some major sources of error that have not as yet been mentioned in this book and to give additional emphasis to a few that have.[1] A knowledge of what the pitfalls are should help the spon-

[1] For an earlier discussion, see H. Kahn and I. Mann, *Ten Common Pitfalls*, The RAND Corporation, RM-1937, July 1957. The inspiration for this chapter and many of the ideas in it came from this work. See also Bernard O. Koopman, "Fallacies in Opera-

sor and the analyst to avoid them, and those who evaluate and use the results of analysis to discover any errors that might be present. A second purpose of the Chapter is to point out that analysis, as applied to military planning and policy, has some fairly definite limitations.

We might begin by remarking the obvious: human beings do make mistakes, and experts do turn out to be wrong. Needless to say, there are few precepts to tell us how to avoid the sorts of errors that stem from ignorance, stupidity, or simple carelessness, and the literature of systems analysis is strewn with examples of what T. H. Huxley once called "the great tragedy of Science – the slaying of a beautiful hypothesis by an ugly fact." Besides the more or less empty counsel that only smart, well-trained, and careful people should work as systems analysts, there is little practical advice one can give. Careful checking and qualitative evaluation of the reasonableness of the results are helpful in discovering the ugly little facts before it is too late. But the possibility of simple mistakes or blunders will inevitably remain.

Is the expert or the professional less likely to make such errors? There is every reason to think so. On the surface, of course, systems analysis seems so reasonable, and so simple in concept, that people are readily led to suppose that the only prerequisite to "doing" it (or using it) is common sense. Since many decisionmaking problems in the real world depend on little else, this is not such an outlandish idea – provided, however, that we temper it by recollecting that common sense, as someone has said, is "the faculty which tells you that the world is flat." Whatever else he may be, the expert is at least the fellow who knows enough to be wary of the obvious and to appreciate the value of the informed hunch, sometimes even his own. And even for the simplest of problems, trained professional judgment is sometimes essential. A case in point from World War I is related by Col. Leonard P. Ayres, then Chief Statistical Officer for the U.S. Army:

> I was at my quarters one stormy night in Chaumont, working hard trying to perfect my French conversation. There was a knock and in came the Adjutant General himself with orders to get transportation and proceed half way across France to the Supreme War Council at Versailles. I did not know what it was all about but I gathered such tables, charts and equipment as I thought might be necessary, took

---

tions Research," *Operations Research*, Vol. 4, No. 4, August 1956, pp. 422–426; C. J. Hitch, *Professor Koopman on Fallacies: A Comment*, The RAND Corporation, P-870, May 1956 (this paper was also published in *Operations Research*, on the pages immediately following the paper by Koopman); C. J. Hitch, *Economics and Military Operations Research*, The RAND Corporation, P-1250, January 1958; C. J. Hitch and R. N. McKean, *The Economics of Defense in the Nuclear Age*, Harvard University Press, Cambridge, Mass., 1960, particularly pp. 120–225; E. S. Quade, *Pitfalls in Military Systems Analysis*, The RAND Corporation, P-2676, November 1962; and E. S. Quade, "Pitfalls in Systems Analysis," in E. S. Quade (ed.), *Analysis for Military Decisions*, Rand McNally & Company, Chicago, 1964, Chapter 16.

a trusted lieutenant with me, and got transportation. Next morning we got to Versailles, which I suppose was the most military place in the world at that time, and I found shortly there were temporarily quartered there the ranking Ordnance officers of all the Allied forces.

The Ordnance officers were there because it was then thought that new artillery, exemplified by the German Big Bertha, would prove to be of genuine importance in the war, and the Allies wanted to undertake immediate production of long range guns of that sort. The Ordnance officers had been ordered to confer on the problem, and to wait at Versailles until there was a dud from Big Bertha and to find out what they could from the dud. They had been waiting for a long time, and finally somebody said that General Pershing had a statistical organization that could answer such questions as the one relating to the amount of time that might elapse before there would be a dud. That is why the Adjutant General had come around personally to see that I got to Versailles.

The problem that they were putting up to me was that of estimating the probability of an event of which there had never been even a single instance. I was inwardly appalled, and yet this was the first time that the Allies had ever called upon the American headquarters for something that they did not have, and it seemed to me impossible to let our headquarters down. I had to give an answer to that question. I asked what evidence they had. They took me into another room where there were long tables covered with white cloth. On the tables were pieces of Big Bertha shells which had been collected whenever such a shell exploded. Every French public functionary, policemen, officials, firemen, postmen, etc. had been ordered to pick up everything that could be found when a Big Bertha shell exploded. All the fragments that came from each burst were put in one pile and labelled with the date, the number, and the location. Most of them were pieces as small as a half dollar, small nearly circular pieces, but there was one long sliver about four inches wide and three feet long, and there were a good many brass pieces.

The tables were covered with these fragments. I examined them, trying to think hard and fast about what to say next, and then I realized that the brass pieces were from the fuses and I had a real quick thought, and I said, "Can you officers tell me if each one of these shells had one or two fuses?" They went into a quick huddle and after a long deliberation they said, "Yes, each shell had two fuses."

Meanwhile I had thought of the next question, which was whether by any type of microscopic or other examination they could ascertain if both these two fuses in each shell had originated in the same arsenal or one from one place and the other from another. That was more difficult. They worked on that all that day and part of the next one, and then they said they had ascertained with substantial conclusiveness that the two fuses came not from the same arsenal or factory, but from separate places of origin. I asked if they had dud rates from other shells with those same types of fuses, and they had. I asked to see them and they showed them to me. What I really had done was to go a long way toward finding out whether the cause of a dud would probably be a unit cause, or whether it would probably have to result from the combination of two independent causations, because it would not be likely that two separately manufactured fuses would fail because of the same defect. Assuming that the causes of failure would be independent I squared the probable dud rate and the result was so infinitesimally small that I took my courage in my hands and told them there was never going to be a dud.

The probability was so minute that it seemed a safe deduction. Mathematically it wasn't safe. Maybe you were only going to get one dud in ten thousand shells, but that one might come today. But I told them they were not going to have a dud in the war and they didn't have a dud during the war.[2]

[2] Col. Leonard P. Ayres, "The Uses of Statistics in War," Address before the Army Industrial College, AIC 195, March 4, 1940, pp. 9–10.

SOME COMMON FALLACIES IN ANALYSIS

Since fallacies, unlike simple blunders, are errors in reasoning, we have more hope for their elimination. Why do we have fallacies? As far as systems analysis is concerned, the main reason is the one suggested earlier: the lack of theory. We do, however, have a certain amount of experience, and that experience, plus common sense and what theory we have, should help us to avoid the more flagrant fallacies.

The fact that a fallacy has been found in a particular study does not necessarily invalidate all the work, for it may be corrected. And the very fact that someone can point out where a systems analysis has gone wrong strongly attests to the value of the approach. It is thus a serious mistake not to make any analysis and the judgments on which it depends explicit. For if they are not, we surrender the three great advantages that the analytic approach has over its competitors – namely, that someone else can examine the work, can evaluate it, and can modify it as new information or insight becomes available.

A number of the more common pitfalls or sources of error are listed below. We will say something about each, beginning with those associated with problem formulation.

Underemphasis on problem formulation
Inflexibility in the face of evidence
Adherence to cherished beliefs
Parochialism
Communication failure
Overconcentration on the model
Excessive attention to detail
Neglect of the question
Incorrect use of the model
Disregard of the limitations
Concentration on statistical uncertainty
Inattention to uncertainties
Use of side issues as criteria
Substitution of a model for the decisionmaker
Neglect of the subjective elements
Failure to reappraise the work

Underemphasis on Problem Formulation

An analysis must begin with problem formulation. A major pitfall is the failure to allocate the total time intelligently, so that a sufficient share of it will be spent in deciding what the problem really is. It is a pitfall to give in to the tendency to "get started" without having devoted a lot of thought to the problem.

In the first systems analysis I worked on I fell into this pitfall. The analysis was being carried out (in 1948) to help set the requirements for a strategic bomber that might become operational in the mid-1950's. One moot question was whether or not to put a tail turret on the bomber, given the premium on saving weight and the known inaccuracies of rearward-firing machine guns at the speeds contemplated. The day I joined the project it was suggested that I work on one aspect of this problem, and attempt to determine how much then current estimates of the probability of destroying an attacking fighter would be lowered if the correlation between aiming points in air-to-air machine gun fire – something always present because the shots are not aimed independently – were taken into account in the calculations. It was an interesting mathematical problem, and I started right to work. Several months later, after we had found a satisfactory approximation and checked it by a Monte Carlo simulation, I saw some test firings of an early version of an air-to-air guided rocket. I was much impressed. Not until some days later, however, did it suddenly dawn on me that by the time the bombers we had under study could become operational, they would rarely, if ever, have an opportunity to fire back at fighters with machine guns. Thinking about the problem as a whole and asking a few questions would very likely have saved all of us a lot of time.

In the haste to get started on model building and computation, a common error is to take the first criterion and measure of effectiveness that suggests itself. For instance, there is a tendency to measure what a system can do rather than what it should do – a tendency to measure, say, the number of bombing sorties carried out or tons of explosives delivered rather than the effect of the bombing or damage on what we are trying to accomplish.

This book has frequently made the point that the difficulties of systems analysis often lie more in deciding what ought to be done than in deciding how to do it. Rather than be guided completely by what the sponsor states is the best approach, a good systems analyst will therefore always insist on formulating his own.

Inflexibility in the Face of Evidence

The statement, early in the analysis, of possible conclusions or recommendations, is occasionally regarded as a pitfall. This itself is a mistake. Once we recognize that analysis is iterative and that a single cycle of formulation, data collection, and model building is unlikely to give the answer, we realize that the pitfall lies not in forming a preconceived or early idea about the solution, but in being unwilling to discard such an idea in the face of new evidence. In other words, we have to be flexible. When someone looks at the work and suggests that we might be wrong, we at least have to entertain the possibility that he might be right. A set of tentative conclusions helps to guide the analysis; it tells us what we are looking for while we are looking. But more important, it offers something concrete for others to probe.

The unwillingness to give up a concept one once thought highly of is well-illustrated by the persistence of horse cavalry in armies of the world until the mid-twentieth century:

> The horse cavalry has had to review its role in war four times since the end of the nineteenth century in the face of four great changes in the science of war: the development of repeating automatic and semi-automatic weapons, the introduction of gasoline and diesel-fueled engines, the invention of the air-borne weapon, and the coming of the nuclear battlefield. Each new challenge to the horse has been, of necessity, seriously considered. Each has demanded a review of doctrine, a change in role and mission.[3]

Each time minor adjustments were made. But the cavalry remained. As late as 1944, statements like the following were still appearing: "Currently we are organizing and training adequate mechanized horse cavalry for field employment."[4] In foreign armies, the lance and the charge survived to World War II and, in the United States, the cavalry was not disbanded until around 1951 (the last Army mule, aside from the West Point mascot, escaped retirement until 1956). And in 1956, the Belgian General Staff seriously suggested that for the kind of dispersed war that low-yield atomic weapons necessarily creates, the horse should be reintroduced into the inventory of weapon systems.[5]

Adherence to Cherished Beliefs

The most serious error likely to be made in problem formulation is to look at an unduly restricted range of alternatives. Although narrowing our range of choice certainly makes the analysis easier, we may pay a high price for the labor we save if some of the excluded alternatives are better than those remaining.

The most frequent cause of failure to look at the full range of alternatives is an "attention bias." This often takes the form of an unconscious adher-

---

[3] Edward L. Katzenbach, Jr., "The Horse Cavalry in the Twentieth Century: A Study in Policy Response," *Public Policy*, Graduate School of Public Administration, Harvard University, 1958, p. 121.

[4] Katzenbach, p. 148.

[5] "Belgians Hit U.S. Concept of Atomic War," *Christian Science Monitor*, August 25, 1956; quoted in Katzenbach, p. 148.

ence to a "party line" or "cherished belief." All organizations foster one to some extent; RAND, the military services, and the DOD are no exception. Experience suggests that Kahn and Mann were right when they called the party line "the most important single reason for the tremendous miscalculations that are made in foreseeing and preparing for technical advances or changes in the strategic situation."[6] The failure to realize the vital interdependence among political purpose, diplomacy, military posture, economics, and technical feasibility is the typical flaw in most practitioners' approach to the analysis of problems of national security. Examples are plentiful – the military planner whose gaze is so fixed on "winning" a local war by military actions that he excludes other considerations; the statesman so convinced that peace can be maintained through deterrence that he completely disregards what might happen should deterrence fail; the weaponeer so fascinated by startling new weapons that he assumes they can of course be used; or the political negotiator who seeks to conciliate the potential enemy at a military cost that is far too great, because he does not trace out the full military implications of his actions.

The history of strategic bombing studies since World War II illustrates the workings of this influence. In World War II, the bombing problem was to penetrate the defenses, bomb accurately, and return. The bomber's concern was with enemy fighters, antiaircraft guns, and missiles – not with enemy bombers. For years, even in studies for time periods long after the Soviets were expected to have nuclear weapons, no serious attention was paid either to the possibility that our bombers might be vulnerable on the ground or that an attack on theirs might be highly lucrative. Requirements and specifications for future bombers hardly considered the problem of surviving the enemy offense. This oversight was not mere stupidity. For instance, in the Navy-SAC controversy over the B-36 in 1949, the Navy questioned the B-36 on every basis it could think of – including the argument that strategic bombing was immoral – but the question of the bomber's vulnerability on the ground did not come up. RAND strategic bombing studies, even after the seriousness of the problem of bomber survival on the ground had been pointed out, continued to concentrate on such questions as speed, altitude, low versus high penetration, supersonic dash, bombing altitude, small versus large planes, and target selection. Thus, in a 1952 comparison of aircraft with missiles for strategic bombing, we ignored any effects of possible differences between the two types of systems in ground survivability – and no one we briefed took us to task for this omission! The Russians and the British took even longer than

[6] H. Kahn and I. Mann, *Ten Common Pitfalls*, The RAND Corporation, RM-1937, July 1957, p. 42.

we did to recognize ground vulnerability, and some people think the French have not yet really absorbed the idea. It took the extensive briefings of a RAND study devoted to proving that base vulnerability was a serious threat to national security to get major attention on the problem.

The party line can be most influential in shaping the study during the early stages. What can happen is that the participants and successive reviewers become aware that some of the alternatives or certain assumptions being considered are frowned on by higher ranking officers. It seems useless, even hazardous, to support such unpopular views strongly, and gradually they may be stressed less and less or even forgotten. In its most extreme form, this influence can, in effect, cause the analyst to lose his independence of view.

Parochialism

A similar pitfall is to expect the man or the organization that created a system to discover its faults. Either may fail completely to take into account some technical notion or fact. An analyst is sometimes shocked to discover that other organizations regard the system he is advocating as controversial at best, or even consider the engineers on whom he depended for technical advice to be completely wrong. This attitude may be due to what has been called the NIH (Not-Invented-Here) factor. But the analyst may discover from others that simple countermeasures, for example, can be devised to render the system almost worthless. We might recall the story of the English invention of "window," or, as we call it here, "chaff." In 1937, in Great Britain, the merits of radar and infrared were being argued. There was talk of shutting down research on infrared because of its vulnerability to countermeasures. It was then that R. V. Jones, a physicist who was working on infrared, came up with the idea of "window" to show (in his words), "They weren't so invulnerable themselves." Watson-Watt, the English radar pioneer, apparently never had quite the heart to urge trials for Jones' device, which seemed so certain to spoil his beautiful invention. Much later, others had the trials carried out. To appraise a system and discover its value, good analysts obligate themselves to consult people with an adverse opinion of the worth of the system, largely because they know how hard it is to get a scientist or engineer to display much ingenuity in tearing down a technically brilliant design that he has been working on for years.

One manifestation of parochialism in analysis occurs when service A proposes a future weapon system with fabulous properties for victory and compares it with the semiobsolete current equipment of service B rather than with the new system that B hopes to introduce.

Communication Failure

Sometimes an error occurs because the analyst fails to communicate effectively with the professional people on whom he must rely. For example, as background for the previously mentioned analysis to choose a next-generation bomber, we made use of a parametric study that covered thousands of possible designs, each of which was then evaluated by means of computer-simulated strategic bombing campaigns. In the comparisons, the number of engines unexpectedly appeared as a most significant factor in bomber survival. Since we felt sure that the aircraft designers had some flexibility as to the number of engines, although this was not a parameter in the study, we went to the people who supplied the designs to determine what penalties would be incurred if the number of engines was increased. We found out that, although they had worked out the total thrust required for each of the designs, they had no rule for, and indeed had not specified, the number of engines. The people who gathered the data for the attrition model had obtained the number of engines by counting those shown on a diagram that indicated the pertinent features of each configuration. It then turned out that the artist who drew the illustrations had decided that, without engines, the diagrams just did not look like aircraft and had simply drawn them in, supplying the number that looked suitable to him for the shape and size of the wing! A possible way to avoid this type of pitfall is to have someone on the analyst's team who is at least a lay expert in all the important fields with which the study is concerned; however, sharp limits of time and money may make this impossible.

Overconcentration on the Model

There are a great many pitfalls associated with models and model building. One of these is for the analyst to become more interested in the model than in the problem itself. Technical people with specific training, knowledge, and capability like to use their talents to the utmost. It is easy for analysts to focus attention on the mechanics of the computation or on the technical relationships in the model rather than on the important questions raised in the study. They may thus find out a great deal about the inferences that can be drawn from the model, but very little about the question they set out to answer.

There is, of course, always the problem of the "conscientious scholar" who approaches his problem with meticulous thoroughness – and, naturally, is careful not to fall into any of the pitfalls listed here – but finds the decision taken before he can report his results.

Excessive Attention to Detail

While there are dangers in oversimplifying the model, in a general sense it pays to be simple. Complicated formulas, or relationships so involved that it is impractical to reduce them to a single expression, are likely to convey no meaning at all, while a simple, though possibly approximate, relation may be easily understood. A major error may invalidate the more complicated expression and, yet, in the general complexity of the formulation, pass unnoticed. In uncomplicated expressions, serious error is apt to become obvious long before the computation is completed, because the relationships may be simple enough to reveal whether or not the behavior of the model is going to be reasonably in accord with intuition. The most convincing analysis is one which the nontechnician can think through.

Neglect of the Question

Another pitfall is to attempt to set up a model that treats every aspect of a complex problem simultaneously. What can happen is that the analyst finds himself criticized because the first model he has selected has left out various facets of the situation being investigated. He is vulnerable to these criticisms if he doesn't realize the importance of the point made earlier about models: *The question being asked, as well as the process being represented, determines the model.* Without attention to the question, he has no rule for guidance as to what to accept or reject; he has no real goals in view and no way to decide what is important and relevant. He can answer criticism only by making the model bigger and more complicated. This may not stop the criticisms, for something must always be left out. The size of the model is then determined not by what is really relevant but by the capacity of the computing machine.

One approach to designing a model is to attempt to reduce the real system to a logical flow diagram. The dangers of this approach are that the model may tend to be too detailed and that components of the real process will be included that contribute nothing to the question to be answered. For this reason, it is advisable to design the model around the questions to be answered, rather than as an imitation of the real world.

Incorrect Use of the Model

One error is to accept as useful output from a model the results of computation that are merely incidental to the question the model was designed to answer. For example, in war-outcome calculations comparing missile systems, a certain missile may show up best when a particular strategy is used. This may be an optimal strategy as far as the model is concerned, but, if the purpose of the model was to make a cost-effectiveness comparison, many considerations important to the way missiles would be employed in actual combat may have been suppressed as not affecting the comparison.

Sometimes a mathematical model indicates a preference for extreme strategies, as in one tactical study which indicated that all offensive aircraft should be used against the enemy air forces for the first few days and then, suddenly, that they should be used in close support of troops for the rest of the campaign. No war has ever been fought in this way, and one should be extremely dubious about such a strategy. This may be the correct solution for the model, but a model cannot reflect all of the smoothing-out factors present in the real world. Such calculations can still be valuable, but modifying circumstances must be considered before they can be offered for operational guidance. Too neat a solution, particularly if it goes against established experience and intuition, should be viewed cautiously.

Disregard of the Limitations

A serious error is to forget, late in the study, limitations that were imposed on the ranges over which some of the approximate relationships used in the model were expected to hold. In early strategic studies, for instance, it was assumed that fissile material would be in short supply. The alternative was to send a "cell" of three, four, or five bombers to a target, only one of them with an A-bomb. After one study, we found another analyst claiming that much better results would be obtained by sending cells of about 750 to 1000 bombers to all the targets in sequence. When I protested that that was absurd, he clinched his argument by telling me he was using the same model we used. It took quite a bit of argument to convince him that, while the model might be satisfactory for a few planes, it was nonsense for a large number. The assumption of the model was that a defense installation would distribute its fire uniformly over all attacking bombers. With this in mind, he had been led to argue that, if the number were made large enough, the fire would be spread so thin that no bomber could accumulate enough hits to be shot down.

Concentration on Statistical Uncertainty

Systems analysis is concerned with problems whose essence is uncertainty. Where the probability of occurrence is more or less objective or calculable, this uncertainty can be handled in the model by Monte Carlo or other methods. The treatment of such uncertainty is a considerable practical problem, however, and a challenge to the analyst. The pitfall for model builders lies in accepting this challenge – to the neglect of the real or unforeseeable uncertainties. These typically involve forms of ignorance that cannot be reduced to probabilities, and their consequences can be devastating.[7] The

[7] Particularly those uncertainties that depend on human caprice. As I think one of Damon Runyon's characters put it: "Nothing what depends on humans is worth odds of more than 8 to 3."

objective in system studies is not to learn how chance can affect a given situation with a specific probability, but to design a system or determine a policy so that any fluctuations are unimportant.

Since a full Monte Carlo investigation may seriously expand the analysis, it is frequently better to carry out first a simple expected-value treatment, deferring a full investigation of fluctuation phenomena until the qualitative aspects of the problem are fully understood. It may then turn out to be unnecessary to perform these more complicated calculations, since consideration of the real uncertainties may have made trivial the effect of any statistical uncertainty.[8]

Inattention to Uncertainties

It is not enough simply to acknowledge that uncertainties exist and to warn that some things have been left out of the analysis because of a lack of information, for such issues may have a critical effect on the conclusions. The user has to come to grips with these issues and he needs to know what their effects will be, how likely they are, when he can expect them, and what he can do about them.

Systems analysis is so striking in its attention to detail and its elaborate calculations that it tends to create the impression that more of the significant factors have been considered than may actually be the case. This impression has enabled systems analysis to get by with an inadequate treatment of future threats on the theory that uncertainties are best taken care of by desensitizing results and including some well-chosen caveats.

In questions dealing with the future, it is ordinarily futile to remove uncertainty by making a best guess. It is essential to do sensitivity analysis rather than depend on "expected values" of key parameters. The DOD presents the matter in these terms:

> As far as the technical and operational parameters are concerned, we have found that the best way to deal with these uncertainties is to explore their limits, and to do our calculations in terms of the range of uncertainty. It is generally useful to begin with a best guess for each of the key parameters and then to introduce into the calculation an optimistic or upper bound estimate, and a pessimistic or lower bound estimate. Although it is usually sensible to design the defense posture primarily on the basis of the best estimates, the prudent decisionmaker will keep asking himself, "Would the outcome be acceptable if the worst possible happened, i.e., if all the pessimistic estimates were borne out?" Carrying three numbers through all of the calculations can increase the workload greatly. For this reason, a certain amount of judgment has to be used as to when the best guesses are satisfactory and when the full range of uncertainty needs to be explored.[9]

---

[8] These matters are discussed in greater detail in Chapters 5, 11, and 12.

[9] A. C. Enthoven, "Operations Research and the Design of the Defense Program," in *Proceedings of the 3rd International Conference on Operations Research*, Dunod, Paris, 1964, pp. 532–533.

There are always uncertainties about context. For example, it may be assumed that the enemy consists of a certain group of countries. We may then want to investigate what would happen if other countries were to join the original group. It is, of course, a matter of judgment how far to carry this type of activity, but it is an error not to give it some attention.

A cherished objective of systems analysis is to have results that are insensitive to what the enemy actually does, primarily because the system chosen will then be effective under any of the conditions considered and therefore be a hedge against the enemy's selection among these choices. This avoids committing oneself from the beginning to a move that is based on predicting what the enemy will do about one's own programmed force. Also, it may enable the analyst to save the effort necessary to make predictions that will be reasonably accurate. We do not deny the desirability of this objective, but rather caution that this method of dealing with uncertainty has some disadvantages. For example, if we take insensitivity to be a property of systems that "work for the full range of extremes" and apply it to determine the over-all strategic posture, we may find we have overcommitted ourselves. This cannot only be costly in resources, but might influence or provoke a larger reaction from the enemy than is consistent with U.S. objectives.

Use of Side Issues as Criteria

A practice that can lead to serious error is suggested by the following statement: "If several alternatives have similar cost and effectiveness and if these results are quite sensitive to the values assigned the inputs, some other basis for decision must be found." This may amount to saying that if, after honest analysis, we are fundamentally uncertain about which of several alternatives is best, the issues should then be resolved on the basis of some specious side criterion not originally judged adequate to discriminate. On the contrary, the point to stress is that the decision must be made on the basis of forthright recognition of the fundamental uncertainty.

Substitution of a Model for the Decisionmaker

The failure to realize the importance of the question in designing the model leads to another pitfall: the belief that there are "universal" models – one model, say, which can answer all questions about a given activity and which therefore can be used to evaluate, without supplemental judgment, a full range of alternatives.[10] For example, it has been proposed a number of

[10] There are, however, models which allow the user to experiment with a wide choice of parameters and assumptions. A number of large-scale computer simulations, such as the family of strategic planning models described by N. C. Dalkey in Chapter 12, are extremely flexible in this respect.

times (even to the extent of writing a study contract) that a general computer model for strategic air war be set up to supply weapon designers with a systematic evaluation of their design concepts and to enable the Department of Defense to evaluate the worth of alternative "design solutions" developed by computing contractors.

One argument for such a model notes that "the choice of assumptions, the forecast of the future, and the methods of analysis have a marked influence on the performance and physical characteristics of the weapon system set forth as preferred or optimal"; therefore, a uniform framework would mean that "the results obtained by the various contractors would be comparable since the effect due to variation in the assumptions they might have chosen to form their models would have been eliminated." This may indeed be the case, but will the end result be desirable? A rigidly specified framework may mitigate one sort of undesirable bias – by making it difficult for an analysis to be used to rationalize conclusions already otherwise derived – but only at the severe risk of introducing other biases.

A fundamental objection is that a uniform framework necessarily conceals or removes by assumption many extremely important uncertainties, and therefore tends to lead to solutions that disregard the value of hedging against those uncertainties. Another is that even if efforts were made to keep the model up to date, this would turn out to be impossible, for the analyst must be able to modify his model in the terminal stages of his study to accommodate information acquired during the early phases. Indeed, in a problem involving the struggle between nations, there are so many factors of shifting importance, and such radical changes are likely in objectives and tactics, that most models are obsolete long before the recommendations from the study can become accepted policy. Moreover, if a model or a mathematical formula were used to indicate which proposal to select, the proposer's emphasis would soon focus on how to make his design look good in terms of this analytic definition, and not on how to make it look good against the enemy – a much harder problem.

C. J. Hitch, at the time Assistant Secretary of Defense (Comptroller), made this last point with an analogy:

> Another kind of problem that might be encountered with an analytically based contract would be "rule beating." An analogy can be found in the case of some of the handicapping rules drawn up by yachting organizations. The intent of these rules is to allow the owners of often greatly dissimilar sailing yachts, basically designed for cruising, to compete against each other on an equitable basis. The rules are generally empirical in nature, and take into account such factors as the dimensions of the hull, the amount of sail area, and so on, resulting in a handicap for each yacht which reflects its theoretical speed. The rule is expressed in terms of a formula which may be rather complex. So long as the competitors are all sailing relatively conventional yachts, the goal of generally equitable competition can be achieved.

However, once such a rule is established, the serious competitor has a considerable incentive to study it very carefully when he is considering a new yacht, or even a new rig for his old yacht. In such an environment, from time to time, there have appeared some fairly unconventional yachts, designed not in the usual way, but in a way specifically tailored to beat the rule. From a practical point of view, these yachts are freaks; nobody would have designed such a thing or wanted to own one save for the existence of the rule. They tend to be undesirable in most ways save that of winning races through the establishment of an unusually favorable handicap. Whenever such freaks start to win most of the races, of course, there is a strong tendency for the rule committee to plug the previously unsuspected loophole in the rule.

To the extent that such rules have loopholes, emphasis is shifted away from beating other yachts towards beating the rule itself. By the same token, setting up a weapons system contract on the basis I have described would not really mean that the contractor will, by definition, be motivated to develop and produce the best possible system. Rather, he will be motivated to develop and produce the system which best meets our analytical definition of the best possible system. To the extent that our definition is incomplete, or subject to unsuspected loopholes, the product may tend to diverge from what we really have in mind. Thus, this sort of contract would be subject to "gaming" on the part of the contractor – either deliberate or unconscious. He may be able to develop a system which meets the necessarily artificial time-cost-effectiveness model beautifully, but which is, in fact, a rather poor weapon system.[11]

Neglect of the Subjective Elements

It is a serious pitfall for the analyst to concentrate so completely on the purely objective and scientific aspects of his analysis that he neglects the subjective elements or fails to handle them with understanding. Quantification is desirable, but it can be overdone; if we insist on a completely quantitative treatment, we may have to simplify the problem so drastically that it loses all realism.

Since the analyst knows his study will be subject to scrutiny, interpretation, and possible further analysis, he should make his subjective judgments known. Trust is essential because, in large part, the client has to take the analysis and recommendation of any study team on faith. The client cannot repeat the study, will very seldom have the time to review it meticulously, and will be influenced by it depending on his belief about how the analyst reached his conclusions. He cannot hope to master the variety of specialized skills that frequently go into a complicated analysis. At best he can acquire enough background to identify really incompetent or patently biased work. But faith in the analyst's purely technical and scientific competence is not sufficient; what is also required is a similar confidence in his judgment. Trust requires disclosure; the client must know either how the analyst has disposed of the subjective elements in the study or if

---

[11] C. J. Hitch, "Cost Considerations and Systems Effectiveness," Address presented at the SAE-ASME-AIAA Aerospace Reliability and Maintainability Conference, Washington, D.C., June 30, 1964.

he has accepted and used the client's judgments. If the analyst does the latter uncritically, then he is not using the full potentialities of analysis.

On the other hand, one danger associated with analysis is that it may be employed by an administrator who is unaware of or unwilling to accept its limitations. When a study is presented confidently, but little attention is called to its deficiencies, the recipients are prone to read too much into it. A weak or careless or busy administrator may ease his job by transferring a portion of his responsibilities to the analyst or a model, and thus fail to give the study the critical scrutiny it requires.

Failure to Reappraise the Work

Administrators sometimes feel that one of the worst characteristics of systems analysis is that the analysts want to make basic changes in a study after the work is half done. As they see it, the result is a great deal of "wasted work" and deadlines that are not going to be met.

It is, of course, quite true that making a major change in a study at a late stage means that much of the early work cannot be used, and even that, because a change may involve a great deal of additional work, deadlines may not be met. For these reasons, some analysts, when they are one-half, two-thirds, or three-fourths through the study, hesitate to pause to evaluate what they have done thus far. A periodic reappraisal is essential, however, because as the study progresses the analyst broadens his understanding of its scope and purpose. Stocktaking that results in junking a major portion of the work indicates that a reappraisal was especially necessary.

Limitations of analysis

Out of context, the pitfalls we have mentioned seem so obvious that one wonders how they could lead to error a first time, let alone be repeated. One has only to examine actual analyses, however, to find that they are still present. Our hope is that as theory and experience develop they will occur less frequently.

One last pitfall is to believe that, if the work is done correctly, systems analysis is unlimited in the quality of the advice it can supply a decisionmaker. This is far from the case. The systems analysis of military problems has many limitations – some inherent in all analysis, some, if not peculiar to military studies, at least more likely to be found there. As a consequence, it is seldom possible to *prove* to a decisionmaker by analysis that he should choose a particular course of action.

To be helpful, systems analysis needs to be clear about what the analysis can do and what must be left to the intuition and judgment of the decisionmaker. Earlier chapters, in attempting to draw this line, have noted many limitations of analysis. We single out four for further comment: (1) analysis is necessarily incomplete; (2) measures of effectiveness are inevitably approximate; (3) ways to predict the future are lacking; and (4) systems analysis falls short of scientific research.

Analysis Is Necessarily Incomplete

Time, money, and other costs obviously place severe limits on how far *any* inquiry can be carried. The very fact that time moves on means that a correct choice at one point will soon be outdated by events and that goals set down at the start may not be final. This is particularly important in military systems analysis, for the decisionmaker can wait only so long for an answer. Other costs are important here, too. For instance, we would like to find out what the Soviets would do if we put an armed Minuteman on Moscow. One way to get this information would be to launch a Minuteman. But while this might be cheap in dollars, the likelihood of other costs precludes at once this type of investigation.

Still more important, however, is the general fact that, even with no limitations of time and money, analysis can never treat all the considerations that may be relevant. Some are too intangible. For example, such qualities of a system as its flexibility, its compatibility with other systems (including some that are yet to be developed), its contributions to national prestige abroad, and its impact on domestic political constraints can, and possibly should, play as important a role in the choice of alternative force postures as any idealized war-outcome calculations. Ways to measure these things even approximately do not exist today and they must be handled subjectively. (And if we find out how to measure them, other political, psychological, and sociological intangibles will still be left.) The analyst can apply his judgment and intuition to these considerations and thus make them part of the study, but the decisionmaker will rightly insist on applying his own.

Measures of Effectiveness Are Inevitably Approximate

The choice of weapons and strategy – in fact, the entire conduct of warfare from initiation to termination – must be governed by the nation's objectives, not solely by military standards of success. But national objectives are often multiple and ill-defined, and sometimes conflicting. Measures of their attainment are likely to be inadequate approximations at best. In Chapter 3 we mentioned a few problems in measuring deterrence. And as L. D. Attaway explained in detail in Chapter 4, difficulties of the same sort occur in almost every attempt to indicate the attainment of objectives. One might measure the value of a missile defense in protecting our cities by the number of enemy warheads it could destroy. But the installation of a near-perfect system might weaken our alliances by appearing to commit us to a fortress America concept or, by making defense appear possible, increase the risk of an early enemy strike. Similarly, in Vietnam, to measure the progress of the war we are forced to use an avalanche of statistical measures – incidents, defections, body counts, weapons lost and captured – all more or less unsatisfactory.

Moreover, we cannot be as confident that our estimates of effectiveness are essentially correct as we are about our cost estimates. One analyst who has studied the problem of estimating casualties suggests that if a pre-World War II estimator had worked analogously to his brother of today, had known his trade exceptionally well, had been knowledgeable about the means by which World War II military actions produced casualties, had known the probabilities associated with each weapon, and could estimate the number of people subject to each weapon – then he would have underestimated the total cost in human lives of the war to the Soviets by a factor of between three and four.

Such an error in the measurement of effectiveness may not be too important if we are comparing two systems that are not radically unlike one another – two ground attack aircraft, say. But at higher levels of optimization – tanks versus aircraft or missiles – gross differences in system effectiveness may be obscured by gross differences in the quality of damage assessment.

The inability to determine good measures of effectiveness is a severe limitation on the usefulness of analysis. Suppose we are seeking to determine the characteristics of future tactical aircraft. If a measure such as the weight of ordnance that can be delivered is used, the aircraft are likely to resemble the B-52 or the C5. Some combination is required that weighs properly *all* the characteristics that are important to the effectiveness of such aircraft – qualities such as speed, bombing accuracy, runway length required for takeoff, invulnerability to ground fire, ferry range, and so forth. Possibilities to use in a comparison of different types of tactical aircraft are the change in the outcome of a projected ground battle or the motion of the forward edge of the battle area. Ultimately an entire profile of measures may be required. For example, in Chapter 21, to study the mix of tactical air forces in Europe, the authors use the "history" of a range of hypothetical wars in which the effectiveness of various force mixes in resisting enemy attacks of various size is compared on the basis of time indices – the time it takes the enemy to penetrate 30 miles, then 100, then 150, then 300, then 500, and the time it takes the defender to gain air parity, then local superiority, then general superiority, then air supremacy. There are fifteen indices in total, the others involving losses and force ratios.

No Satisfactory Way Exists to Predict the Future

Systems analysis lacks any good methods of predicting a single future in terms of which we can work out the best system or determine an optimal policy. Consequently, we must consider a range of possible futures or contingencies. In any one of these we may be able to designate a preferred course of action, but we have no way to determine one for the entire range of possibilities. We can design a force structure for a particular war in a particular place, but we have no surefire way to work out a structure that is good for the entire spectrum of future wars in all the places they may occur.

Consequently, defense planning is rich in the kind of analysis that tells what damage could be done to the United States given a particular enemy force structure (or, to put it another way, what the enemy would require to achieve a given destruction); but it is poor in the kinds of analyses that evaluate how we will actually stand in relation to our potential enemies in years to come.

Systems Analysis Falls Short of Scientific Research

No matter how we strive to maintain standards of scientific inquiry or how closely we attempt to follow scientific methods, we cannot turn military systems analysis into an exact science. For one thing, there exists no way to verify our models except in rare circumstances.

Except for this inability to verify, systems analysis may still look like a purely rational approach to decisionmaking – a coldly objective, scientific method, free of preconceived ideas and partisan bias, in which the terms used are defined exactly and the conclusions reached depend on formal logico-mathematical reasoning.

But it is not so. Human judgment is used in designing the analysis; in deciding what alternatives to consider, what factors are relevant, what interrelations between these factors to model, and what numerical values to choose; and in analyzing and interpreting the results of the analysis. The terminology may be inherently vague, and the reasoning may be informal. In short, since judgment and intuition are fallible, caution and reservation on the part of both the analyst and the decisionmaker are necessary to avoid errors or misconceptions that could bias or even negate the implications of the analysis.

Chapter 20

THE CHANGING ENVIRONMENT FOR SYSTEMS ANALYSIS

JAMES R. SCHLESINGER

This Chapter emphasizes the growing requirement for care in the design of systems analysis, and for ingenuity and flexibility in systems design. It shows that changes in the strategic, technical, and political environment are difficult to encompass in the stereotyped analysis. And it acts as a reminder of certain points already made: how the optimal allocation of resources may depend on nonquantifiable, and sometimes even unknowable, considerations; how studies conducted at one point in time may become irrelevant as objectives and circumstances change; how the need for ingenuity, for flexibility, for hedging, and yet for timely decisions has grown over the years; how the desire for great precision in systems studies is likely to be self-defeating; and how the growth of complementarities among force components has affected cost analyses.

INTRODUCTION

The purpose of this Chapter is to spell out the increased complexity of decisionmaking in the sixties and its implications, not only for the decisionmaker, but for the analyst as well. Whatever alters the character of decisionmaking will, at a different level of choice, influence the analyst. The consequence of the alterations in the environment is to generate a *growing* requirement for imagination in the design of studies or analyses and for ingenuity and flexibility in the program for the design of systems. This Chapter attempts, specifically, to indicate the trends that are intensifying the demands on the analyst, to illustrate these general trends by reference to specific analytical and decision problems confronting the Air Force and the DOD, and to draw some inferences regarding future analytical work.

To indicate the scope of the discussion, it is perhaps advisable to say a few words at the outset regarding the relationship between system *study* and system *design*. No precise line can be drawn between the two, for their conceptual bases and roles overlap. In what follows, we will use "studies" or "analyses" interchangeably to refer to the broad role of outlining and evaluating the range of plausible contingencies and the range of measures – including plans, equipment, and operational concepts – designed to cope with such contingencies. Others have used the term "system design" in much the same way – that is, to emphasize that the analyst should be concerned with more than merely analyzing the alternatives presented to him. He should instead be striving to design or invent new alternatives which will more adequately satisfy our objectives than existing ones do. To an extent, we shall also be considering system design in this sense. But system design also includes the more circumscribed role of designing specific weapon systems in light of information provided by the broader analyses. System design in this sense is "hardware design." It may be viewed as the appropriate fruit of an expanded R&D program. H. Rosenzweig has already discussed the major issues bearing on technology, technological uncertainty, and research and development.[1] Largely for this reason, we can concentrate on the problem of system studies, emphasizing the selection, operational concepts, and integration of weapon systems – and the guidance provided to those who design such systems.

Major environmental changes

Major changes that have occurred in the strategic, political, and technical environment since the developmental years of systems analysis have intensified the sophistication we require of the art. In particular, four specific factors, operating together, have accentuated the demands upon the analyst. We will see, as we examine them individually, that none of them is wholly new, but growing weight must be assigned to each, since jointly they add to the challenge to the imagination and insight of the analyst.

Increased Perception of Political Fluidity

In the past, the assumption of Soviet malevolence, accompanied by estimates of future Soviet capabilities based on production possibilities, led to the designation of specified threats in particular time periods. The challenge to the analyst was to design a broad system to deal with the assumed threat. This approach was employed in 1950 shortly after the first Soviet atomic test.[2] At that time, the year 1954 was specified as one of "maximum peril" on the hypothesis that by then the Soviets would have produced atomic weapons in sufficient number to neutralize the U.S. atomic deterrent. By that year, it was argued, conventional forces would be required to prevent a thrust of Soviet power. Whatever the merits of the contemplated posture, the argument for it was temporarily eclipsed by the "new look" decisions of 1953–1954.

---

[1] See Chapter 6.

[2] See P. Y. Hammond's study, "NSC-68: Prologue to Rearmament," in W. R. Schilling, P. Y. Hammond, and G. H. Snyder, *Strategy, Politics, and Defense Budgets*, Columbia University Press, New York, 1962.

Again, in the "missile gap" controversies following the Soviet development of the ICBM in 1957, projections were developed indicating 1961–1962 as the period of maximum danger for a Soviet strike, possibly – or even probably – from the blue. The degree of risk was properly regarded as dependent on the U.S. posture. The main objective of the analyst was to devise means for countering a clearly specified threat – one which implied the "worst possible" consequences. In such an environment, the "minimax rule" compelled attention for guiding analysis. Once again, the existence of an acknowledged and specific threat simplified the task of the analyst.

It would be improper to suggest that a technique of specifying and responding to a single dominant threat is simply wrong. As a simplifying assumption in periods of revolutionary changes in military technology, it has its uses. Indeed, if once again the nation were confronted with another such revolutionary change, we might well return to using this analytical device. But consider how different our situation is today. Instead of having to deal with a dominant threat in a specific time period, we deal with *a spectrum of vaguely perceived and more modest threats which may develop at some indefinite time in the future*. No longer do we feel that the Soviets can negate our second-strike capability without signs of buildup – and the absence of a dominant threat cancels out the "minimax rule" as a guide to action. We are less inclined to view the Soviets as either implacable foes or rational game-opponents or to attempt to anticipate their actions on such a basis. No longer do we seem to place much confidence in our ability to predict what the Soviets will do. As a result, we can no longer concentrate our resources on countering a single maximum threat. Instead, we must allocate resources so that we can deal with many (preferably all) of the vaguely perceived threats within a broad spectrum.

Greater Sophistication Regarding the Character of Nuclear War

A second factor, interlocking with the first, that intensifies demands upon analysts is the growing sophistication on both sides regarding the character of nuclear war. We will go into this subject in more detail later; here we might simply note the major changes. First, we now recognize both a number of conflicting objectives and major uncertainty as to which of these objectives would be dominant in nuclear war. Thus, much greater attention must be devoted to the criterion problem – a subject L. D. Attaway considered in Chapter 4. In the fifties, analyses of possible nuclear exchanges did not get much beyond two-sided spasm wars.[3] In both the

[3] The dominant concern was to assure that, if a Soviet attack came, the war would be at least two-sided – and, in view of the character of the existing forces, it was then the appropriate concern.

first-strike and second-strike variants, the objective was clear-cut: destruction of the maximum number of enemy targets, either cities or highly vulnerable strategic air forces. The targets were known, immobile, and soft. Destruction was to be achieved without worrying about collateral damage (which was, at that time, a "bonus" effect). Damage limitation for the United States emerged principally through the reduction of Soviet offensive capabilities either by pre-emption or quick retaliation subsequent to the Soviet first strike.

Retrospectively viewing these early-vintage studies of broad strategic systems, we are probably struck by the relative simplicity of the problem, as it then was seen. Only a single type of weapon system was involved. Penetration was a manageable problem. The focus of attention was on survivability, but it was the short-lived survivability necessary to hit back in a spasm-war strike. Thus, in the early fifties, the relative advantages of ZI-basing versus overseas basing versus basing in the United States could correctly become a major issue in studies of the two main contingencies.

By current standards, the defects of this outlook are numerous – even if we confine ourselves to analysis of central war with the Soviets as the main foe and disregard complexities introduced into strategic planning by the rise of China. Forces and analysis have advanced. Conceptions regarding both objectives and strategy, if not wrong then, are now obsolete. Manipulation of a single offensive weapon system as the main variable for both deterrence and warfighting is now wholly inappropriate. The image of how nuclear war might come about – in massive initial strikes with little or no warning – must now be modified. As a consequence, a sharp line can no longer be drawn between strategic forces and general purpose forces. Today we are concerned with damage limitation, and with combinations of offensive and defensive systems to attain that end. A variety of strategic offensive systems exists which potentially may contribute to our strike capabilities, but the optimal choice will depend upon the circumstances.

Partly because of the inherent difficulties if the Soviets take sensible countermeasures, partly because of the absence of really hard thinking on the part of the services, partly because of high-level policy decisions, it is now generally accepted that the option of a highly successful disarming strike is not open to either side. A principal consequence of this view is the reinforced emphasis on avoiding collateral damage to the Soviet Union. If the Soviet offensive forces capable of inflicting drastic damage on the United States cannot be eliminated, the counsel of wisdom suggests that we provide every incentive to the Soviets to exercise restraint in the use of their surviving forces. This implies keeping the Soviets aware of how much remains at risk, if they behave rashly. This concern about collateral damage means that we must now give careful attention to balancing objectives to which we gave no thought in earlier systems studies – for example, knocking out a hardened target with a weapon of the smallest possible yield. A half-decade ago, we were indifferent to the size of weapons to be dropped on small targets.

Two major consequences of this growing awareness of the complexities surrounding nuclear war should be underscored.

The first stems from the fact, mentioned earlier, that emphasis has shifted from the operating characteristics of individual weapon systems to the combination of numerous systems into an integrated package. The stress has shifted to the complementarities among weapon systems, and *a major goal in trade-off analysis is to improve compatibility between systems*, even if some otherwise desirable characteristics must be sacrificed. This trend is epitomized most revealingly, perhaps, in the recent decision to combine Package I and Package II (strategic offensive and defensive forces) for analytical purposes.

These changes stressing complementarities will require major adjustments in the Air Force and also in work done for the Air Force. In a sense, we will have to pay the penalty for past success and for developing concepts which dealt with readily manipulable pieces of a problem. The weapon-system concept was pioneered at RAND and in the Air Force. In the past, it has been both possible and appropriate for the weapon system to hold the center of the analytical stage. But we must now change our thinking. In a sense, we must now start to grapple with the kind of intricate problem of integration that the Army, for example, has continuously faced – at least implicitly. For the Army, the meaning of the weapon-system concept has always been rather fuzzy. Center-stage has been something like the infantry division, which ties together a number of separate capabilities. What meaning could there be, for example, in a "howitzer weapon system" or an "armored car weapon system"? In the strategic field, at least, the Air Force has been lucky in the past in having relatively clean-cut analytical devices at hand. But this increases the difficulty of adjusting to a package-oriented environment.

The second consequence that we must now recognize involves a related set of complicated factors, which stem from our changed image of how nuclear war may come about. Given the prospective strategic balance, with the potential for devastation embodied in the forces that would survive a disarming attack, it becomes very hard to envisage nuclear war being initiated suddenly with all-out strikes. If it were to come, it is most likely to come in a sequence of escalating steps from a lower-level confrontation. This implies the need for careful study of how best to mesh general purpose forces and strategic forces. Strategic forces will either serve to control or fail to control the process of escalation – "keep the lid on," in the current parlance. The less advantageous the strategic balance is to the United States, the bolder the enemy may be in any specific crisis. On the other hand, limited war forces, including selected nuclear forces, may through their existence or through their employment serve to control conditions which could escalate to central war. Thus, limited war forces – with or without firebreaks – are part of the mechanism of deterrence of central war, and the complementarities and trade-offs between the two types of forces must be carefully analyzed. If crises and potential crises are the seedbed of central war, and if effective crisis management constitutes a principal means for reducing the risk of central war and for obtaining settlements on terms favorable to us, then some of the confidence generated in the past by systems studies must perforce disappear. Crises involve so many unpredictable elements – boldness, resolve, and determination in the pursuit of one's objectives; rapid and unforeseen adaptation or improvisation of military capabilities – that neatness and precision, if obtained in systems studies, will not be consistent with the messiness of real-life conditions.

Both the increased stress on packages, complementarities, and mission trade-offs (under conditions in which we hope that central war, if it comes, will be characterized by restraint), and the increased tie between strategic and limited war capabilities (stressing the recognition that central war, if it comes, will come via escalation) diminish our ability to get a quantitative handle on strategic problems. The role of assumptions in providing a royal road to quantitative conclusions has increased by something like an order of magnitude. While varying degrees of confidence will be placed in such assumptions, the problem of analysis is markedly different from what it was in the fifties, when contingencies could be mapped out in advance. Whatever the confidence placed in assumptions, the probability that they will be wrong is very high.

Increased Emphasis on Highly Specialized Weapon Systems

The demands on the analyst, particularly in force-structure determination in an environment in which mission trade-offs have become critical, is heightened by the highly specialized nature of many modern weapon systems. In part, specialized systems may be required because of our altered objectives in nuclear war. Given a desire to avoid collateral damage, the option of going to higher levels of violence to achieve target destruction disappears. One must attain higher performance levels in target destruction. The end of reliance on big yields may imply a variety of highly specialized delivery vehicles.

Perhaps more importantly, specialized weapon systems are also needed to counter enemy advances or to exploit enemy vulnerabilities. But because of their inflexibility, such weapon systems will be required in the force structure only if the enemy adopts certain courses of action rather than others. A highly specialized system, by definition, invests major resources in a specific kind of capability. On a cost-effectiveness basis, however, the allocation of resources for a highly specialized purpose is warranted only if the enemy chooses to procure and deploy just those capabilities for which the highly specialized system is a countermeasure.

The significance of the growing variety of delivery vehicles and concepts is perhaps best illustrated by the fact that a decade ago there was only one delivery vehicle: the bomber. Before the order-of-magnitude improvement in air defense capabilities, the principal problem was to get the aircraft over the target. A systems analysis concentrated on trade-offs among range, weight, payload, and speed, and paid some additional attention to basing concept, alert status, vulnerability and the like. However, with the arrival of surface-to-air missiles, the picture began to change. Against extensive air defenses, even with defense suppression, high-altitude attack looked less promising. To circumvent Soviet SAMs, low-altitude penetration became the accepted concept. In principle, bombers optimized for low-altitude operations became attractive. At the same time, long-range missiles, free of any initial problems of penetration, were being deployed, and they had major cost advantages over bombers for most tasks involving known immobile targets.

Yet the possibility of effective ABM defenses has lurked in the background. Given the improvement in low-altitude air defense, major difficulties in penetration could develop with existing U.S. delivery vehicles. A case may exist for developing, as a hedge, advanced systems like SLAM, designed to circumvent such defenses through low-altitude penetration at Mach 3. But that case does not extend to procurement and deployment – until such a time that Soviet action makes so specialized and costly a system an attractive buy. If Soviet defenses do not improve to such a degree, other less costly measures will suffice – improved penetration aids, for example. The point is that much of U.S. R&D activity should be devoted to developing specialized capabilities designed to counter Soviet developments which would exploit vulnerabilities in existing U.S. systems. Yet the Soviets cannot acquire capabilities to exploit all our vulnerabilities; they will have to choose. Thus, many of our own specialized systems need never be procured. On cost-effectiveness grounds, appropriate actions for us depend upon those routes the Soviets actually choose to follow. And since force-structure decisions, more so than R&D decisions, are critically dependent upon intelligence, the broader the menu of capabilities from which we must choose, the more vital good intelligence becomes in analysis.

One additional consideration must be added. Development of specialized systems implies, almost by definition, that there are fewer hedges against the failures of *close substitutes*. In the fifties, the F-101, the F-104, and the F-102/106 programs were in some sense substitutes. The opportunities for transfer of subsystems from one program to another were sizeable. But in recent years the increased specialization of systems and subsystems reduces this kind of hedge. Increased specialization means less opportunity for partial overlaps among programs. This imposes greater demands on the R&D program, a fact that leads us to our last point, which concerns the growing financial strains in R&D.

Rising Costs of R&D

That costs of developing military systems since the early fifties have been rising rapidly, perhaps exponentially, may be taken as a datum. Opinions vary regarding causes and possible cures, but the fact itself is beyond dispute. Given relatively stable budgets, either the number and variety of systems both developed and procured will fall, or else a number of systems successfully designed will not be carried through the full and costly development cycle as it is now known.

At the same time that the supply of new weapons is under downward pressure, the demand is rising for varied weapon development to hedge against uncertainty. As mentioned earlier, we live in a period in which the strategic balance is likely to change slowly and in no clearly predictable manner with respect to new weapons. Yet, though the direction of advance is not clearly charted, instability and change do lurk in the background. We must use our R&D resources to counter a number of potential Soviet threats.

Happily, the possibility does exist for directing the R&D more toward hedging against a large number of possible surprises and less toward developing a smaller number of operational systems. There are two contrasting approaches that can be taken to hedge against uncertainty. The first is to have in development a number of complete systems – one of which, as a threat crystallizes and a need is perceived, has qualities which make it adaptable to a new mission. This is the now traditional approach to aircraft development. It characterized bomber and fighter development during the fifties. It was comforting to discover, for example, that the B-52 did possess a low-altitude capability or that the F-104 could be adapted to the role of an attack aircraft.

The second approach is quite different. The stress is on a well-stocked R&D menu, with numerous specialized projects which can rapidly be moved into the procurement-development stage, if the need arises. The focus of the program is shifted away from *full systems* development to exploratory and advanced development *stages*. The goal is to create, in effect, a shelf of advanced weapon hedges. The key concepts here are technical building-blocks, preliminary compatibility studies, and system design. This may be called an option-creating and option-preserving strategy for R&D. It contrasts with a strategy in which the major effort on the scientific-technical base represented mainly feedback from the objective of full systems development. It involves recognition that successful development does not necessarily involve procurement, and that procurement and development will probably not follow from successful development in the majority of cases. A major problem with this strategy, of course, is that the willingness to cut off a successful program goes against the grain for both the technologists and the organizations responsible for its development.

If we are prudent in allocating our energies, there seems to be little reason to wonder which is the appropriate kind of hedge strategy to pursue in the near future. Given the rising cost of systems, the falling supply of new weapon systems, the growth of specialized potential threats which existing systems are unlikely to prove sufficiently flexible to counter, and the need for more specialized capabilities to counter specific threats, it appears that the appropriate means for hedging against surprises is through an enhanced R&D program, in which individual projects are austerely conducted – a program designed to create and preserve a multitude of options. As the Soviet Union gives indications of pursuing particular lines of attack, we could move with moderate speed to counter those actions. We must be aware that such a strategy involves the quick response of the American economy, when production and deployment prove necessary. But in light of the proved flexibility of U.S. industry and technology and the historical sluggishness of the Soviet economy, we can have a measure of confidence that, in the final race for completely operational forces, we would come out ahead.

Yet we should be aware of the greater challenge this strategy represents for system design and systems analysis. The designer must deal more than he would like with preliminary work on incomplete systems. The analyst, as he looks to the future, must deal with more or less hypothetical forces on which it is extremely difficult to get even a rough quantitative handle.

Two contemporary analytical problems

Up to this point, the discussion has run largely in terms of generalities – generalities which help to explain how the character of analytical studies has been altered since the mid-fifties. But as one major purpose of this book is to provide some guidance for those who are, or might be, working on questions of current and future concern – particularly as these questions involve the objectives and criteria problem – let us turn to two specific analytical and decision problems that may demonstrate the relevance of the general discussion up to this point. The problems which we shall consider are (1) the optimal resource allocation for damage limitation, and (2) the choice of a specific offensive force-mix and the limits this choice may impose on strategic options in subsequent time periods. The former is intended to illustrate the points made earlier regarding our growing sophistication about nuclear war and our changed perception of the threat. The latter is intended to illustrate what has been said regarding the problem of covering a broad spectrum of threats in light of the increasing cost and decreasing flexibility of new weapon systems.

Criteria for Resource Allocation for Damage Limitation

The first problem is, of course, the subject of the on-going DOD studies on optimizing Packages I and II (strategic offensive and defensive forces). We could criticize several aspects of the existing studies – in particular, the deficiencies of parametric analysis for long-range force-structure planning in light of the absence of time-phasing and the impossibility of identifying or analyzing critical decision points. But we might do better to concentrate attention on a single aspect: the dependence of the analysis on a subjective parameter. This is a question we should view in light of our earlier discussion of (1) the altered objectives in nuclear war, (2) the spectrum of possible central war scenarios and our uncertainty regarding which is the relevant one, (3) the altered image regarding the initiation of nuclear war, and (4) the much enhanced emphasis on damage limitation, which may require our working on Soviet intent as well as capabilities. And given this background, our conclusion is likely to be that *the optimal allocation will depend upon assumptions regarding a highly subjective parameter: specifically, the probability and duration of a period of mutual restraint and city avoidance in nuclear war.* In short, this question is a case in which what was previously called "the royal road to quantitative conclusions" must be based upon some rather questionable initial assumptions. Either explicitly or implicitly (that is, haphazardly), some estimate of the probability of city avoidance will enter into the analysis and determine the results. But since this highly subjective element will influence both resource allocation and strategic choice, it should be considered explicitly – rather than be ignored in the quest for firm quantitative conclusions.

Broadly speaking, the current studies attempt to reveal, for various budget levels, the optimal point for damage limitation on a trade-off curve (of constant damage) between strategic offensive forces, on the one hand, and optimized civil defense and terminal defense, on the other. (See Fig. 20.1.) But the optimal point is highly dependent on the assumption mentioned earlier regarding the duration of a period of mutual restraint – in that for a war which is primarily counterforce, it will be advisable to invest relatively greater resources in expanding, diversifying, and protecting our offensive capabilities.

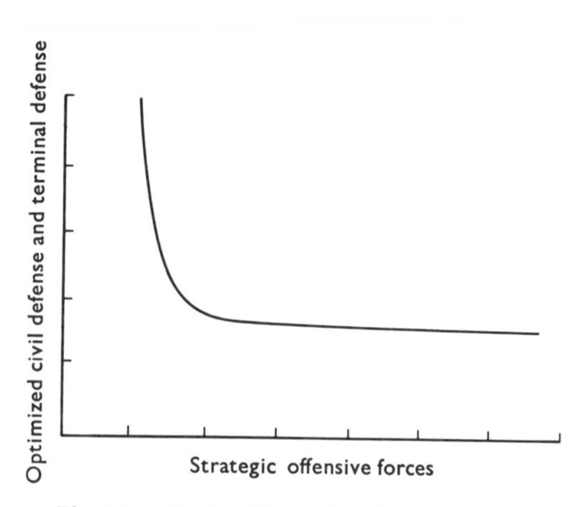

Fig. 20.1 – Trade-off curve for damage limiting

Let us consider two extreme and hypothetical cases, and explore their implications for resource allocations. In the first case, Mr. Brezhnev and his successor, if any, as well as the American President, repeatedly emphasize in their public statements that nuclear war is a terrible thing: that it would be disastrous were it to come; that if it does come, the loss of life must be held down; and that "our" side would never *initiate* a strike at enemy cities, but reserve its "invulnerable" forces for retaliation should the foe strike cities first. Such declarations, if made repeatedly, would certainly influence our view as to the nature of nuclear war and the preparations we should make for it.

In the second case, we feel quite sure that the bulk of the Soviet missile force is *pointed at our cities*, and that the decision to launch will follow immediately upon any substantial U.S. strike – either because the missiles are, in effect, wired for an automatic response or because the authority to fire descends automatically to lower command levels when U.S. warheads impact on Soviet soil.

Clearly, under these two sets of hypothetical conditions, we would assign very different subjective probabilities to the existence and extent of a period of mutual city avoidance. But it should also be clear that, given these alternative probabilities, *we are dealing with two different trade-off functions for our strategic offensive and strategic defensive capabilities.* Thus, in Fig. 20.2, Curve A represents a trade-off function in the case of spasm or near-spasm war. For damage limitation in such a war, our strategic

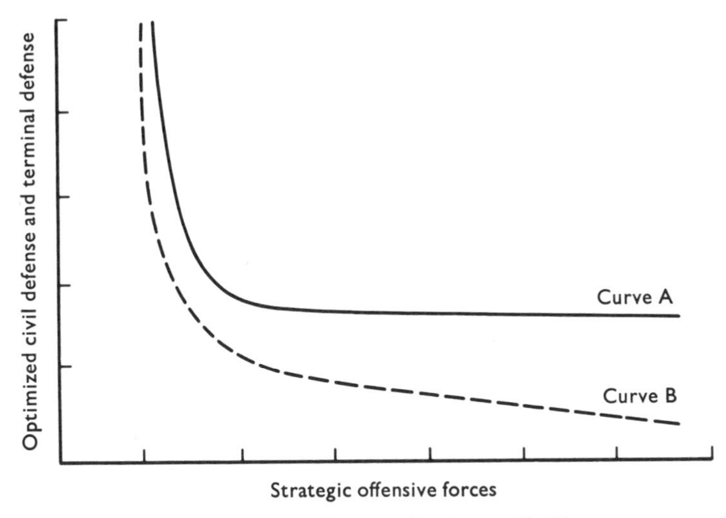

Fig. 20.2 – Trade-off curves for damage limiting

offensive forces, designed to deal with time-urgent targets, would consist largely of missiles. The force would tend to be limited, since very large additional expenditures on offensive capabilities would not buy much in the way of damage limitation. This is indicated by the elbow in Curve A. On the other hand, the payoff to additional defensive forces would be moderately high, so that relatively modest outlays for defensive capabilities would be the equivalent of very large expenditures on offensive capabilities. By contrast, Curve B indicates the trade-off function on the assumption that there will be an indefinite period of city avoidance. The entire curve is shifted downward and to the left, indicating that limiting damage to a *given* level can be obtained with lower outlays on Packages I and II. Under these conditions, the returns for additional outlays on strategic offensive forces may be moderately high. By contrast, beyond a limited initial investment in defensive capabilities, very large additional outlays on such capabilities may buy comparatively little – and relatively modest outlays on offense will be the equivalent of very large outlays on defense.

Why does this sharp divergency exist? The answer is that defensive capabilities (aside from fallout shelters) perform a somewhat different and more limited function than do offensive capabilities. This difference in function is obscured in many existing studies, which examine optimization in terms of alternative war outcomes based upon potential damage at a single point in time rather than provide a time-sequential analysis of the war that recognizes the possibility of a period of mutual restraint. To illustrate this difference in the nature of the two force-sets, let us turn to Fig. 20.3. As the arrows at the left of the Figure suggest, both the offensive and defensive capabilities of the United States serve to reduce the Soviet *potential* for damage. But there is a difference which could prove to be very important. While both sets of forces from the very beginning do limit potential damage, *the United States could employ its strategic offensive forces immediately at the outbreak of war in order to alter the character of the Soviet threat*. The longer the period of restraint, the more extended is the *intra-war* opportunity to alter that threat. Moreover, the longer the period, the more options may be open to us to make such an attempt. By contrast, the defensive capabilities are, in a sense, "withheld." They perform their "active" role only in the relatively brief period required for a Soviet strike against cities. *A period of restraint does not result, therefore, in an expansion of the list of interesting defensive options.*

Thus, if restraint is preserved for a period, the actual employment of defensive capabilities to blunt the Soviet attack would occur only after what may be a considerable lag – during which the strategic offensive forces could be employed to perform their function of reducing the possible

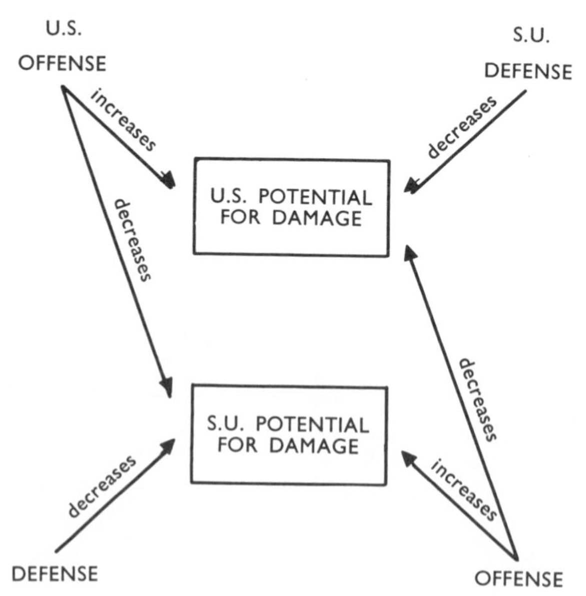

Fig. 20.3 – Measuring war outcomes

weight of the attack that the Soviets must ultimately decide whether or not to launch.[4]

There is a need, therefore, to consider carefully how long such a lag might last, and to include the best probability estimates among the parameters used to optimize force allocation. But this important factor is neglected in many studies, which more or less implicitly assume near-spasm war. The reason for this is that only in the context of a near-spasm war can the strategic offensive and strategic defensive forces be compared simply and without qualification. Once the possibility of restraint is introduced and the contrasting functions of the two categories of forces are underscored, the complexity of the calculations is increased many times. And this complexity occurs not only in the calculations of systems effectiveness. One must ask himself searching questions regarding what portion of the population he is willing to risk (for a limited but indeterminate time period) in order to provide greater capabilities for reducing long-run Soviet damage potential. To such a question there is no objective answer. Yet, if one avoids such questions, the result may be that the force structure is optimized for dealing with what may be the least likely type of central war.

To illustrate the way such calculations may influence the optimal force structure, let us consider three hypothetical systems – distinguished in accordance with the speed of effective reaction. System A (say, missiles) can react immediately to destroy Soviet damage-inflicting capabilities. System B (say, reconnaissance-strike capabilities) reacts more slowly, its maximum effectiveness occurring from twelve hours to two weeks after the outbreak of war. System C (say, ASW capabilities) requires several weeks or even months to accomplish its mission. To limit damage in a near-spasm war, one would want to rely primarily on System A. But procurement would be relatively limited, with the balance of funds going into defensive capabilities because the *marginal cost* of killing additional Soviet offensive capabilities would rise rapidly for this system. However, if there is a lengthy period of restraint, Systems B and C may become attractive for reducing Soviet damage potential. If the period of restraint lasts for a week, for example, the marginal cost of destroying surviving Soviet land-based missiles through reconnaissance-strike capabilities may be relatively moderate – and this option *could* become interesting. If the war goes on for months, ASW capabilities designed to seek out and destroy Soviet missile-launching submarines might be very interesting. In a brief

[4] There is the added possibility, depending on the hardware characteristics and the command and control arrangements of the Soviet forces, that they could also be used to degrade Soviet targeting capabilities.

period, the marginal cost of destroying Soviet SLBMs could be infinite. But, over an extended period, it might be moderate enough to be highly attractive, especially when it is remembered that knocking out a submarine represents a bargain in terms of missiles destroyed. In an extended counterforce war, all of the enemy's capabilities can be made vulnerable.

The moral of this story is that where enough time is available, a slow reacting system may be relatively cheap in terms of the *marginal costs* of destroying additional enemy capabilities. The high marginal cost of damage limitation through strategic offensive forces applicable in a near-spasm war may cease to be relevant if an extended period of restraint occurs.

As a simple example of these points, let us examine several situations, defined by the data in Table 20.1, in which the Soviet forces consist entirely of missiles. There are 800 missiles, of which 200 are elusive targets that can be discovered only after some time has passed. If the war is essentially over after an initial exchange, the use of strategic offensive forces to reduce enemy damage potential becomes too costly. The marginal cost of taking out enemy capabilities may be very high in relation to saving the lives of, say, a million people. Enough enemy missiles (350) survive, in any event, so that one should invest heavily in defensive capabilities. If, however, the war goes on for several months, one may be able, through the use of time-consuming offensive systems, to reduce enemy forces to the point that the surviving missiles (55) represent a much more modest threat to civil society. From the standpoint of damage limitation, the optimal mix in this case is skimpier defensive preparations and far heavier investment in strategic offensive forces – a solution that means little more than that the allocation of resources depends on the ratio in the final showdown between enemy vehicles and one's own lucrative – that is, civil – targets. If one cannot reduce this ratio substantially through extended counterforce operations, then the payoff of heavy outlays on defensive measures will be much greater than if one can.

TABLE 20.1

Effects of different allocations of strategic forces

Type of Missile Attacked	Initial Force Size	Size after Initial Strike	Size after One Week	Size after Several Months
Targetable	600	150	40	25
Not initially targetable	200	200	200	30
Total	800	350	240	55

To demonstrate that for optimal resource allocation we are interested in the ratio, in the final showdown, between enemy vehicles and one's own civil targets, let us turn to Fig. 20.4. This Figure expresses, once again, our major theme that an extended period for employing offensive forces may permit us to alter the character of the threat – and perhaps to do so at relatively low cost by the use of techniques (essentially look-shoot-look tactics) which would not be available in a near-spasm context. The Figure

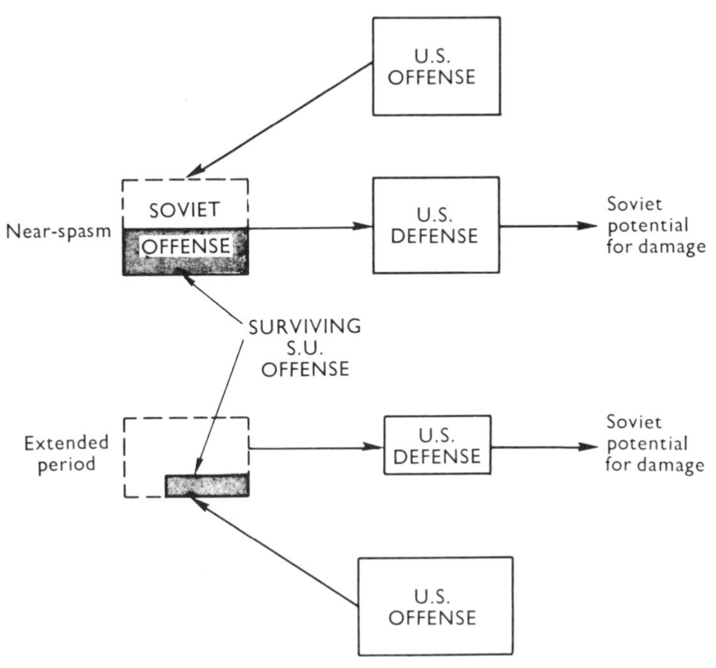

Fig. 20.4 – Variation in optimal resource allocation depending on assumed character of war

indicates (1) that the measure in which we are ultimately interested in studies of damage limitation is the Soviet potential for damage, and (2) that both offensive and defensive capabilities affect this variable, the offense by directly reducing Soviet offensive capabilities, the defense by blunting the effects of an attack, so that what is finally filtered through is Soviet potential for damage. The questions are: How much to U.S. offense, how much to defense? In the near-spasm context, the United States may be unable to make a very substantial dent in Soviet offense, and the optimal strategy would be to invest heavily in defense. Where, on the other hand, there is an extended period, U.S. offense, through a variety of measures, may reduce Soviet offense to a very low level – so low that much less should be allocated to defense to blunt the now much-reduced Soviet attack. The size of the boxes in Fig. 20.4 indicates that in the second case an entirely different allocation may be appropriate for the United States – one in which much more is invested in offense and far less in defense.

In leaving this example concerning optimal resource allocation, we should perhaps note that the studies of damage limitation in the near-spasm context are unquestionably useful in providing an *initial* basis for analysis. From them one can speculate on the sequence of interactions as each side responds to the perceived outcomes by altering its intentions and capabilities. However, from the foregoing discussion, we can conclude that raising the issue of complementarities between missions enormously complicates analytical work and raises questions regarding the confidence that we can place in the results. To design an analysis which points to a single and unequivocal set of conclusions regarding strategic-forces resource allocation is well-nigh impossible. The point that should be remembered is that the ultimate decision regarding resource allocation must rest on nonquantifiable or subjectively quantifiable elements and that it cannot rest solely on presumably quantifiable technical data. Thus, an essentially unknowable parameter becomes critical in determining the ultimate results.

Constraining Future Strategic Options by an Early-on Force-Structure Decision

Let us turn now to our second illustration, which involves a problem in choice with which the DOD is continuously struggling, that is, determining the composition of U.S. missile forces. The purpose of this example is to underscore the desirability of maintaining flexibility in planning in order to cope with a gradually unfolding threat environment dominated by uncertainty. It was suggested earlier that the way to retain flexibility in such an environment is through an aggressive R&D program designed to develop multiple options and through avoidance of force-structure decisions until such decisions are forced upon us by the flow of events. The present example stresses the advisability of delaying major force-structure commitments until long lead-time elements force a decision. The case for such a decisionmaking pattern is quite strong *when one has moderate confidence that delay will permit the resolution of major uncertainties regarding the future strategic and technological environment*. This kind of flexibility may be contrasted with the premature foreclosing of strategic options implicit in commitments which are made at too early a date.

To indicate the advisability of such a decisionmaking pattern, let us examine the particular set of decisions, made early in 1961, which determined the character of our missile forces. Needless to say, the purpose in going back to these earlier decisions is not to indulge in some pointless second-guessing, but to learn what we can for the future.

The decisions made in March 1961 were advertised as part of a "quick and dirty look" at the force-structure program inherited from the Eisenhower administration. Major conclusions were to expand greatly the projected Minuteman (and Polaris) force and to reduce the size of, and the emphasis upon, the Titan II force. The effect of these decisions was to determine that the intermediate-run U.S. forces would be overwhelmingly composed of small payload missiles. Two background aspects of these decisions should be kept in mind. First, they were made before the new intelligence then becoming available had been fully absorbed. While fears of a major "missile gap" were being dissipated, we were still unaware how great our strategic superiority was. Consequently, substantial emphasis remained on a quick buildup of a second-strike capability. Second, there existed certain political pressures for making changes (particularly in light of the preceding campaign) that would dramatize and highlight the shift away from the policies of the preceding administration. The atmosphere was one calling for decisiveness in a period of presumed crisis, as is perhaps suggested by the phrase "quick and dirty look." In any historically fair assessment of these decisions, these considerations must be kept in mind.

Nevertheless, we should ask ourselves a question: What can we as analysts learn in retrospect from these decisions? It now is clear that major difficulties existed in formulating the early-on cost-effectiveness studies which served as their basis. On the one hand, there were major uncertainties regarding the size and character of prospective Soviet forces and also regarding the strategic concept that we would adopt. In addition, major technical uncertainties regarding both Minuteman and Titan II remained unresolved. As a consequence of these deficiencies in information, it was inevitable that only the crudest observations could be made regarding the effectiveness component of the decisionmaking schema. Cost considerations, therefore, became dominant. Yet, even here, because of the unresolved technical problems, not much confidence could be placed in the cost calculations. As it turned out, these calculations were strongly biased against Titan II because of the drastic underestimation of missile operations and maintenance costs. This favored the missile with the lower initial capital cost, that is, Minuteman.

The upshot was that, as a result of calculations based mainly on cost, decisions were made, in effect, against large payload missiles and for small payload missiles. At the time, there was, to be sure, a developing emphasis on the desirability of avoiding collateral damage through the use of weapons of lower yield (which may have been associated with a stress on small payload vehicles), but it seems fair to say that cost was the main consideration pushing in the direction of a force composed of small payload missiles. Had it been necessary to make that decision at that time, of course, it would have been equally necessary to have based it on whatever information was then available. But we can say with reasonable confidence that there was then no compelling reason why it had to be made. It could have been delayed, and – in retrospect – we can see several still unresolved strategic and technical issues which suggest why it should have been delayed. Let us examine six of these issues.

Counterforce Strategy and the Prospective Hardening of Soviet Missiles. Perhaps the most important has been the elaboration of the counterforce strategy in its controlled response variant, with its emphasis on initial targeting of military targets exclusively and the attempt to avoid major damage to the fabric of Soviet civil society. In addition, although Soviet ICBMs were then soft, there was the prospect, later emphasized by Secretary McNamara in Congressional testimony, that in the future, ICBMs would be hardened. The extent of hardening and, *a fortiori*, the degree of hardness could not be anticipated with any precision. Particularly as the Soviets hardened their missiles, the United States might require higher-yield weapons to destroy Soviet capabilities. The degree to which our own CEPs could be lowered was unknown and, consequently, higher-yield weapons might be needed to substitute for targeting inaccuracies. Moreover, if the Soviets failed to press the development and procurement of missile-armed submarines, the pressure upon us to avoid the use of high-yield weapons out of concern for collateral damage would be much reduced, because we still might be able to achieve a major disarming blow. In light of the still unknown parameters, the prospective Soviet moves toward hardening should have led us to emphasize the large-payload hedge rather than commit ourselves so early to a force composed largely of small-payload missiles.

Possible Soviet ABM Deployment. Since we were aware of the possibility that the Soviets might deploy an ABM system, the implications of such an eventuality for a U.S. missile force composed primarily of small-payload missiles might well have been considered. Re-entry vehicles with small-yield weapons appear to be particularly vulnerable to ABM systems. Large-payload vehicles represented a hedge against Soviet ABM deployment in that the re-entry body could be toughened up, much higher-yield weapons could be employed, and a wide assortment of penetration aids could be packed into the vehicle.[5] In short, the commitment to a force composed primarily of small-payload missiles could be regarded as an inadequate hedge against the possibility of the Soviet ABM system.

[5] See *The New York Times*, June 20, 1966, pp. 1, 9.

Test Moratorium and Test Ban. In early 1961, it could not be assumed with any confidence that the Soviets would break the test moratorium during the summer of 1961, thereby permitting our own test series and possible improvements in yield-to-weight ratios. In subsequent reviews, the possibility of the test ban treaty, which materialized in the summer of 1963, should have been kept in mind. The treaty now inhibits our ability to reduce the size and weight of warheads in Minuteman or to increase the yield with a warhead of given size. With given weight and size constraints in the re-entry vehicle, a reduced ability to vary the physical size of the weapon implies a lessened possibility of doing such things as toughening up the re-entry vehicle or packing in additional penetration aids.

Limitations of Numbers. Limiting the number of missiles available to both sides has been discussed at Geneva. Moreover, even in the absence of a formal agreement, some implicit bargaining has taken place between the two camps with the intention of holding numbers down. If numbers are held down, for whatever reason, much larger capabilities are provided by a force composed of large-payload as opposed to small-payload missiles. The large-payload vehicle represents a hedge to offset the effects of the likely inclination to hold numbers down.

Multiple Warhead Options. For quite obvious reasons, the possibility of a missile's carrying multiple warheads, each individually delivered, very much increases the utility of a large-payload vehicle.[6]

Questionable Systems Reliability. The fact that the technical characteristics of missile systems and subsystems were still unknown in 1961 points to one final possible advantage of the large-payload vehicle. In such a vehicle, if reliability problems were encountered, subsystems such as guidance packages could have been placed in parallel, thereby reducing the risk of unreliability – and possibly economizing on operation and maintenance costs.

The ultimate influence of any one of these considerations could have been such as to make it advisable to press forward with large-payload vehicles; yet decisions made earlier, largely on the basis of cost considerations which were crude in themselves, had already inclined the United States in the direction of small-payload missiles. The entire episode illustrates the need for flexibility, for hedging, and yet for *timely* decisions respecting force structure. It also demonstrates what should be obvious, how cost-effectiveness studies conducted at one point in time may become irrelevant as strategic objectives and circumstances change. The point here, however, is that the decision did not have to be made so soon.

[6] Cf. *The New York Times* story cited earlier.

When the external environment permits delay, and when one has moderate confidence that major uncertainties will be largely resolved, delaying the decision is likely to be the wisest course of action.

These conclusions seem inescapable, but they still leave room for the observation that, in fact, the Minuteman decision has not worked out too badly. This is due particularly to the slow buildup, hardening, and dispersal of Soviet forces and to the slow advance on ABM systems. The favorable resolution of technical problems and the brief resumption of testing by the United States have been helpful. But the point is, if we have been right, we have been right because of developments that could not have been predicted with high confidence. Whatever analysts may say, it is undoubtedly better to have been right for invalid reasons than wrong for the right reasons. Yet, since we cannot count on such good fortune in all cases, we are well advised to see what we can learn from earlier experiences. One lesson this example demonstrates is the desirability of maintaining options and of putting off critical force-structure decisions until forced to make them by long lead-time items. Another is that the dominant role of uncertainties in this example (like the role of assumptions regarding highly subjective parameters in the previous examples) indicates that undue expectations regarding precision in systems studies is likely to be self-defeating.

Some final inferences

The purpose of these last few paragraphs is to draw some inferences from the discussion that should be helpful in future analytical work. We can group these inferences under six headings.

Uncertainties: State of the World, Objectives, and Strategies

The first issue to mention is the inevitable one: uncertainty. Two points in particular should be emphasized here: (a) we are now more aware of the uncertainties that have always existed in the environment, and (b) in many relevant ways that environment has become more uncertain. This heightened awareness of uncertainty is reflected in our perceptions of the nature of the enemy, the character of nuclear war and how it may be initiated, and the future array of military capabilities and the possibility that they may be coherent. But these uncertainties regarding the environment, objectives, and strategy vastly complicate the decisionmaking schema. One view of the decision process, taken from formal decision theory, has attracted some attention among systems analysts. The procedure is to assign subjective probabilities to possible states of the world, array them accordingly along one axis, examine strategies or alternative lines of action along another axis, and then make a choice among them by means of some decision rule.

This is a neat and intellectually elegant way to structure the problem, but, if improperly understood, may create more problems than it solves. While a useful first approximation, such a model is inadequate in at least two respects. First, the assignment of probabilities to perceived possible states of the world will inevitably be misleading, because the chances are very great that the state of the world that does materialize will be one which was not perceived in advance. Second, both the optimal strategy and the strategy finally chosen are likely to be different from those which were arrayed in the payoff matrix. In assessing decision theory, it is important that we keep in mind the distinction that exists between *risk* and *uncertainty*. For risk, anticipation is possible and appropriate calculations, even if subjective, can be made. By contrast, how uncertainty will be resolved is impossible to foresee, and its existence will partially destroy the relevance of all advance calculations.[7]

Disparate Approaches to Analysis

This matter of uncertainty raises the question of how to approach analysis. In the past, both within and without RAND, there have been two disparate points of view. A first group, whom we might call the contingency *planners*,[8] has felt some confidence in our ability to chart in advance successful policies for the unknown future. Their method has been to designate the probable states of the world and to design a system which can deal adequately with each of them. A second group, whom we might describe as *contingency* planners,[9] has tended to emphasize the uncertainties and our inability to predict the future. Those who hold this view have consequently stressed the need for sequential decisionmaking, for improvisation, for hedging, and for adaptability. The increase in uncertainties should tend to make us view with greater sympathies the approach of the latter group.

Developing and Selecting U.S. Military Capabilities

We face a wide spectrum of threats, but we cannot tell which, if any, will actually materialize. Given the cost and specialization of our own weapon systems, we cannot afford to procure all the systems necessary to deal with such a multitude of threats. We may actually buy fewer systems, each of which is designed to deal with a relatively limited threat. The effectiveness

[7] Note Albert Madansky's discussion of these questions in Chapter 5.

[8] That is, those who believe that the array and character of future contingencies can be specified in advance, and that *detailed advance planning can be done to deal with whichever one does occur*.

[9] That is, those who believe that future developments will have a large element of the unforeseen, that contingencies cannot be specified precisely in advance, and that *whatever planning one does must be done so that it may be adapted to the contingent and the unforeseen*.

of any given U.S. system depends on what is in the Soviet force structure. Many major force-structure decisions will have to be delayed until we have clear evidence of the direction in which the Soviets will proceed. On the other hand, although specialized weapon systems deal with narrow threats, we must protect ourselves against a broad range of threats. The chief way to do this is through a wide-ranging, austerely-conducted R&D program, in which it is fully recognized that many successful developments will not lead to procurement.

Although the chief way of building flexibility into the future force structure should be an R&D program which provides a rich menu, we ought not to neglect the possibility of building flexibility into individual systems. In a continuously changing strategic environment, it must be kept clearly in mind that the choice of a weapon system does not simply optimize – it also constrains the choice of future strategy. The adaptation of strategy should not be unduly limited by the selection of weapon systems through the choice of criteria which inherently reflect a single set of strategic conditions.[10]

The Problem of Cherished Beliefs

A determination to stress adaptability and the avoidance of premature commitments in the future implies that we must be on our guard against the cherished beliefs that are carried forward from previous conditions and previous battles. As E. S. Quade implied in Chapter 19, this admonition no doubt represents a counsel of perfection, one that sounds naive when extended toward frail human nature as it must perform in a bureaucratic environment. We may be certain, however, that counsels of perfection are necessary in providing warning flags for the kind of error into which we fall through seduction rather than through bungling.

This matter can be illustrated by referring once again to decision theory and to one of its basic tenets. In principle, our choice of force structure and strategy should be dependent upon, and subsequent to, our estimates of the probabilities pertaining to various states of the world. In practice, however, this is rarely the case. Partly, this may be ascribed to the long lead-time associated with the purchase and deployment of weapon systems – which implies that our strategic choices must be made well in advance of any hard knowledge about the state of the world. Much more important, however, in imposing obstacles to logical choices are bureaucratic pressures and the fact that in most human beings there is a proclivity to decide a question more or less on the basis of intuition, and then to adjust one's assessment to the state of the world accordingly. Rather than the

[10] In this regard, note particularly the comments of L. D. Attaway (Chapter 4), H. Rosenzweig (Chapter 6), and Roger Levien (Chapter 15).

state of the world determining one's strategy, as in the model, the assessment of the state of nature is not even arrived at independently, but all too frequently is merely a reflection of strategic choice.

This is how analysis is distorted by – or, more properly, is made to reflect – preconceptions. If we are to achieve true adaptability and suppress cherished but obsolete beliefs, we shall all have to try, both individually and organizationally, to make our assumptions explicit.

Complementarities and Mission Trade-offs

Recognition of complementarities among major missions means that the problems of over-all system design have become increasingly intricate and that more attention must be paid to them. More emphasis must be placed on force integration; less attention can be concentrated on the individual weapon system. As a result, the opportunity for traditional systems analysis – in the sense of analysis to assist a simple choice between several given systems for accomplishing a single objective – has diminished. This shift implies that in analytical work, choice becomes more dependent upon parameters which are only implicitly or subjectively quantifiable – and which may even be unknowable. Under these circumstances, it is probably preferable to have an acknowledged imprecision in systems studies rather than a spurious precision.

Quantitative Precision

Finally, we might observe that the imprecision in analytical results, at least in the quantitative sense, stems not only from the growing role of mission trade-off studies which rest on parameters at best only subjectively quantifiable, but also from the passing sharpness of the distinction between capabilities for central war and limited confrontations, from the absence of dominant threats, and from the general growth of uncertainty. But this may imply that in military systems analysis we are passing through a great transition. Previously, there may have been an overemphasis on intuition, but now that the battle for the recognition of quantitative studies has been won, the current problem may be an overemphasis on those objects of analysis that can be readily quantified. The stress on the quantitative is not obsolete, but now that its importance has been recognized, we should be increasingly aware that it does not represent the whole story. We should be more inclined, perhaps, to recognize the element of art in systems analysis – and to stress what the best practitioners have always known: that judgment and intuition (in handling quantitative considerations, to be sure) are the critical inputs.

Chapter 21

THE TRADE-OFF STUDY REVISITED

L. H. WEGNER and M. G. WEINER

This Chapter attempts to illustrate certain major conceptual and methodological problems of systems analysis by reconsidering the example of trade-offs between ground and air forces presented in Chapter 2. One object of this discussion is to suggest how better results might be obtained. The fundamental aim, however, is to illustrate that a sound evaluation of any analysis depends on an understanding of its structure and methodology, as well as the assumptions and data used.

Introduction

The highly simplified trade-off study presented in Chapter 2 purported to show that of two alternative mixes of groundpower and airpower used in a non-nuclear limited war, the second was capable of producing essentially the same result in combat as the other, but at a saving of nearly $8 billion. Although this rather dramatic, if not downright suspicious, result flowed directly from the premises spelled out in the course of the example, the Chapter concluded by pointing out that the key to appraising that analysis – or any other – is as likely to be found by examining its basic structure to see what is included or excluded as it is by examining what assumptions are made.

It is true, of course, that a great deal might be learned by focusing on the assumptions alone. Indeed, it can be shown that, given even the same criterion, ground rules, and cost data used in Chapter 2, but calculating the effectiveness of the airpower in a slightly different though equally reasonable way, and then crediting the friendly ground forces with improved effectiveness, we would find that the date at which enemy and friendly ground forces are at the same strength comes later, that the composition of the equal-effectiveness mixes changes, and that the difference in cost between *these* alternatives drops to $.8 billion. Changes in a very few numbers thus change the conclusion of the original analysis by a factor of almost 10, and lead to results that can hardly be as significant in making a force-structure decision.

But this is only one approach.

Revisiting the trade-off analysis presented in Chapter 2 offers the opportunity to discuss, as part of the conclusion of this book, the alternative of approaching an analysis not through its assumptions or results, but through its structure. Moreover, it permits us a chance to illustrate, in some detail, what a systems analysis is likely to involve. Of course, earlier Chapters have repeatedly made the point, and rightly, that because systems analysis is still at least as much an art as it is a science, no single example can adequately reveal its full character – its strengths and its weaknesses. Perhaps the best we can do is to examine a problem that is broad in scope, rich in complexities (some inexpressible in quantitative terms), directly relevant to major questions of national security, and thus of interest to decisionmakers on the highest level. Even though our discussion of the trade-off of airpower and groundpower in Chapter 2, and in what follows, is necessarily wrapped in hypothesis and uncertainty, it surely satisfies these conditions.

Enough has been said in these paragraphs to suggest that, in fact, something is seriously amiss in the analysis in Chapter 2. And enough has probably been said in the last nineteen Chapters to show most readers what it is – or how to find it. Our intent here, therefore, is simply to begin again. Directly or indirectly, the flaws in Chapter 2 should thereby present themselves.

ESTABLISHING A POINT OF VIEW

Basically, then, we want to do two things in this Chapter: Define an approach to making a trade-off analysis, and then illustrate the application of this approach. We can do both by examining, as in Chapter 2, the question of comparing the military effectiveness of different mixes of tactical air and land combat forces in tactical non-nuclear operations. Our interest, of course, is solely in how one force mix might be compared with another, and not in whether any of the mixes we will consider will actually exist at the time our hypothetical conflict takes place.

The analyst's first obligation, if not his first task, is to recognize that several approaches to a trade-off analysis may be possible, each of which can be supported rationally. He knows, of course, that none of them can be free of uncertainties or the difficulties these uncertainties imply. There are, for example, the uncertainties introduced by the sheer diversity and complexity of tactical operations, which encompass political, technological and economic as well as military considerations. Moreover, the variety of possible conflict situations that may develop; questions about the size, disposition, and effectiveness of enemy forces; questions about the size and nature of the forces of our allies; the very fact that neither national policies nor the capabilities of military forces remain static in or out of battle – all of these and other factors combine to the analyst's disadvantage.

In view of these considerations, the analyst has to make decisions about the scope of the analysis. To compare the military effectiveness of different mixes of air and land combat forces in a hypothetical campaign some years hence, we need to decide, first, on a level of analysis. Should we limit our attention to, say, the ability of these mixes to attack specific targets? Should we take a more comprehensive point of view? What criteria apply on each level? What measures of effectiveness? Second, having decided on a level of analysis, we have to specify the force mixes to be compared. What forces will be available and should be considered? How can we estimate the cost of these forces? Third, we should specify the nature of the threat. What contingency might call these forces into play? Is there a range of possible threats? Fourth, we want to define the model we intend to use to study the possible forces and threats on the various levels. What are its restrictions? What are its capabilities?

Now, the basic purpose of comparing the combat capabilities of combat forces is to determine the effectiveness of different combinations or mixes of forces in implementing national policy. But the extent to which forces can be mixed is limited by the necessity for balanced forces, capable of appropriate responses in not one crisis, but across a spectrum of different military situations. A complete substitution of the forces of one service for those of another is unreasonable: it would contradict the history of warfare, which demonstrates an increasing interdependence among the services, and it would overturn the present posture, organization, and employment of general purpose forces, which for many situations are crucial. Thus, between the present force posture on the one hand and the requirements for balanced forces on the other lies the area in which trade-offs may be considered. And within this very broad area, trade-offs can indeed be considered on a number of different levels. Let us examine three of them, and something of the basic methods of analysis that might be used on each.

Levels of analysis and criteria

Level 1. Trade-offs Between Different Forces to Accomplish the Same Specific Task

Analysis on this level – the "task" level – would aim at assessing the effectiveness of various combinations of air-delivered and ground-delivered weapons in achieving a variety of particular missions, as, for example, destroying an enemy artillery emplacement or denying the enemy a hill position. At its simplest, the analysis might proceed by first defining a series of such tasks, from which a list of targets involved in accomplishing the task could be identified. A matrix could then be drawn, as in Fig. 21.1, which would relate mixes of weapons to targets, according to a measure like the number of rounds or bombs required to achieve a specified level of damage. Data on the cost of producing this damage could then be developed. In turn, the cost data and the results appearing in the matrix could be used to provide either an "equal-effectiveness, different-cost" comparison or a "different-effectiveness, equal-cost" comparison, against which the various mixes could be weighed.

Task 1	Mix of Forces				
	1	2	3	4	. . . n
Target 1					
T - 2					
T - 3					
. . . T - n					

Fig. 21.1 – Basic weapon-target matrix

One way in which we might make such an analysis more valid would be by introducing a "distance measure." A list of targets likely to be found at various distances from the forward edge of the battle area (hereafter, FEBA) could be constructed. The addition of distance would introduce different types of fire missions, such as close fire support for targets near the FEBA, close support for those farther away, interdiction for those still farther away, and so on, as indicated by the matrix of Fig. 21.2. The same procedure as in the simpler analysis could be carried out, and the relative cost and effectiveness of different mixes of air- and ground-delivered weapons could be established. But the addition of a distance measure would provide several new insights. Of these, the most useful would be the suggestion of a scale that would indicate at one end the unique capabilities of ground-delivered weapons; at the other end, those of air-delivered weapons; and, between these points, various mixes of the two. For example, safety considerations for ground forces could define the minimum distance from the FEBA at which air-delivered weapons are acceptable. Similarly, artillery ranges could define the maximum distance from the FEBA at which artillery can be used. Between the two

Mission / Target	Mix of Forces			
	1	2	3	. . . n
Close fire support (0 – 3 km)				
Target 1				
T – 2				
T – 3				

Close support (3 – 12 km)				
T – 1				
T – 2				
T – 3				

Interdiction (12 – 75 km)				
T – 1				
T – 2				
T – 3				

Deep interdiction (over 75 km)				
T – 1				
T – 2				
T – 3				

Fig. 21.2 – Weapon-target matrix with a distance measure added

extremes, cost and other considerations might be used to define the more efficient forms of delivering ordnance.[1]

Analysis at the task level can be carried at least one additional step by adding a time dimension. A series of target lists could be drawn up, with each list representing a target structure at a different point in time. Thus, a hypothetical development of the conflict could be depicted by a changing target structure. In this approach, the changing target structure could be derived from previously played war games or exercises.

In sum, methods *can* be created for analyzing possible trade-offs between mixes of tactical air and land combat forces in terms of their relative capabilities to accomplish specific tasks. But an analysis on this level would have several serious drawbacks. For one thing, the utility of defining "specific" tasks and analyzing the relative effectiveness of different mixes would be limited, unless such "situational" factors as terrain, tactics, or intelligence were introduced. But even if they were introduced, task analysis would still be limited because the effect on the total conflict situation of accomplishing these tasks would still have been left out of account. Thus, by themselves, the capabilities of force mixes for achieving specific missions could not be used in assessing the total utility of the forces.

Moreover, it is probably not reasonable to construct a detailed two-sided game situation for use only in a task analysis. Note that an analysis on this level involves essentially only one of the criteria of trade-offs, the potential of different force mixes to destroy enemy targets. But to apply even this one criterion would involve great effort, since an adequate evaluation of the effectiveness of different types of weapons against different targets can be made only when many characteristics of the weapons and targets are incorporated in the analysis. Since this information would also be a significant part of more comprehensive analyses, the construction and play of a game on the task level might better be deferred in their favor.

Level 2. Trade-offs Between Different Forces in the Same Situation

The step from Level 1 to Level 2 – the "situational" level – brings into the analysis the setting or context in which the military operations are conducted. This incorporates the objective of the military operation, the purpose for which the forces have been committed to combat, and some consideration of specific policy, economic, and strategic goals, as well as a

[1] Obviously, not too much should be made of this scale, since it would underrate the flexibilities that have been built into military forces. Land combat forces might, in theory and in practice, attack enemy airfields beyond artillery range by the use of air-delivered assault units. Air forces might stop the movement of enemy ground forces by interdiction of the supplies, reserves, and lines of communication of the enemy.

range of purely military goals. On the task level, the connection between national policy and the destruction of an enemy tank may safely be left out of account. On the situational level, where we combine many such tasks and examine their influence on the course of the conflict, this relationship is more important.

Consequently, the analysis of trade-offs on the situational level requires an approach different from that used on the task level. For one thing, it necessitates the use of definite (albeit hypothetical) conflict situations with specific military objectives, since military situations are never independent of military objectives. Thus, a scenario for a specific military situation and its accompanying objective has to be defined. The situation will usually incorporate joint and combined operations of different services, and therefore provide a framework within which the critical interdependence or balance of military forces can be analyzed in some depth. Moreover, the use of specific situations requires the inclusion of a host of interacting factors necessary to trade-off choices – geography, time, enemy actions, attrition, logistics support, and so on. Including such factors will make the analysis more comprehensive. On the other hand, it will also tend to decrease the amount of certainty that can be attached to the conclusions.

There is another difference. On the task level, it is possible to be somewhat confident that the results, although limited, will have a reasonable validity for a usefully long time. After all, such specific military tasks as attacking artillery positions will be part of most foreseeable military operations. On the situational level, the ability to define a military situation that may arise in the future, and the manner in which we will respond to it, involves a good deal of judgment. How might the situation develop? What combat forces would be employed? How would they be employed? How would policy considerations, nuclear options, and other factors influence the conflict? As we have seen, these are just a few of the questions that can appear on this level of analysis. Clearly, therefore, such analyses are infeasible within any reasonable limits of time and effort unless judgment is used to restrict the possibilities that might characterize the situation. In short, the analyst's critical conceptual problem in conducting trade-off analyses on this level is to define an appropriate situation and to identify and define within the situation those factors that contribute to a useful comparison of force mixes.

Figure 21.3 presents a list of basic criteria that could be used at the situational level of trade-off analysis. All of them appear to be directly relevant to the comparison of different mixes of tactical air and land combat forces. We have already discussed how data might be developed so that the first criterion, destructive potential, would be appropriate to the analy-

Level		CRITERIA		Mix of Forces 1	2	3	. . n
Level 1 / Level 2	I	Destructive Potential	How well can the force mix destroy targets?				
Level 2	II	Responsiveness	How rapidly can the force mix be ready for military actions?				
Level 2	III	Deployability	How rapidly can the force mix move to the theater?				
Level 2	IV	Mobility	How rapidly can the force mix move in the theater?				
Level 2	V	Supportability	How effectively can the force mix be supported and maintained?				
Level 2	VI	Survivability	How vulnerable is the force mix to enemy actions?				
Level 2	VII	Flexibility	How many different postures or capabilities can the force mix employ?				
Level 2	VIII	Controllability	How responsive is the force mix to command requirements?				
Level 2	IX	Complementarity	How well does the force mix complement the forces of our allies?				

Fig. 21.3 – Trade-off criteria on Levels 1 and 2

sis. Let us consider briefly what might be done to provide measures of effectiveness for each of the others.

II. Responsiveness. To estimate how rapidly the force mix can be ready for military action, the analyst needs to determine the status of air and ground forces on both sides at some time before the hypothetical conflict is assumed to occur. With that information, he could then develop the requirements – in time, dollars, manpower, equipment, and so on – necessary to bring the various force mixes fully into action. These requirements would thus indicate the "cost" of achieving a specified level of readiness and provide one possible measure of responsiveness.

III. Deployability. The simplest measure for this criterion is the time required to move the ready forces of each of the force mixes from their ZI and overseas positions to the theater of operations. Estimates could be made for different conditions of available airlift, base posture, sealift, prepositioning, and so on.

IV. Mobility. The analyst can represent movement in the theater in at least two ways: movement from the peacetime posture to a military posture appropriate to the conflict situation, and movement of forces during combat operations. The latter can be viewed in terms of the time required to bring destructive potential to bear on the enemy at various times during the course of the conflict. Differences between mixes of tactical air and land combat forces in destructive potential and the rapidity with which it can be brought to bear should be balanced against the ability to maintain this destructive potential over time. Thus, in its usual meaning, "mobility" could include both movement and non-movement or "stayability" – that is, the ability to maintain destructive potential over time.

V. Supportability. The criterion of supportability involves several considerations. These include the ability to *support* the forces in the theater from the ZI or stocks, and the ability to *maintain* the forces within the theater. The former involves the amounts of materiel, personnel, and carriers needed to conduct combat operations, as well as the time required to provide them. The latter involves the ability to maintain and service the forces in the theater and requires the analyst to estimate the numbers and types of personnel, skills, equipment, and other requirements necessary to repair and service aircraft, to replace land combat personnel and equipment losses, and so on.

VI. Survivability. The capabilities to support the forces and to maintain their mobility is dependent on the losses and damage which the forces suffer. Thus, the survivability of the force mixes is an important test of their combat capability. Survivability can be represented in terms of attrition – that is, direct combat losses in such categories as personnel, aircraft, and equipment. It should also include losses of support equipment and limitations in movement due to enemy actions against depots, lines of communication, and support vehicles.

VII. Flexibility. The conflict situation should provide an opportunity to examine the ability of different force mixes to modify their combat capabilities as the situation demands. Can the air component operate from different basing postures? Can the ground component operate with different lines of communication? Can the forces develop different combat organizations or procedures to meet specific circumstances? These and other characteristics of force flexibility – such as the ability to move to nuclear operations in both offense and defense – are among the more difficult standards to measure, since they are highly dependent on the particular conflict situation. Nevertheless, in the comparison of different air-ground mixes they can play an important role.

VIII. Controllability. Related to the flexibility of the force mixes is their controllability. This can be represented by the timeliness with which the force mix can respond to such command requirements as to change the type or location of the combat operation, or make the transition to nuclear weapons. Controllability can be distinguished arbitrarily from flexibility in terms of the time required to respond to command requirements.

IX. Complementarity. In any conflict situation, the size and nature of a nation's commitment will be determined in part by the capabilities that its allies possess to meet aggression. Since these commitments are intended to supplement each other, complementarity can be represented in the analysis of different force mixes by the extent to which the different mixes possess the types of forces that round out those of the allies.

From this very rough description of criteria, it should be apparent that the model used to examine force trade-offs would be most useful if it represented combat operations that change over time; geography, bases, and supply routes; attrition to air and ground forces; logistics and supply of the combat forces; weather conditions; force deployments; the size, composition, and employment of allied forces; command and control; different contingencies, both military and political, that might characterize the conflict situation; and, of course, the alternative force mixes themselves. It is particularly important that the model provide the opportunity to examine variations in contingencies within the same military situation – this can be accomplished with different scenarios – and variations in the manner in which the forces are employed to achieve the same objective. In this way, no force mix would be penalized by being considered in too narrow a framework. Later we shall describe the model that will be used in our illustration.

To sum up the situational level, we should consider these points. Trade-offs between different force mixes can be examined at this level, and will include criteria that cannot be included at the task level. Such analyses will involve greater complexity, judgment, and uncertainty, but will also permit broader comparisons of the effectiveness of different force mixes. The addition of different contingencies within the situation and different force employment policies appropriate to the specific force mixes will provide a more comprehensive basis for choosing between alternative force mixes. The results of a situational analysis, however, can be considered appropriate only to the situation analyzed. If there are major differences in the capabilities of the tactical air and land combat forces between the force mixes, the comparisons should be extended to other situations. The approach of Level 3 includes this aspect of trade-off analyses.

Level 3. Trade-offs Between Different Forces to Implement National Policy

National policy is a dynamic process that must consider a variety of actual and possible military situations and contingencies. The tactical force posture, therefore, cannot be exclusively determined by the ability to respond to any one threat, even if one predominates. To repeat something said earlier: The requirement that tactical military forces be capable of employment in a variety of different military situations necessitates that the different mixes of tactical air and land combat forces be compared *in* those situations. But on Level 3 – the "policy" level of trade-off analyses – judgment becomes central. What situations, with what priorities and what weighting of importance, should be examined?

Although the main emphasis in establishing and maintaining a military posture rests on selecting the best posture commensurate with military requirements to respond effectively to a spectrum of military conflicts, related considerations of national policy also influence the choice among alternative force mixes. These considerations include a host of domestic and international issues, such as the relation of the tactical forces to the strategic forces, the mobilization capacity required to augment the combat forces, the impact of the forces on the gold flow, and the national and international political responses the military forces might inspire. These issues are frequently difficult to define precisely, and, in many cases, impossible to measure numerically. Nonetheless, their role may be crucial in force-posture choices.

Figure 21.4 extends the list of criteria presented earlier by adding some that are pertinent to the evaluation of force mixes on the policy level. Of these criteria, some are directly related to the multi-situational capability necessary for military purposes, and some to the broader issues of national policy. Let us look at each.

X. Versatility. In part, the policy level can be considered as defining a series of varied situations in different areas of the world with concomitant differences in geography, weather, force size, logistics capabilities, and so on. Versatility – that is, the range of different military or politico-military situations in which the force mixes can be used efficiently – thus becomes the major criterion of trade-off analysis on this level. For the analyst, this multi-situational criterion would involve comparing the "utility" of the force mixes in one situation with their utility in others. The measure or measures of utility would include all of the measures used at the situational level. In contrast to the following four criteria, versatility is thus likely to lend itself to quantitative estimates.[2]

XI. Deterrent Capability. To estimate the extent to which each force mix contributes to the deterrence of aggressive action by the enemy at different levels is a complex problem. It involves many aspects, such as enemy "risk calculations," and the magnitude and timeliness of response. As such, deterrent capability is one of the criteria which depends heavily on intelligence information, political appraisals, and other considerations with large components of judgment.

XII. Expandability. In attempting to determine the extent to which each force mix permits additional mobilization of forces that contribute to its effectiveness – and this is what we mean by "expandability" – the analyst's task is to test such maxims

[2] Of course, even on the situational level, not every factor can be quantified. Some, like leadership or morale, cannot be measured objectively, although in many cases rankings or orderings can be made. In any case, however, a limited or nonquantitative estimate seems preferable to excluding the criteria completely.

	CRITERIA		Mix of Forces			
			1	2	3	. . n
I	Destructive Potential	How well can the force mix destroy targets?				
II	Responsiveness	How rapidly can the force mix be ready for military action?				
III	Deployability	How rapidly can the force mix move to different theaters?				
IV	Mobility	How rapidly can the force mix move in the theaters?				
V	Supportability	How effectively can the force mix be supported and maintained?				
VI	Survivability	How vulnerable is the force mix to enemy actions?				
VII	Flexibility	How many different postures or capabilities can the force mix employ?				
VIII	Controllability	How responsive is the force mix to command requirements?				
IX	Complementarity	How well does the force mix complement the forces of our allies?				
X	Versatility	How effective is the force mix in a variety of military and politico-military situations and crises?				
XI	Deterrent Capability	How much does the force mix contribute to our ability to deter aggression?				
XII	Expandability	How fast can additional capability be mobilized for the force mix?				
XIII	National Acceptability	How readily will the force mix be accepted domestically?				
XIV	International Acceptability	How readily will the force mix be accepted by other nations?				

Fig. 21.4 – Trade-off criteria on all three levels

as this: force mixes involving skills that require an elaborate training base cannot be expanded as rapidly as those with less stringent requirements. The analyst can, for example, assess the cost of having standby production capability available in each force mix.

XIII. National Acceptability. This criterion expresses the extent to which there are differences between the mixes in relation to the totality of considerations of the nation's policy, economics, technology, production, manpower, and so on. Among the questions the analyst would want to investigate are the impact of each mix on the gold flow problem, on the existing crises, and on national attitudes toward military expenditures.

XIV. International Acceptability. Here the analyst is concerned to estimate the differences, if any, between the mixes insofar as the various attitudes and postures of the nation's allies, potential enemies, and non-aligned nations are concerned. Although these considerations are certainly not ultimate determinants of military posture, they contribute to its form and nature. The contribution may be direct (for example, through control of the availability of bases or lines of communication) or indirect (for example, through policy reactions).

How significant these policy-level criteria are, and the extent to which they should be included in any trade-off analysis, are open questions. Without doubt, methods for using them are limited, uncertainty is great, and judgment is crucial. But it is important that the issues they raise be recognized in the creation of any major trade-off analysis. Whether or not they should or can be incorporated in the analysis itself depends, in part, on how comprehensive an analysis is undertaken. In the illustration that concludes this chapter, we have purposely restricted our analysis to the situational level in order to avoid some of these difficulties.

An illustration

So far, we have said nothing about cost calculations, the model, or the range of contingencies. These matters we will take up in the context of the illustration, highlighting some of the practical problems involved and letting the earlier discussions of theory[3] stand without additional comment. Since the primary purpose of this example is to indicate that analytic tools for comparing different force mixes can, in fact, be developed, it should be perfectly clear from the start that what follows is intended solely *as* an example. For this reason, what it does or does not accomplish is much less significant than how it goes about accomplishing it. The model, the force mixes, and the results could be different in point of fact (and indeed would be, if the example were not hypothetical), but the methodology would remain basically the same.

The Force Mixes and Their Cost

Three force mixes will be considered. Mix I is a hypothetical force presumed to exist at the time; it consists of 24 wings of tactical air forces and 16

[3] Particularly in Chapters 3, 7, 10, 14, and 16.

divisions of ground forces. Using the cost of this mix as a base, we can define two other equal-cost alternatives:

> Mix II, which is Mix I less 2 divisions and plus 8 wings of relatively low performance, inexpensive aircraft. (The cost of adding the 8 wings is used to determine the number of divisions to be subtracted.)
>
> Mix III, which is Mix I plus 4 divisions and less 4 wings of jet aircraft. (These 4 wings are taken arbitrarily as the number to be subtracted; the money thus freed is used to purchase the additional divisions.)

Because our interest is in outlining a method of analysis, the actual cost calculations required in order to derive these mixes need not be entered into here. But it might be useful to say something about their complexity. Since the basic problem is to estimate the resource changes due to aircraft additions and deletions, translate these changes into costs, and then translate these costs into ground divisions, the first step is to estimate the cost of the first eight wings of aircraft (Mix II) and the last four wings of jet aircraft (Mix III). For the added aircraft, we would want to discover the costs of their RDT&E, initial investment, and annual operation (for either 5 or 10 years). A good description of how this is done is given in Chapter 9. For the four wings that are subtracted, only the costs of initial investment and annual operation need be considered.

Cost-sensitivity analysis is appropriate at several points in these calculations of aircraft costs, but if we limit our attention to just one – base operating support (BOS) costs – the difficulties may be clear. Base operating costs are of two types: constant costs, which are associated with the base itself, and variable costs, which depend on the level of base activity. In Mix II, the relatively inexpensive aircraft system costs turn out to be sensitive to the inclusion of the constant value, which reflects the assumption that new bases would be built for these aircraft. But would it be necessary to build new airbases? The analyst must come to some conclusion. Here we assume that they would not. Similarly, Mix III subtracts four wings from the air forces. To include the BOS constant value in the calculations would be to assume that bases would be closed. If this were done, the base investment costs would be unrecoverable, and only the annual operating costs would be included in the total system costs. It follows that Mix III costs are insensitive to the inclusion of the BOS constant value. Should this value then be excluded from the aircraft system costs in computing the number of divisions? Here we assume that it should.

Estimating division costs is at least as complicated. The major problem is simply to achieve a consistent cost analysis for the air systems and the divisions, and to include only direct costs and those indirect costs that change measurably with variations in force size. To achieve this consistency in deriving the size of the divisions to be added or subtracted in the present example, it becomes necessary to adjust certain cost categories and add new ones. This done, the next problem is to recompute the investment and operating costs for individual divisions under a variety of costing assumptions, so as to develop the costs of complete divisions of the proper size, type, and number. With this information, we are then in a position to see what forces we can buy in Mixes II and III.

Again, cost-sensitivity analysis is indispensable, even if, as in the tests made for this example, the division costs turn out to be insensitive to the variables examined. For example, one variable is initial training of the division personnel. If we compute this cost at both 100 per cent and 50 per cent of the cost of *full* training, in order to determine how sensitive the system costs are to the inheritance value of trained personnel, we find that variations in initial training costs are not significant. The explanation is that, while training represents a moderate portion of investment costs, its impact as far as total costs are concerned is diminished because the operating costs are much larger. This is true whether five or ten years of annual operation are assumed. The point to bear in mind, however, is that it is as necessary to discover such insensitivities as it is to discover the sensitivities.

The Threat

Having specified the force mixes we intend to compare, a next step is to define the situations they are assumed to face. In this example, we will analyze four hypothetical cases – by no means an exhaustive list of possibilities. These four have been chosen solely because they provide some variations for evaluating our three force mixes.

Case 1: *An Intermediate Red Attack*. Here we assume a situation of high tension. The Red forces begin movements to their attack positions on D-3. Blue forces, alerted by the Red movements, adopt their forward defense positions. Forces in the ZI are alerted, and preparations for their deployment to the theater are begun. The Red plan is to attack with a small number of assault divisions and a large part of his air strength, and to commit additional divisions and aircraft if necessary. Blue forces in position at the time of attack consist of fewer divisions and aircraft than Red has available, and the total of Blue divisions and aircraft that could be committed within the first 90 days are assumed to be lower than the total Red forces committed.

Case 2: *A Limited Red Attack*. This Case also assumes a situation of tension. As in Case 1, the Red forces begin moving to their attack positions on D-3. Again, Blue forces are alerted and adopt their forward defense positions. Forces in the ZI are alerted, and preparations for deploying them to the combat theater are begun. The initial Red attack force is the same as in Case 1. Red, however, plans a more limited reinforcement: he commits fewer total divisions to the operation, and plans to introduce them at a slower rate. Blue's response is the same as Case 1. For Red and Blue, the air commitments are the same as in Case 1.

Case 3: A Major Red Attack. Case 3 is identical to Case 1, except that we assume an initial Red commitment of 60 divisions, which is greater than the initial commitment in Case 1 but the same as the size of the Red forces in the example presented in Chapter 2. The total Red commitment and the augmentation rate is the same as in Case 1. The Blue ground commitment and the Blue and Red air commitments are the same for this Case as they were in Case 1.

Case 4: An Intermediate Red Attack with Preemptive Air Strike. Case 4 and Case 1 are identical in forces committed. In this Case, however, Red initiates operations by an air strike against Blue's airfields, defenses, and aircraft prior to beginning ground operations. Red ground forces start their attack with limited close air support for the first few days.

To simplify the force mix comparisons, we can narrow the list of important influences that we might otherwise want to consider by introducing, for each of these four cases, the following additional assumptions: (1) The major Red attack occurs on one front; (2) neither side is engaged in any major conflicts elsewhere in the world at the time; and (3) strategic balance exists between the sides. The immediate effect of these assumptions (and others made earlier, such as that all the conflicts would be non-nuclear) is to clear away some problems relevant primarily on the policy level. Their deeper effect, of course, is to limit the usefulness of the analytic results, since we are now ignoring some questions a decisionmaker in the real world might well want to have answered.

The Model

The model (TAGS-II) we shall use for the force mix evaluation is a substantially modified version of the Theater Air-Ground Study (TAGS) computer model developed by The RAND Corporation in the early fifties for studies of tactical forces.[4] As modified, it is a two-sided campaign model that incorporates the following major conflict elements for both Red and Blue:

---

[4] See C. P. Siska, L. A. Giamboni, and J. R. Lind, *Analytic Formulation of a Theater Air-Ground Warfare System (1953 Techniques)*, The RAND Corporation, RM-1338-PR (DDC No. AD 86022), September 1954, and J. R. Brom, *Narrative Description of an Analytic Theater Air-Ground Warfare System*, The RAND Corporation, RM-1428-PR (DDC No. AD 86709), February 1955.

Initial aircraft inventories in the theater for each of three types: a high-payload, high-performance type; a medium-payload type; and a low-payload type

Initial land combat forces in the theater

Air augmentation forces

Ground augmentation forces

Airfields (subject to attack)

Airfields (not subject to attack, that is, in sanctuary)

Aircraft shelters

Air missions

1. Counter-airfield
2. Interdiction
3. Close support
4. Air defense
5. Counterair defense (radars, control centers, etc.)
6. Counter-SAM (surface-to-air missiles)

Allocation policy at different times during conflict for above missions

Ground missions

1. Offensive
2. Defensive
3. Holding

Theater stock levels

Consumption rates for air units in combat

Consumption rates for ground units in combat

Line of communication (LOC) capacities

Capacities required for moving ground units

SAM inventories

SAM augmentation

Antiaircraft artillery

Terrain

The model, which is shown schematically in Fig. 21.5, involves 300 parameters. Of these, approximately 200 are used for intermediate calculations. Of the remainder, 11 describe characteristics of the initial forces; 22 describe augmentation, repair, and supply; 25 describe force employ-

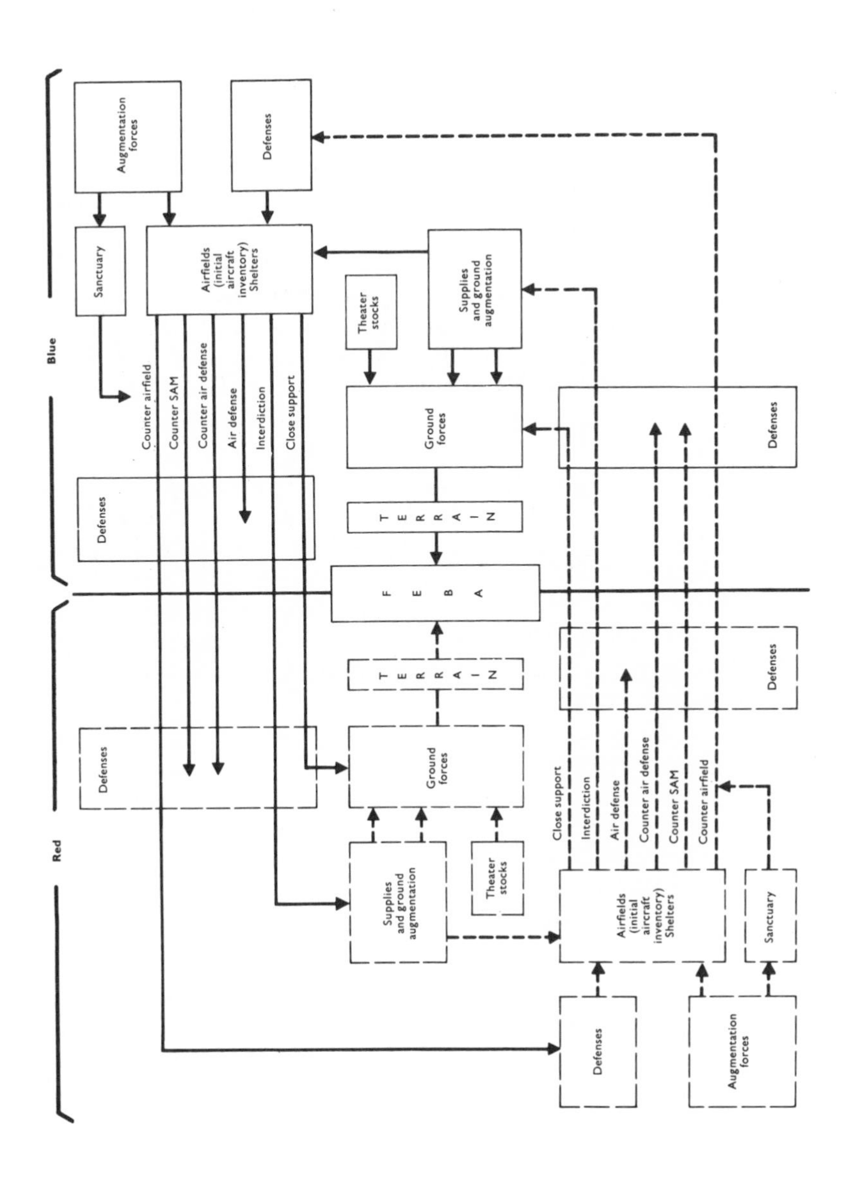

Fig. 21.5 – Structure of the TAGS-II model

ment; 17 describe offensive operations; 21 describe defensive operations; and 6 describe other ground operations. Values for all the parameters can be fixed for each computer run, can be preset to change on any War Day during the run, or can be set to change when the value of any other parameter reaches a particular point. These are important features, since they allow us to alter such things as air and ground augmentation rates to meet changed conditions.

To evaluate the performance of our three force mixes, the model incorporates specific measures for each of the nine criteria relevant on the situational level of analysis.[5] The first, *destructive potential*, is represented in various ways. In the counter-airfield mission, offensive strikes are made against aircraft parked on airfields or in shelters. Airfield defenses are attacked by defense suppression aircraft immediately before the primary strike aircraft arrive.[6] In the interdiction mission, which is carried out solely within the theater, aircraft disrupt the flow of men and materiel by cutting rail lines and destroying bridges on the main transportation routes. In the close air support mission, casualties are produced among ground combat personnel, and the movement of troops is restricted. In the counter–SAM mission, area-deployed SAMs are destroyed in a rollback operation that clears corridors for subsequent deep penetration by aircraft on other missions. In the counterair defense mission, targets such as air defense radars, command centers, and high-altitude SAMs are destroyed, thus forcing the air defense aircraft into a combat patrol mode of operation. From the ground, aircraft are destroyed by means of antiaircraft fire and SAMs. The ground combat is modeled quite simply, in the sense that casualties are calculated on the basis of planning factors derived from statistical records of World War II and the Korean war. A measure is also obtained of the rate and degree to which the actual commitment of ground divisions approaches the planned commitment.

Responsiveness and *deployability* are included in the model by assumption. That is, the ready state of the tactical air and land combat forces in each of our three mixes, and their deployability, are introduced in terms of the time required for the forces to reach the combat theater. Input values for these characteristics are derived for this example on the assumption that each side has a ZI that is outside the tactical theater environs and thus invulnerable to enemy tactical aircraft.

Mobility is represented in the model in a limited manner. Maximum rates of movement based on terrain and other factors are included and

---

[5] See pp. 396–397.

[6] Although aircraft losses through counter-airfield attacks are tallied, the effects of damaged airfield facilities on subsequent operations are not represented.

are modified in light of the combat situation.[7] Similarly, aircraft sorties for different types of aircraft and missions – close support, interdiction, air defense, counter-SAM, and so on – are also included. But differences in mobility between types of land combat divisions, alternative basing postures for aircraft, or details of the tactical deployments in the combat situation are not included. It should be noted generally that, for use in the model, all the ground forces are considered as homogeneous division slices; that is to say, no distinction is made between armored, mechanized, or infantry divisions.

Supportability of both the air and land combat forces is represented in terms of the gross supply requirements for their deployment and combat operations. Such characteristics as the size of theater stocks, the capacity of lines of communication, the daily consumption rate, and the effects of extending lines of communications are included. For the purposes of the example, we assume that aircraft replacements, ground force reinforcements, and supplies are drawn from the ZI. The theater itself contains the ground forces, the supply lines (primarily a rail network), and the tactical airfields.

Survivability is represented in several different ways – basically, as the obverse of the results obtained under *destructive potential*. Thus, losses in tactical air and land combat forces from the ground combat situation, from SAMs, from antiaircraft artillery, from air defense interceptor aircraft, and so on are included. The protection afforded by aircraft shelters and "sanctuaries" is taken into account. Losses in supplies and reductions in the capacity of lines of communication because of interdiction or because of their extension as the FEBA moves are also incorporated.

Flexibility and *controllability* of the forces in the different mixes are represented in limited detail. Different allocations of air strikes are provided for, as are changes in these allocations during the course of the conflict. Various basing postures for aircraft, various arrangements of SAM defenses, and various patterns of movement of the land combat forces – including the changes produced by arrivals of land force augmentations – are included. But command relationships, differences in flexibility and control, and rules for making a transition to nuclear operations are not included.

Examples of Output

Among the outputs of the TAGS-II model are the following:

[7] The FEBA is assumed to move as a unit. The velocity and direction of this movement are calculated as the average movement of the entire theater front, which, in turn, depends upon the ground strengths of each side and the number of aircraft sorties that strike close support targets.

Position of the FEBA in miles, plus or minus, from its original position

Number of Blue and Red divisions in combat

Number of Blue and Red aircraft, total and by type of aircraft

Number of SAMs in combat

Supply capacity available to each side

Number of divisions lost

Number of aircraft lost

Number of SAMs lost

Number of close support sorties

Number of interdiction sorties

Number of counter-airfield sorties

Number of air defense sorties

Number of counterair defense sorties

Number of counter-SAM sorties

Losses due to non-combat factors

Losses due to enemy AAA

Losses due to enemy air defense

Losses due to enemy counter-airfield attacks

Losses due to enemy airfield defenses and SAMs

The list could be extended without great difficulty. But for present purposes, let us focus attention on the results if we take our four cases and the different force mixes and use the model outputs to make explicit comparisons of only a few relevant indices.

One major indicator of the capability of a force mix in combat is the progress of the ground battle. And one over-all measure of the progress of the ground battle is the first item on the preceding list: the movement of the FEBA. Although the TAGS-II model can present this information in various ways, we will use only the following five indices:

INDEX 1: The day on which Red ground forces penetrate approximately 30 miles from their forward position.

INDEX 2: The day on which Red ground forces penetrate approximately 100 miles from their forward position.

INDEX 3: The day on which Red ground forces penetrate approximately 150 miles.

INDEX 4: The day on which Red ground forces penetrate approximately 300 miles.

INDEX 5: The day on which Red ground forces penetrate approximately 500 miles from the original position of the FEBA.

Another indicator of the ground battle is the number of divisions lost. Considering only Blue's losses, we can call out an additional three indices:

INDEX 6: The day on which Blue's total ground losses equal approximately 10 per cent of his initial strength.
INDEX 7: The day on which Blue's total ground losses equal approximately 20 per cent of his initial strength.
INDEX 8: The day on which Blue's total ground losses equal approximately 33 per cent of his initial strength.

The results of air-to-air action, antiaircraft artillery, and SAMs can be indicated by a number of values, among them these seven:

INDEX 9: The first day on which air parity is achieved. "Air parity" is arbitrarily defined here as the point at which Blue aircraft in combat are equal to the Red aircraft in combat.
INDEX 10: The first day that "local air superiority" is achieved. This is arbitrarily defined as the point at which Blue achieves a 2:1 ratio of aircraft in combat over Red aircraft in combat.
INDEX 11: The day on which "limited air superiority" is achieved – that is, the day on which the ratio of Blue to Red combat aircraft is 5:1.
INDEX 12: The day on which "air supremacy" is achieved – that is, the day on which the ratio of Blue to Red combat aircraft is 10:1.
INDEX 13: The day on which the Blue aircraft inventory is approximately two-thirds of total strength. We can also include here the ratio of Blue aircraft to Red aircraft on that day.
INDEX 14: The day on which the Blue aircraft inventory is approximately one-half of total strength. We also include the ratio of Blue aircraft to Red aircraft on that day.
INDEX 15: The day on which the Blue aircraft inventory is approximately one-third of total strength. We also include the ratio of Blue to Red aircraft on that day.

The results for these 15 index values are presented in Table 21.1.[8]

Discussion of Results

The figures in Table 21.1 indicate that, between the cases, each force mix varies somewhat in limiting the impact and rate of enemy action. Within the cases there are only small differences in the effectiveness of the three mixes, especially as the ground battle is concerned. On balance, both within and between the cases, it would appear that the "more air, less ground" mix, Mix II, is the most successful mix of the three. Why this is

---

[8] Since Table 21.1 presents the first comparison of the relative combat capabilities of the three force mixes, it may be appropriate at this point to re-emphasize that these results are not intended to demonstrate the value of any one mix, but rather of *a way of comparing* the mixes. To avoid any misinterpretation, we have presented no actual values in the Table, but have instead indicated only the magnitude and directions of change from the Case 1, Mix I, situation. (That is, the capital letters in the Case 1, Mix I column should be taken to represent a particular numerical result; the numbers shown in the other columns are to be read as the difference, plus or minus, from the appropriate base value.)

TABLE 21.1

Force mix comparisons against 15 index values (days)

INDEX NO.	DESCRIPTION OF INDEX	CASE 1 (Intermediate Red Attack)			CASE 2 (Limited Red Attack)			CASE 3 (Major Red Attack)			CASE 4 (Intermediate Red Attack, with Preemptive Air Strike)		
		Mix I	Mix II	Mix III	Mix I	Mix II	Mix III	Mix I	Mix II	Mix III	Mix I	Mix II	Mix III
1	Red ground forces penetrate 30 miles	A	0	0	+1	+1	+1	−2	−2	−2	0	0	0
2	Red ground forces penetrate 100 miles	B	0	0	+2	+3	+2	−2	−2	−2	0	0	0
3	Red ground forces penetrate 150 miles	C	0	0	+6	+6	+5	−2	−2	−2	0	0	0
4	Red ground forces penetrate 300 miles	D	+4	−1	+20	+26	+21	−7	−6	−6	−4	−3	−1
5	Red ground forces penetrate 500 miles	E	+5	+2	+26	+27	+27	−14	−12	−10	−11	−8	−9
6	10% of Blue divisions lost	F	0	0	0	0	0	0	0	0	+2	+2	+2
7	20% of Blue divisions lost	G	0	−1	0	0	−1	0	0	−1	+2	+2	+1
8	33% of Blue divisions lost	H	+1	−5	0	+1	−5	0	+1	−5	+1	+2	−4
9	Air parity (1:1)	I	−1	0	0	−1	0	0	−1	0	+5	+3	+16
10	Blue local air superiority (2:1)	J	−2	+4	0	−2	+4	0	−2	+4	+6	+3	+12
11	Blue limited air superiority (5:1)	K	−5	+7	0	−5	+7	0	−5	+7	+13	+6	+24
12	Blue air supremacy (10:1)	L	−4	+9	0	−4	+9	0	−4	+15	+14	+8	+25
13	Blue acft inventory=2/3	M	+1	0	0	+1	0	0	+1	0	—	—	—
	Force ratio, B : R=	N	+0.1	−0.1	0	+0.1	−0.1	0	+0.1	−0.1	—	—	—
14	Blue acft inventory=1/2	O	0	−1	0	0	−1	0	0	−1	−3	−3	−4
	Force ratio, B : R=	P	+0.3	−0.2	0	+0.3	−0.2	0	+0.1	−0.2	−0.6	−0.5	−0.6
15	Blue acft inventory=1/3	Q	+6	−3	0	+6	−3	0	+6	−3	−13	0	−27
	Force ratio, B : R=	R	+1.9	−1.0	0	+1.9	−1.0	0	+1.9	−1.0	−2.3	−0.6	−2.8

so can be seen, perhaps, by looking at the results in Case 2, the contingency in which Mix II's relative superiority seems most clearly indicated.

Case 2, it will be recalled, involved a limited Red attack. It assumed a relatively small initial Red land force and a slow rate of augmentation. In other particulars, including the size of Blue's ground and air forces, the opposing sides were identical to those assumed in Case 1.

Figure 21.6 illustrates the movement of the FEBA in Case 2. Each mix of Blue forces is considerably more successful in slowing the Red advance that it is in the other cases, but Mix II is slightly more effective. The explanation seems to lie in its ability to generate a large number of close air support and interdiction sorties once the enemy air threat has been substan-

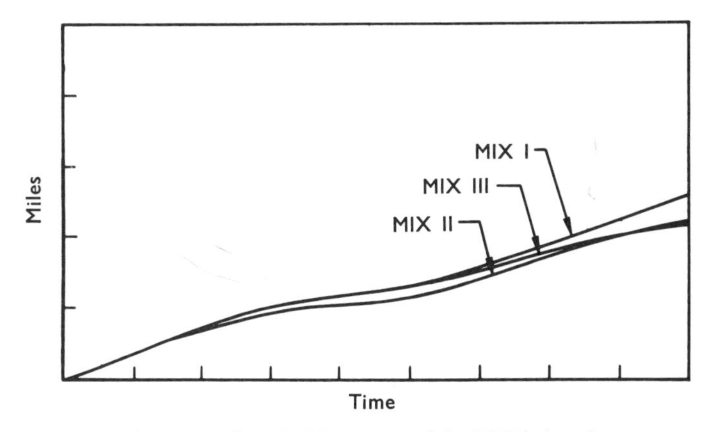

Fig. 21.6 – Case 2: Movement of the FEBA (west)

tially reduced. This is indicated in Fig. 21.7, which presents the cumulative number of sorties flown by each mix. The consequences of this capability are clear: The increased sortie rate not only helps Mix II to lower its own ground losses (Fig. 21.8), and to increase (though very slightly) the rate at which Red aircraft are destroyed in the early phases of the campaign (Fig. 21.9), but it also means that more Red divisions are defeated, or the same number are defeated sooner (Fig. 21.10). Moreover, Mix II seems to perform better than its alternatives in slowing the rate of the Red ground commitment (Fig. 21.11) and in permitting the smallest number of Red divisions to survive at the front (Fig. 21.12).

On the other hand, the increased sortie rate of Mix II entails a cost to Blue: greater exposure to enemy air, SAMs, and AAA. Thus, as indicated in Fig. 21.13, a greater number of aircraft is lost by Mix II.

From the results of this illustrative "situational" analysis, it is tempting, but inappropriate, to compare the alternative force mixes. We have already considered at length the assumptions that have gone into this

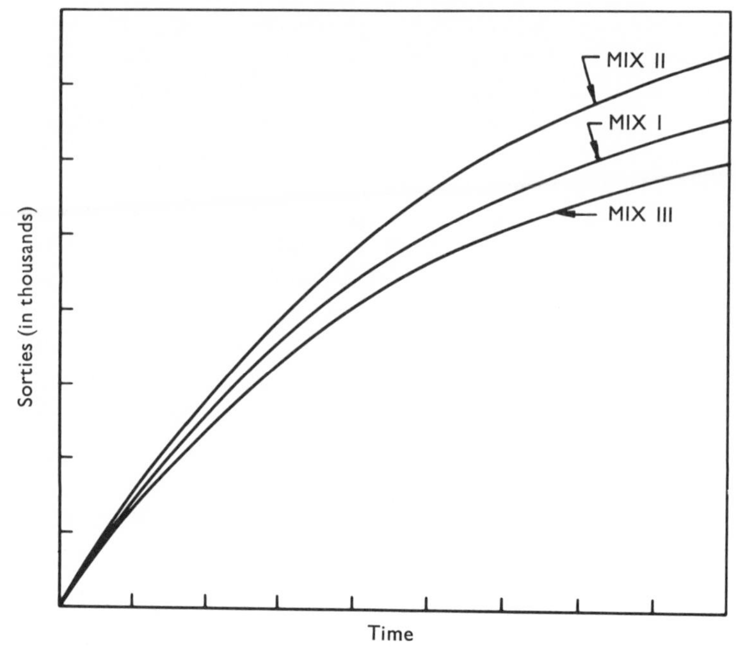

Fig. 21.7 – Case 2: Blue air sorties

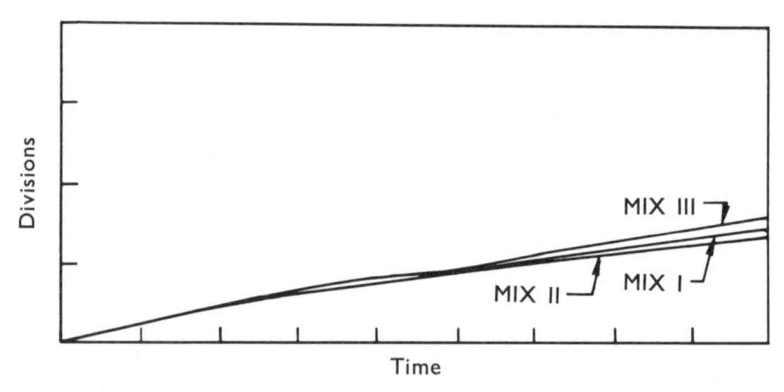

Fig. 21.8 – Case 2: Blue division losses

example. Simply because we now have a few curves and a table of indices does not mean that the assumptions or the uncertainties have magically disappeared. To draw any conclusions, we would need, first, an improved development of the model, the inputs, the costing, the details of the scenarios, the criteria, and the measures of effectiveness. We would need to examine a greater number of situations. We would need a larger variety of sensitivity tests – not merely for reducing the uncertainty in the analysis, but also for identifying the significant parameters and assumptions of the

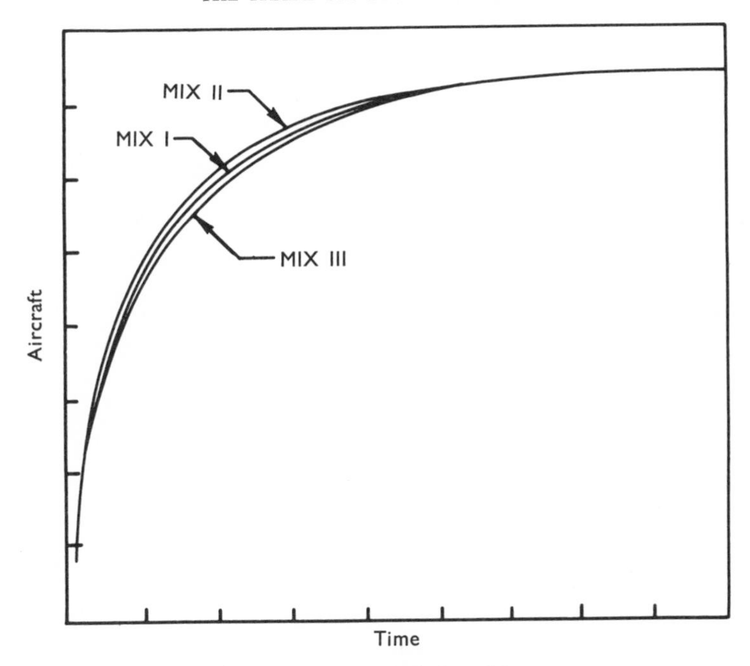

Fig. 21.9 – Case 2: Red aircraft losses

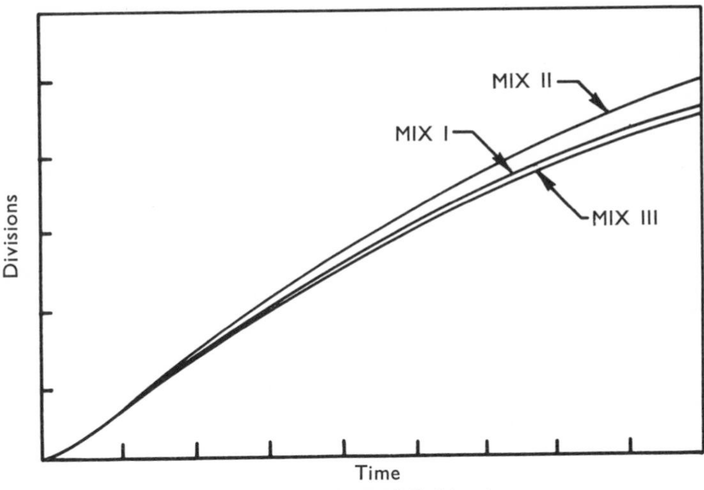

Fig. 21.10 – Case 2: Red division losses

analysis. Without these, the results shown above allow us to say little about the relative capability of our different force mixes.

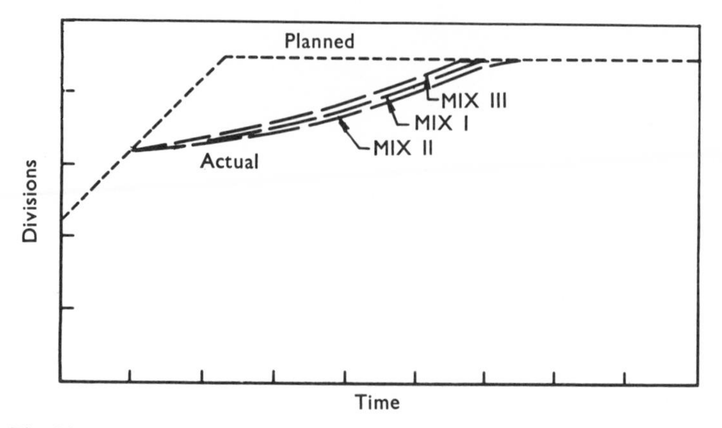

Fig. 21.11 – Case 2: Planned versus actual Red ground force commitments

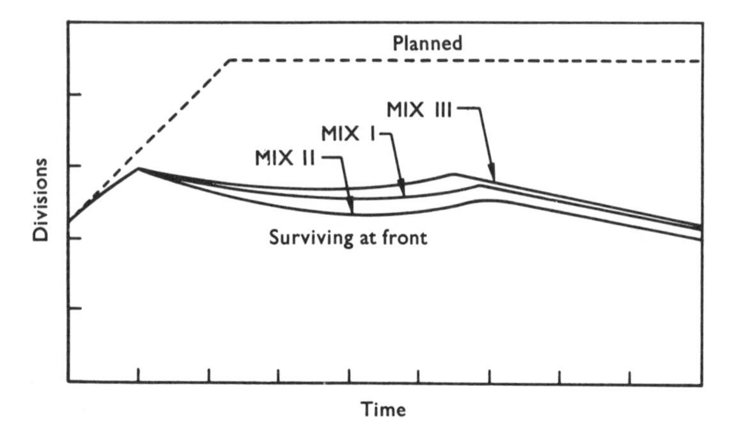

Fig. 21.12 – Case 2: Planned Red ground force commitments, and divisions surviving at the front

If, however, we stand back from the details of the example, and reconsider generally the conceptual and methodological problems of producing – through analysis – a basis for comparing alternative force mixes, several broad conclusions present themselves. We might mention five of them.

1. It is unlikely that any single criterion is adequate to compare the relative effectiveness of different force mixes. For the purposes of this discussion, we introduced nine criteria; others could be developed. Moreover, for each of the criteria, no single measure of the relative effectiveness of the mixes seems possible. Although the over-all measure of effectiveness

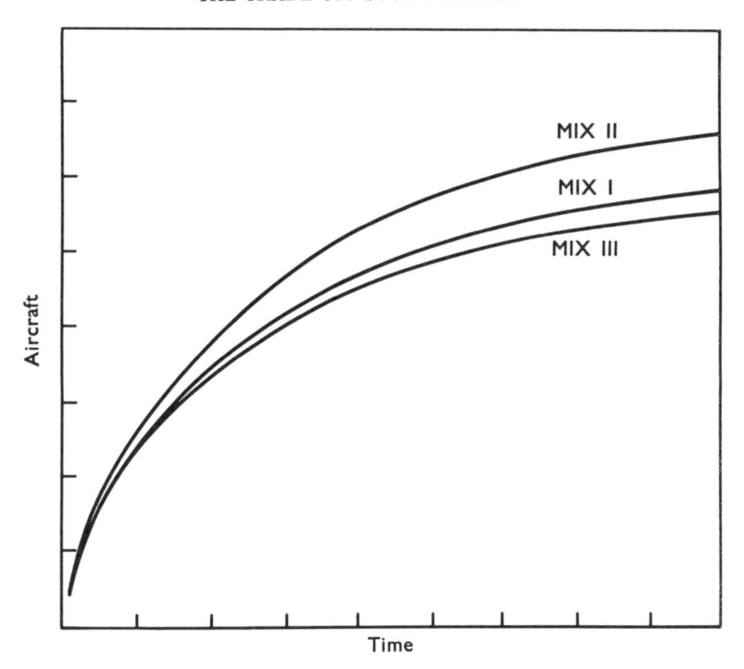

Fig. 21.13 – Case 2: Blue aircraft losses

used in the example was the movement of the FEBA, this measure showed essentially no difference between the mixes. The use of other measures, such as combat losses, rates of loss, and the time required to gain control of the air, did reveal differences between the mixes. Multiple criteria and multiple measures of effectiveness thus seem necessary in any major force trade-off study to account for the different capabilities of the forces.

2. It is likely that the utility of any force mix will depend on the specific conflict situation. In the illustration, the mixes performed differently in each of the four cases. Within each case, the differences were less pronounced, and no one mix dominated the others in all measures. For any major trade-off study, therefore, it will be necessary to define the situations that are most reasonable or credible in order to "weight" the significance of the results.[9] Defining such situations will require both analysis and judgment.

3. Within any trade-off analysis, there will be important parameters and criteria that cannot be handled quantitatively or by formal analytic

[9] For example, it might be unreasonable to use a given situation for evaluating the effectiveness of non-nuclear force mixes if it assumes that they will be met with a strategic nuclear response.

techniques. Leadership, morale, the relative controllability of the mixes – none of these was incorporated in the example. Qualitative analysis outside the formal model may be required in such cases.

4. For some trade-off problems, the first step may be to establish the level of the forces required to achieve the military objectives. Then we could vary the force mixes to determine their effectiveness. In the example, none of the mixes was capable of halting the Red advance; the most we could learn, therefore, was the relative effectiveness of the mixes in slowing the advance or in gaining time to implement other options (such as a nuclear response). To establish the utility of different mixes for obtaining a favorable military outcome *in the four conflict situations we postulated*, we would have had to make an initial "requirements" investigation.

5. The results of any trade-off study will be sensitive to the assumptions made in the cost analysis and in the effectiveness analysis. To determine which of the assumptions or parameter values have the greatest effect on the outcome, a variety of sensitivity tests will be needed. An important by-product of such testing can be to identify the factors to which the results are insensitive; these can then be omitted from further consideration.

A FINAL NOTE

No matter how much time and effort are spent in identifying and defining criteria, developing hypothetical conflict situations, generating various force mixes, or incorporating assessments of other factors important in trade-off analyses on the situational level, the resulting representation of the real world is certain to be imperfect. This sort of imperfection is by no means limited to trade-off analyses. As other authors in this book have made clear, it is true of any model used in systems analysis. Indeed, even the models used by the exact scientist, which are part of a well-confirmed body of scientific knowledge, may involve this same imperfection and have to be improved through experimentation. The systems analyst, to whom experimentation in national conflict is not available and who has no well-established theory for the phenomena he is dealing with, must construct a model as best he can. As insights accrue from working with the model (the nearest thing we have to experimentation) and more information becomes available, the existing model can be improved or replaced by a more representative model. The goal – and, in some cases, the result – is a model that is fully adequate to handle the questions we are studying.

This process of refining our models through approximation has been born of necessity. Nevertheless, as the basis of operations research, model building has met with considerable success in industry in coping with the problem of economic choice. It is much more difficult, unfortunately, to make really adequate models of military conflict. The model we have just discussed, for instance, places its stress on such characteristics as weapon effectiveness, gross firepower, and vulnerability, and tends to de-emphasize the human factor – how men are likely to perform, whether at the broad policy and strategy level or at the level of small combat actions – since it is difficult to represent this within an analytic structure. More important, perhaps, is that not only are many of the model's existing elements and their interactions imperfectly understood, but they concern a future time period, and thus introduce the serious problem of predicting new or altered elements and interactions.

Consequently, we stress the importance that should be given to the structure of the analysis, whether it be in the form of a computer model or a political assessment. Unless that structure represents the salient – and relevant – aspects of the real world as best we understand them, we cannot have great confidence in the resulting predictions. Sheer size or complexity are not guarantees that a model represents the real world in as valid a way as our knowledge permits. To the extent that any model forces us to make explicit the elements of the situation that we are considering and imposes on us the discipline of clarifying the structure we are using, thus establishing unambiguous, intersubjective communication about the problem under consideration, we progress toward greater validity. As a result, the insights and recommendations that stem from this process have a better chance of being appropriate than do those that are produced without the use of an explicit model.

Chapter 22

BY WAY OF SUMMARY

E. S. QUADE

This Chapter lists a number of precepts to guide the analysts, reviews the general principles of analysis, emphasizes certain points about which little or nothing has been said elsewhere in the book, comments briefly on the dangers of relying too heavily on any of the various types of analysis, and, finally, attempts to indicate something of the future of systems analysis.

In general, this book has dealt with the problem of what can and what should be done by analysts, engineers, and scientists – including social as well as physical scientists – to help people who must make decisions in the face of real uncertainty. It is this uncertainty that makes the problems difficult. A long-range military problem is comparable, for example, to the problems of the owner of a racing stable who wants to win a horse race to be run many years hence, on a track not yet built, between horses not yet born. To make matters worse, the possibility exists that when the race is finally run the rules may have been changed, the track length altered, and the horses replaced by greyhounds. Yet, in spite of such uncertainty, analyisis may be able to help.

This book has tried to demonstrate the necessity of abstraction and show its nature in dealing with any complex problem of the real world; to explain, in simple language, what a model of a problem is; and then to show, by examples, the usefulness which explicit models can have, despite their inability to be realistic in all details. Alternative methods of handling questions of policy – methods that do not involve explicit analysis – also necessarily involve models. But these models, because they are implicit, are more likely to be dangerously inadequate. Systems analysis, in contrast to these alternatives, provides its answers by processes that are accessible to critical examination, capable of duplication by others, and more or less readily modified as new information becomes available. As has been pointed out, however, much more is involved than the collection of information and its manipulation in mathematical models. Indeed, whatever techniques are used, asking the right questions, inventing ingenious alternatives, and skillfully interpreting the results of the computations and relating them to the many nonquantifiable factors are all part of the analytic process. These steps in the process may prove more helpful in decisionmaking than thousands of machine computations or a thorough knowledge of sophisticated mathematical techniques.

As remarked earlier, systems analysis, while it is becoming more scientific, is still to some extent an art. An art is hard to summarize. Rather than attempt a comprehensive review of the earlier chapters, we might best conclude this book simply by noting and saying a few additional words about some of the major questions that have concerned us in these pages. For convenience, we might begin by considering two lists of principles – lists that may or may not be mutually exclusive. The first (shown in Fig. 22.1), we can call "Precepts for the Systems Analyst"; the second (shown in Fig. 22.2), "Principles of Good Analysis." Should items appear to be on the wrong list, it need not matter, since, after all, they pretend to be nothing more than heuristics or sophisticated rules of thumb. Our hope is that the cases in which they are sound advice far outnumber those in which they are not.

PRECEPTS FOR THE SYSTEMS ANALYST

1. *Pay attention to problem formulation.* A large share of the effort by the leaders of a project (sometimes more than 50 per cent) must be invested in thinking about the problem, exploring its proper breadth, trying to discover the objectives of the systems or operations under consideration, and searching out good criteria for choice. It is useful to know as much as possible about the background of the problem – where it came from, why it is important, and what decision it is going to assist. Problem formulation,

1. Pay major attention to problem formulation.
2. Keep the analysis systems oriented.
3. Never exclude alternatives without analysis.
4. Set forth hypotheses early.
5. Let the question, not the phenomena alone, shape the model.
6. Emphasize the question, not the model.
7. Avoid overemphasizing mathematics and computing.
8. Analyze the enemy's strategies and tactics.
9. Treat the uncertainties explicitly.
10. Postpone detail.
11. Suboptimize with care.
12. Do what you can.

Fig. 22.1 – Precepts for the systems analyst

as we have seen, necessarily involves a great deal of judgment and intuition about the actual subject, scope, detail, precision, and assumptions of a problem, and it is, therefore, as much a part of analysis as the rest of the study.

2. *Keep the analysis systems oriented.* Emphasis should be placed on the simultaneous consideration of all the factors relevant to the decision under study. To simplify the problem by looking just at its components and deliberately neglecting their interactions is to miss the point: the effort should be to extend the system boundaries as far as necessary to determine which interdependencies are important and then to study the total system. The effectiveness, costs, and risks associated with each of the various alternatives must be compared in a context that includes everything important to all of them. This is an ideal, of course. The practical always limits the optimal, and piecemeal analysis and approximate criteria must inevitably be employed. But it is desirable at least to *think* about the problem as a whole.

3. *Never exclude alternatives arbitrarily or without analysis.* The traditional, conventional, or plausible way of carrying out a task may not be the only way. It is as important to look for ideas and new alternatives as it is to look for facts and relationships that help to explain those known initially. Unless we have ideas and alternatives, there is nothing to analyze or choose between. It is important not to exclude alternatives merely because they run contrary to past practice, or superficially seem impractical, or do not fit in with organization policy. It may be time to change the "party line."

4. *Set forth hypotheses early.* Analysis is an iterative process. Any investigation that attempts to make use of scientific methods may require many cycles or passes at the problem. In a real-world problem, the first set of assumptions and the first model are unlikely to do more than help to decide how to continue. If an assumption dominates the results, then it must be re-examined. Setting forth hypotheses and possible conclusions early in the study is essential in guiding the analysis that follows, but the analyst must stand ready to discard his early notions about the solution in the face of later evidence.

5. *Let the question, not the phenomena alone, determine the model.* For most systems and situations there are many possible representations; but the appropriate model depends on the question being asked. No one model can handle all the questions about a given activity.

6. *Emphasize the question, not the model.* In all analysis, the use of models is inevitable. Because problems of military planning are complicated, the models are frequently detailed and elaborate, requiring the investment of many man-hours to formulate, program, and apply. No matter how challenging the model is or how rich in opportunities to create new

techniques, the analyst must not allow himself to become more interested in it than in the question he set out to answer.

7. *Avoid overemphasizing mathematics and computing.* Mathematics and computing machines, while extremely useful, are limited in the aid they can give in broad policy questions. A computer can help only with problems that the analyst knows conceptually how to solve by himself. If the problem can be completely formulated mathematically, techniques such as dynamic programming, game theory, and Monte Carlo are the means of obtaining the best solution. In most analyses aimed at policymaking, however, computations with models and machines are more often valuable because they aid intuition and understanding than because of the results they provide.

8. *Analyze the enemy's strategies and tactics.* The enemy is not inert. In conflict situations, the effect of the enemy's decisions on ours, and vice versa, must be taken into account explicitly, and allowance must be made for the possibility that he may change his plans as a result of our analysis. This requires, as part of a complete systems analysis, an analysis of the enemy's systems, operations, and strategies. And, if possible, we would like this to be done from the enemy's point of view.

9. *Treat the uncertainties explicitly.* The analysis should be carried out with the uncertainties in mind. In real-world problems, uncertainties that cannot be removed by further analysis always exist. This uncertainty must be explicitly treated and the analysis reported with full recognition that the uncertainty exists. The goal is to find an alternative that will work well in the most likely contingencies and might even give some sort of a reasonably satisfactory performance in a number of others. To find precisely how much better one alternative will perform than another in a single contingency is seldom very helpful.

10. *Postpone detail.* Detailed treatment usually should come late in the study. There it can be important in discovering misconceptions and mistakes. Early in the study, it is generally a mistake to spend much time on relatively well-understood details or complicated models. For turning up ideas, a rough treatment of many models is better than a careful and detailed treatment of one.

11. *Suboptimize with care.* Suboptimization is usually necessary to avoid working with models that are excessively large or aggregative. But there are dangers in suboptimizing – for example, plausible criteria for lower-level choices can easily be inconsistent with those on higher levels. To prevent serious error, it is necessary to ponder higher-level criteria and to make certain that the criterion selected for suboptimization is not inconsistent with them.

12. *Do what you can.* Inquiry can seldom be complete. If not the difficul-

ties of the problem, limitations on time or money see to that. The decisionmakers responsible for action must get along without the additional analysis, that, given time and money and infinite wisdom, could be done. By their judgment, intuition, experience, and what they learn from others, they must take into account the considerations the analysis could not or did not have time to deal with.

Partial analysis is therefore far better than no analysis. Even if a study can produce no convincing comparison of the alternatives because, say, no satisfactory measure of effectiveness exists, the preparation of an exhaustive list of alternatives, together with an assessment of their potential feasibility, attainments, and costs, can make the effort well worthwhile. Indeed, the fact that analysis is incomplete is the main reason why one can think the analysis is very good and yet find much to criticize.

One of the most valuable means of assisting a policymaker is a demonstration of the extent to which his various options can be implemented for equal dollar expenditures. This may not tell him which alternative is superior or whether any of them are worth drawing resources from other uses, but the knowledge of the costs of his actions should help him to a better decision.

Principles of good analysis

Let us now consider some fairly general principles or considerations that we have come to associate with good analysis (see Fig. 22.2).

1. Efficient use of expert judgment is the essence of analysis.
2. Choice of the right objectives is essential.
3. Sensitivity testing is important.
4. The design of alternatives is as important as their analysis.
5. Interdisciplinary teams are usually necessary.
6. The analysis of questions of R&D should not emphasize optimization.
7. For broad questions, comparisons for a single contingency are not enough.
8. Partial answers to relevant questions are more useful than full answers to empty questions.
9. Estimates of cost are essential to a choice among alternatives.
10. The decisionmaker by his actions can compensate to an extent for partial analysis.
11. A good new idea is worth a thousand evaluations.

Fig. 22.2 – Principles of good analysis

1. *The essence of systems analysis is the efficient use of expert judgment.* Analysis does not free us from reliance on judgment. Indeed, the reason systems analysis is a successful technique is that it is able to make systematic and efficient use of the judgment that specialists or experts in the field of interest can apply to the problem. This is done by constructing a model appropriate to the situation; since the model introduces a precise structure and terminology that serves as an effective means of communication, the experts can make their judgments in a concrete context, and, through feedback, arrive at a clearer understanding of the subject matter and the problem.

2. *The choice of objectives is crucial.* It is more important to discover the right objective than it is to find a perfect optimization procedure; the choice of the wrong objective means that the wrong problem is being attempted, while an imperfect optimization merely means that only a partially satisfactory alternative may be chosen. In other words, the discovery of what ought to be done has to come first; we can then try to find out how to do it. Satisfactory objectives are not independent of the means to obtain them, but we should delay the search for some idea of the cost and difficulty of attaining them until we know what they are.

3. *Sensitivity testing is important.* Ordinarily there is no unique, "best" set of assumptions, but a variety of possibilities, each of which has some basis for support. A good systems study will include sensitivity tests on the assumptions in order to find out which ones really affect the outcome and to what extent. This enables the analyst to determine where further investigation of assumptions is needed and to call the attention of the decisionmaker to possible dangers that might be present.

4. *Systems analysis should try to create as well as to eliminate alternatives.* The invention of new alternatives can be much more valuable than the exhaustive comparison of given alternatives, none of which may be very satisfactory. The job of the systems analyst is thus not only analysis but design. His analysis should suggest new alternatives or changes in given ones that will make the system or operation that is eventually preferred more satisfactory.

5. *The type of problem systems analysis is designed to handle usually calls for an interdisciplinary team consisting of persons with a variety of knowledge and skills.* This is not merely because a complex problem is likely to involve many diverse factors which cannot be handled by a single discipline. An even more important reason is that a problem looks different to, say, an economist, an engineer, a sociologist, and a military professional, and their different approaches to the problem contribute to finding a solution.

6. *Although analysis is as applicable to problems of research and development as it is to other problems of choice, it is absurd to concentrate technical*

competence and military expertise on the problem of recommending for development the system that will be most effective in the environment expected to exist at some future time. For in spite of our efforts, the future will remain broad and uncertain, and the specific system designated is very likely to come into being in a completely different environment. The real problem is to provide a menu of alternatives to confront a spectrum of future events. This is not only a more tractable problem, but a far more helpful one. Analysis is likely to be more useful in pointing out where further information would be particularly valuable and how to get it than in evaluating or specifying the "best" system. But analysis alone cannot locate all the theoretical and empirical knowledge required; experiment and development cannot be replaced by systems analysis.

7. *In broad policy questions, an alternative cannot be preferred merely because it is the lowest-cost choice in a single contingency.* Even at best, such a comparison can reflect only the most probable future circumstance. A preferred alternative should also go a long way toward achieving the objectives in a less probable or even in many improbable situations – and we would even like it to offer a good chance to attain many of the objectives of lower priority that are always present.

8. *Real questions are full of uncertainty; it is more useful to try to get some sort of an answer to important questions than to try to get good answers to questions that have been so reduced by assumption that they are no longer really meaningful.*

9. *Estimates of cost are essential to a choice among alternatives.* It is important to cost alternatives in some appropriate sense, however approximate, prior to choice. Even though effectiveness may be impossible to measure satisfactorily, differences may be apparent, and a decisionmaker can make a better decision if he at least knows the cost implications of his actions.

10. *The decisionmaker can compensate to some extent for uncertainty, and the analysis can advise him how to do this.* Difficulties such as contradictory objectives, the necessity for dependence on subjective judgment, the uncertain implications of costs over time, inaccurate or missing data, and the freedom of action of the enemy may preclude full optimization. The decisionmaker can protect himself to some extent against serious error due to these difficulties by temporizing decisions – decisions to postpone action, decisions for mixed rather than pure solutions, or decisions for parallel efforts. Advising on such actions is an appropriate role of analysis, and the knowledge that they are possible can make the job easier.

11. *A good new idea may be worth a thousand evaluations.* In an analysis aimed at policy-making, an investigation of the relevance of the many factors and contingencies affecting the decision is likely to be more useful

than any narrow optimization achieved by sophisticated analytic techniques. A good new idea – technical, operational, or what have you – may be worth a thousand elaborate optimizations, for it may offer a means of overcoming our difficulties.

NATURE OF THE DECISIONMAKER AND HIS RESPONSIBILITY

Taking due account of these principles and precepts for systems analysis requires, among other things already discussed, a fairly accurate notion of who the user of analysis is. In this book, as in the last sentence, we have tended to speak of the decisionmaker as if he were isolated. In the real world, this view of the decisionmaker is an obvious fiction; for even when the decision is not made by an organization, the individual decisionmaker is usually imbedded in an organization, and his decisions are colored by that organization – so much so that, as one RAND analyst, W. M. Jones, has put it: "One should not think of solving a strategic problem as such – say, recommending the requirements for a next manned strategic bomber. One should rather think of solving an organizational problem in which the analysis of the requirements for the manned bomber plays an important role." The costs to an organization of changing its policies and procedures may be large; in considering a range of alternatives, these costs should also be taken into account. Organizational decisionmaking is a complex activity and difficult to analyze, but one thing can be said: Successful analysis is characterized by a continual interaction between the analysts and the decisionmakers, not just a formal presentation when the analyst thinks that the study is over.

The man who expects to act on the basis of someone else's analysis needs more than confidence in the technical qualifications of the analyst to assure himself that the work is competent. To act wisely he should, at the very least, understand the important and fundamental principles of analysis the work involves. Since he does not ordinarily have the time or the training necessary to work through all the aspects of the analysis, he must, if he wants to understand those that he cannot accept intuitively, resort to questioning the analyst. With these questions he can attempt to make sure that the study has been conducted according to generally accepted principles of good analysis, that the analyst has properly communicated the necessary doubts, and that, if the conclusions are not acceptable to common-sense reasoning, he understands why.

It is easy to exaggerate the degree of assistance that analysis can offer a policymaker. Using value judgments, imprecise knowledge, and intuitive estimates of enemy intent, gleaned from specialists or from the policymaker himself, a study can do little more than assess the implications of choosing one alternative over another. This assessment can help the decisionmaker make a better decision than he would otherwise make. But the man who has the responsibility must interpret the information he is presented with in the light of his own knowledge, values, and estimates, and assess other implications himself. The decision must become his own.

To illustrate this point, we can draw an analogy between the decisionmaker using a study team for advice and a medical doctor using a clinical laboratory.

Suppose, for example, that the doctor is trying to decide whether to send his patient to a surgeon to have his stomach resected or to treat him medically for a gastric ulcer. This decision will be influenced by several considerations. One is the technical findings of the laboratory crews. Like the military decisionmaker, the doctor may or may not be able to carry out these investigations himself, but it would probably not be economic for him to do so. He would depend, therefore, on laboratory reports, some of which will be on cold slips of paper without comment or nuance – numbers alone. Other reports might be written out at length, or involve discussions with the doctor and an examination of X-ray plates under the guidance of the technician.

The doctor himself would, of course, make his own observations or analyses. Some of these he would put in the form of written notes; others he would simply hold in his head. He would also consult his experience, recollecting impressions of the risks and possibilities of success with various treatments. Some of these impressions would come from medical reports.

In the end, the doctor, like the military decisionmaker, must make a judgment based on whatever facts or analyses he has. This judgment will represent the ultimate synthesis the doctor makes of the numerical tests, the written out but relatively diffuse notes, the unrecorded conversations with technicians, and his own introspection. It cannot be a mere calculation but must be made on intuitive grounds. Sometimes one consideration will be important enough to make the final decision fairly easy, but generally he just will not *know* what to do. He could do more analysis, sometimes even risk the patient's life in order to guard it – call for a liver puncture or other dangerous procedures – but his inquiry will never be complete. His judgment, like that of the military decisionmaker, must be made with uncertainties in mind. Analysis will have helped him by answering some of his questions, sharpening his intuition, and broadening his basis for judgment. But in very few cases should we expect it to *prove* that a particular course of action is best.

Some dangers of analysis

How can we summarize any danger that there might be in relying on systems analysis, or any similar approach to defense decisions? For one thing, as other authors in this book have remarked, many factors fundamental to national defense problems are not subject to rigorous, quantitative analysis. They may, therefore, be neglected, deliberately set aside, or improperly weighted in the analysis itself or in the decision based on such analysis. The danger also exists that an analysis will appear so scientific and quantitative on the surface that it may be assigned a validity not justified by the many subjective judgments it involves. In other words, we may be so mesmerized by the beauty and precision of the numbers that we overlook the simplifications made to achieve this precision, neglect analysis of the qualitative factors, and overemphasize the importance of idealized calculations in the decision process. But better analysis and careful attention to where analysis ends and judgment begins should help to reduce these dangers.

Even if we are aware of these dangers, systems analysis may sometimes still look like a purely rational approach to decisionmaking, coldly objective and scientific. As we have seen many times in this book, it is not. But, in contrast to other aids to decisionmaking, systems analysis extracts everything possible from scientific methods, and its virtues are the virtues of those methods. *Furthermore, its limitations are shared by its alternatives.* And if we exclude intuitive judgment, then, in a sense, these alternatives are also analysis, but poorer analysis, less explicit, less systematic, and less quantitative.

The future of systems analysis

Finally, what can we say about the future of systems analysis in the study of problems of defense and national security?

Resistance by the military to its use in broad problems of strategy is gradually breaking down. With regard to systematic, quantitative analysis in general, the military is in an evolutionary flux, like industry in the throes of automation. Military planning and strategy have always involved more art than science; what is happening is that the art form is changing from an *ad hoc*, seat-of-the-pants approach based on intuition and experience to one based on analysis *and supported by* intuition and experience. The military itself has changed. In the old Air Force, for example, a man flew, or fixed aircraft that flew. Today, the ratio of staff to pilots is increasing to the point where the force appears to be turning completely into staff and support personnel. With this change the computer is becoming increasingly significant – as an automaton, a process controller, a trouble-shooting technician, a complex information processor, and as a decision aid. And yet the computer is not more than a tool to expedite analysis; even in the narrowest military decisions, considerations not subject to any sort of quantitative analysis can always be present. Thus, while the military is turning increasingly to modern analytic methods, its big decisions have not been, nor are they ever likely to become, the *automatic* consequence of a computer program, of cost-effectiveness analysis, operations research, or any application of mathematical models or of systems analysis.

For complex questions, involving force posture and composition or the strategy to achieve the objectives of foreign policy, intuitive, subjective, even *ad hoc* study schemes must continue to be used – but they will no doubt be supplemented more and more often by systems analysis. As Alain Enthoven has put it:

> Ultimately all policies are made and all weapon systems are chosen on the basis of judgments. There is no other way and there never will be. The question is whether those judgments have to be made in the fog of inadequate and inaccurate data, unclear and undefined issues, and a welter of conflicting personal opinions, or whether they can be made on the basis of adequate, reliable information, relevant experience, and clearly drawn issues.[1]

In recognition of the profound need for clarity and informed judgment, systems analysis will no doubt see a greater use of scenarios, gaming, and techniques for the systematic employment of experts, along with a growing use of quantitative analysis for problems where it is appropriate.

Moreover, new approaches and techniques are being proposed constantly. Many of these are primarily mathematical in nature, but increasing attention is being devoted to systematic methods for taking into account the various organizational, political, and social factors heretofore so poorly understood but often so critical to national security problems. In the computer field, the trend toward a better union of man and machine through personalized, on-line, time-sharing systems that use natural language and graphical input and output, and store submodels on discs, will be a great boon to the systems analyst.[2] It will give him the capability to change his program instantly, to experiment, and to perform numerous excursions, parametric investigations, and sensitivity analyses. In mathematics, new advances in game theory are giving us insight into the many-person and nonzero-sum situations of conflict and cooperation. And, more importantly, for questions not amenable to quantitative treatment, new techniques for the direct use of expertise are giving us a way to grasp these difficult-to-treat aspects of our problems.

At the very least, analysis can provide a way to choose the numerical quantities related to a weapon system so that they are logically consistent

[1] Alain C. Enthoven, "Choosing Strategies and Selecting Weapon Systems," *United States Naval Institute Proceedings*, Vol. 90, No. 1, January 1964, p. 151.

[2] Developments like the RAND Tablet and JOSS point in this direction. (See T. O. Ellis and W. L. Sibley, *On the Development of Equitable Graphic I/O*, The RAND Corporation, P-3415, July 1966, and C. L. Baker, *JOSS: Introduction to a Helpful Assistant*, The RAND Corporation, RM-5058-PR, July 1966.)

with each other, with an assumed objective, and with the analyst's expectation of the future. Systems analysis strives to do more, however, than simply supply solutions that correctly follow from sets of arbitrarily chosen assumptions in narrow problems. It aspires to help the decision-maker find solutions that experience will confirm in the broadest of problems. This goal, as many of the chapters in this book have made clear, is still far from being attained. But a greater understanding of the nature and roles of systems analysis promises to bring it closer.

SELECTED BIBLIOGRAPHY

Ackoff, R. L. (ed.). *Scientific Method*, Wiley, New York, 1962.

Ackoff, R. L. (ed.). *Progress in Operations Research*, Vol. I, Wiley, New York, 1961.

Ackoff, R. L., and P. Rivett. *A Manager's Guide to Operations Research*, Wiley, New York, 1963.

Adelson, Marvin (ed.). "Planning Education for the Future," *American Behavioral Scientist*, Vol. 10, No. 7, March 1967, pp. 1–31.

Ansoff, H. Igor. "A Quasi-Analytic Method for Long-Range Planning," Paper presented at the First Symposium on Corporate Long-Range Planning, College on Planning, The Institute of Management Sciences, Chicago, Illinois, June 6, 1959.

Asher, Harold. *Cost-Quantity Relationships in the Airframe Industry*, The RAND Corporation, R-291 (DDC No. AD 105540), July 1956.

Averch, Harvey, and M. M. Lavin. *Simulation of Decisionmaking in Crises: Three Manual Gaming Experiments*, The RAND Corporation, RM-4202-PR (DDC No. AD 605476), August 1964.

Baker, C. L. *JOSS: Introduction to a Helpful Assistant*, The RAND Corporation, RM-5058-PR (DDC No. AD 636993), July 1966.

Banister, A. W. "The Case for Cold War Gaming in the Military Services," *Air University Review*, Vol. XVIII, No. 5, July–August 1967, pp. 49–52.

Bell, Chauncey F., and Milton Kamins. *Determining Economic Quantities of Maintenance Resources: A Minuteman Application*, The RAND Corporation, RM-3308-PR (DDC No. AD 407200), January 1963.

Bell, Chauncey F., and T. C. Smith. *The Oxnard Base Maintenance Management Improvement Program*, The RAND Corporation, RM-3370-PR (DDC No. AD 292909), November 1962.

Berkovitz, L. D., and M. Dresher. *A Game Theory Analysis of Tactical Air War*, The RAND Corporation, P-1592, January 1959. (Also published in *Operations Research*, Vol. 7, 1959, pp. 599–620.)

Berman, E. B. *Toward a New Weapon System Analysis*, The RAND Corporation, P-1493 (DDC No. AD 224124), September 23, 1958.

Bickner, R. E. *The Changing Relationship between the Air Force and the Aerospace Industry*, The RAND Corporation, RM-4101-PR (DDC No. AD 608579), July 1964.

Blackett, P. M. S. "Operational Research," *Advancement of Science*, Vol. 5, No. 17, 1948, pp. 26–38.

Blackwell, David, and M. A. Girshick. *Theory of Games and Statistical Decisions*, Wiley, New York, 1954.

Boldyreff, A. W. *Systems Engineering*, The RAND Corporation, P-537 (DDC No. AD 422829), June 16, 1954.

Boren, H. E., Jr. *Individual Weapon System Computer Cost Model*, The RAND Corporation, RM-4165-PR (DDC No. AD 603005), July 1964.

Brodie, Bernard. *Scientific Progress and Political Science*, The RAND Corporation, P-968 (DDC No. AD 605109), November 30, 1956.

Brown, B., and O. Helmer. *Improving the Reliability of Estimates Obtained from a Consensus of Experts*, The RAND Corporation, P-2986 (DDC No. AD 606970), September 1964.

Bureau of the Budget Library. *Program Analysis Techniques: A Selected Bibliography*, Bureau of the Budget, Washington, D. C. (rev. ed.), 1966.

Campbell, Robert H. *A Methodological Study of Expert Utilization in Business Forecasting*, Unpublished doctoral dissertation, University of California, Los Angeles, 1966.

Caywood, T. E., and C. J. Thomas. "Applications of Game Theory in Fighter versus Bomber Combat," *Journal of the Operations Research Society of America*, Vol. 3, 1955, pp. 402–411.

Charnes, A., and W. W. Cooper. *Management Models*, Vols. I and II, Wiley, New York, 1961.

Chernoff, Hermann, and Lincoln E. Moses. *Elementary Decision Theory*, Wiley, New York, 1959.

Churchman, C. W., R. L. Ackoff, and E. L. Arnoff. *An Introduction to Operations Research*, Wiley, New York, 1957.

Clapp, R. E., and H. E. Boren, Jr. *MISCOM: An Individual Weapon System Computer Cost Model*, The RAND Corporation, RM-5142-1-PR (DDC No. AD 647975), February 1967, and Addendum (DDC No. AD 650235), March 1967.

Clement, G. H. *Weapons System Philosophy*, The RAND Corporation, P-880 (DDC No. AD 605051), July 27, 1956.

Cohen, John. *Chance, Skill, and Luck*, Penguin Books, Inc., Baltimore, Md., 1960.

Dalkey, N. C. *Families of Models*, The RAND Corporation, P-3198 (DDC No. AD 620777), August 1965.

Dalkey, N. C. *Games and Simulations*, The RAND Corporation, P-2901 (DDC No. AD 601138), April 1964.

Dalkey, N. C., and O. Helmer. *An Experimental Application of the Delphi Method to the Use of Experts*, The RAND Corporation, RM-727-PR (Abr.), July 1962. (Also published in *Management Science*, Vol. 9, No. 3, April 1963, pp. 458–467.)

Daniels, Everett J., and John B. Lathrop. "Strengthening the Cost-Effectiveness Criterion for Major System Decisions," Paper presented at the Joint National Meeting of the Operations Research Society of America and The Institute of Management Sciences, Minneapolis, Minn., October 7–9, 1964.

Dantzig, George B. *Linear Programming and Extensions*, The RAND Corporation, R-366 (DDC No. AD 418366), August 1963. (Published commercially by Princeton Univ. Press, Princeton, N. J., 1963.)

Davis, Harold. *The Use of Cost-Effectiveness Studies in Military Decision-Making*, Operations Analysis, Hq USAF, Operations Analysis Paper No. 1, November 1961.

Davis, M., and M. Verhulst (eds.). *Operational Research in Practice*, Macmillan (Pergamon), New York, 1958.

Dienemann, P. F. *Estimating Cost Uncertainty Using Monte Carlo Techniques*, The RAND Corporation, RM-4854-PR (DDC No. AD 629082), January 1966.

Dienemann, P. F., and G. C. Sumner. *Estimating Aircraft Base Maintenance Personnel*, The RAND Corporation, RM-4748-PR (DDC No. AD 626087), October 1965.

DonVito, P. A. *Annotated Bibliography on Systems Cost Analysis*, The RAND Corporation, RM-4848-1-PR (DDC No. AD 810910), March 1967.

Dordick, H. S. *Maintainability: A Primer in Designing for Profit*, The RAND Corporation, P-2886 (DDC No. AD 601080), May 1964.

Downs, A. *Inside Bureaucracy*, The RAND Corporation, P-2963 (DDC No. AD 604847), August 1964. (Published commercially in A. Downs, *Inside Bureaucracy*, Little, Brown, Boston, 1967.)

Dresher, M. *Games of Strategy: Theory and Applications*, The RAND Corporation, R-360 (DDC No. AD 257899), May 1961. (Published commercially by Prentice-Hall, Englewood Cliffs, N. J., 1961.)

Dresher, M. *Some Military Applications of the Theory of Games*, The RAND Corporation, P-1849, December 10, 1959.

Ellis, T. O., and W. L. Sibley. *On the Development of Equitable I/O*, The RAND Corporation, P-3415 (DDC No. AD 637781), July 1966.

Enthoven, A. C. "Choosing Strategies and Selecting Weapon Systems," *United States Naval Institute Proceedings*, Vol. 90, No. 1, Whole No. 731, January 1964, pp. 151–158.

Enthoven, A. C. "Operations Research and the Design of the Defense Program," *Proceedings of the Third International Conference on Operations Research*, Dunod, Paris, 1964, pp. 531–538.

Feeney, G. J., J. W. Petersen, and C. C. Sherbrooke. *An Aggregate Base Stockage Policy for Recoverable Spare Parts*, The RAND Corporation, RM-3644-PR (DDC No. AD 408943), May 1963.

Fisher, G. H. *The Analytical Bases of Systems Analyses*, The RAND Corporation, P-3363 (DDC No. AD 634512), May 1966.

Fisher, G. H. *The World of Program Budgeting*, The RAND Corporation, P-3361 (DDC No. AD 633069), May 1966.

Fisher, G. H. *The Role of Cost-Utility Analysis in Program Budgeting*, The RAND Corporation, RM-4279-RC (DDC No. AD 608055), September 1964.

Fisher, G. H. *Some Comments on Program Budgeting in the Department of Defense*, The RAND Corporation, P-2721 (DDC No. AD 402137), March 1963.

Fisher, G. H., and the Staff of the RAND Cost Analysis Department. "Costing Methods," in E. S. Quade (ed.), *Analysis for Military Decisions*, The RAND Corporation, R-387-PR (DDC No. AD 453887), November 1964, Chap. 15.

Fort, D. M. *Systems Analysis as an Aid in Air Transportation Planning*, The RAND Corporation, P-3293-1 (DDC No. AD 629769), March 1966.

Frye, Alton. "Politics – The First Dimension of Space," *Journal of Conflict Resolution*, Vol. 10, No. 1, March 1966, pp. 103–112.

Frye, Alton. *Space Arms Control: Trends, Concepts, Prospects*, The RAND Corporation, P-2873 (DDC No. AD 438436), February 1964. (Also published in *Bulletin of the Atomic Scientists*, Vol. 21, No. 4, April 1965, pp. 30–34.)

Frye, Alton. *The Proposal for a Joint Lunar Expedition: Background and Prospects*, The RAND Corporation, P-2808, January 1964. (Also published in *Air Force/Space Digest*, Vol. 47, No. 11, November 1964, pp. 62–68.)

Frye, Alton. "Our Gamble in Space: The Military Danger," *The Atlantic*, Vol. 202, No. 2, August 1963, pp. 46-50.

Geisler, M. A. *Development of Man-Machine Simulation Techniques*, The RAND Corporation, P-1945 (DDC No. AD 224314), March 17, 1960.

Geisler, M. A. *Simulation Techniques*, The RAND Corporation, P-1808 (DDC No. AD 224158), September 4, 1959.

George, Alexander L. *Propaganda Analysis: A Study of Inferences Made from Nazi Propaganda in World War II*, A RAND Corporation Study, Harper and Row, New York, 1959.

Glennan, T. K., Jr. *An Economist Looks at R&D Management*, The RAND Corporation, P-2819 (DDC No. AD 429982), November 1963.

Glennan, T. K., Jr. *Some Suggested Changes in Research and Development Strategy and Their Implications for Contracting*, The RAND Corporation, P-2717 (DDC No. AD 402992), March 1963.

Goldhamer, Herbert. *Human Factors in Systems Analysis*, The RAND Corporation, RM-388 (DDC No. ATI 78026), April 15, 1950.

Goldhamer, Herbert, and Hans Speier. *Some Observations on Political Gaming*, The RAND Corporation, P-1679-RC (DDC No. AD 261649), April 30, 1959.

Goldsen, J. M. (ed.). *International Political Implications of Activities in Outer Space: A Report of a Conference, October 22-23, 1959*, The RAND Corporation, R-362-RC, May 5, 1960.

Goode, H. H., and R. E. Machol. *System Engineering*, McGraw–Hill, New York, 1957.

Hall, A. D. *A Methodology for Systems Engineering*, Van Nostrand, Princeton, N. J., 1962.

Hawthorne, G. B., Jr. "Digital Simulation and Modeling," *Datamation*, Vol. 10, No. 10, October 1964, pp. 25–29.

Haywood, O. G., Jr. "Military Decision and the Mathematical Theory of Games," *Air University Quarterly Review*, Vol. 4, 1950.

Helmer, O. *Analysis of the Future: The Delphi Method*, The RAND Corporation, P-3558 (DDC No. AD 649640), March 1967.

Helmer, O. *The Use of the Delphi Technique in Problems of Educational Innovations*, The RAND Corporation, P-3499 (DDC No. AD 644591), December 1966.

Helmer, O. *Social Technology*, Basic Books, New York, 1966.

Helmer, O. *Social Technology*, The RAND Corporation, P-3063 (DDC No. AD 462520), February 1965. (Published commercially in O. Helmer, *Social Technology*, Basic Books, New York, 1966.)

Helmer, O. *The Systematic Use of Expert Judgment in Operations Research*, The RAND Corporation, P-2795 (DDC No. AD 417901), September 1963.

Helmer, O., and T. J. Gordon. *Report on a Long-Range Forecasting Study*, The RAND Corporation, P-2982 (DDC No. AD 607777), September 1964. (Published commercially in O. Helmer, *Social Technology*, Basic Books, New York, 1966).

Helmer, O., and E. S. Quade. "An Approach to the Study of a Developing Economy by Operational Gaming," in *Recherche Opérationnelle et Problèmes du Tiers Monde*, Dunod, Paris, 1964, pp. 43–57.

Helmer, O., and N. Rescher. *On the Epistemology of the Inexact Sciences*, The RAND Corporation, R-353 (DDC No. AD 236439), February 1960. (Also printed in *Management Science*, Vol. 6, No. 1, October 1959, pp. 25–52.)

Hertz, David B., and Roger T. Eddison (eds.). *Progress in Operations Research*, Vol. II, Wiley, New York, 1964.

Hirschman, A. O., and C. E. Lindblom. *Economic Development, Research and Development, Policymaking: Some Converging Views*, The RAND Corporation, P-1982 (DDC No. AD 224286), May 4, 1960.

Hitch, C. J. *Decision-making for Defense*, Univ. of California Press, Berkeley and Los Angeles, 1965.

Hitch, C. J. "Cost Considerations and Systems Effectiveness," Address presented at the SAE-ASME-AIAA Aerospace Reliability and Maintainability Conference, Washington, D.C., June 30, 1964.

Hitch, C. J. "Cost/Effectiveness," Address presented at the Thirteenth Military Operations Research Symposium, Washington, D.C., April 29, 1964.

Hitch, C. J. "Programming's Role in Defense," Address before the First International Meeting of the Western Section of the Operations Research Society of America, Honolulu, Hawaii, September 1964; quoted , in part, in *Aviation Week and Space Technology*, Vol. 81, No. 15, October 12, 1964, p. 17.

Hitch, C. J. *Uncertainties in Operations Research*, The RAND Corporation, P-1959, April 1960. (Also published in *Operations Research*, Vol. 8, July-August 1960, pp. 437–445.)

Hitch, C. J. *Economics and Military Operations Research*, The RAND Corporation, P-1250 (DDC No. AD 224037), January 1958. (Also published in *The Review of Economics and Statistics*, Vol. 40, No. 3, August 1958, pp. 199–209.)

Hitch, C. J. *Professor Koopman on Fallacies: A Comment*, The RAND Corporation, P-870 (DDC No. AD 605045), May 21, 1956. (Also published in *Operations Research*, Vol. 4, No. 4, August 1956, pp. 426–430.)

Hitch, C. J., and R. N. McKean. *The Economics of Defense in the Nuclear Age*, The RAND Corporation, R-346 (DDC No. AD 243098), March 1960. (Published commercially by Harvard Univ. Press, Cambridge, Mass., 1960.)

Hoag, M. W. "The Relevance of Costs," in E. S. Quade (ed.), *Analysis for Military Decisions*, The RAND Corporation, R-387-PR (DDC No. AD 453887), November 1964, Chapter 6.

Hoag, M. W. *Nuclear Strategic Options and European Force Participation*, The RAND Corporation, P-2594-2, July 1963.

Hoag, M. W. *An Introduction to Systems Analysis*, The RAND Corporation, RM-1678 (DDC No. AD 101071), April 1956.

Hollingdale, S. H. (ed.). *Digital Simulation in Operational Research*, The English Universities Press, London, 1967.

Hoos, Ida R. "A Critique on the Application of Systems Analysis to Social Problems," Address presented at the Thirteenth Annual Meeting of the American Astronautical Society, Dallas, Texas, May 2, 1967.

Jantsch, Erich. *Technological Forecasting in Perspective*, Organization for Economic Cooperation and Development, Paris, October 1966.

Jones, W. M. *Factional Debates and National Commitments: The Multidimensional Scenario*, The RAND Corporation, RM-5259-ISA (DDC No. AD 812381), March 1967.

Jones, W. M., M. B. Shapiro, and N. Z. Shapiro. *The Flight Operations Planner*, The RAND Corporation, RM-2415 (DDC No. AD 227738), July 16, 1959.

Jones, W. M., M. B. Shapiro, and N. Z. Shapiro. *Introduction to the Automatic Flight Planner*, The RAND Corporation, RM-2147 (DDC No. AD 150691), April 10, 1958.

Kahn H., and I. Mann. *Ten Common Pitfalls*, The RAND Corporation, RM-1937 (DDC No. AD 133035), July 1957.

Kahn, H. *War Gaming*, The RAND Corporation, P-1167 (DDC No. AD 606531) July 1957.

Kahn, H. *Techniques of Systems Analysis*, The RAND Corporation, RM-1829-1 (DDC No. AD 133012), rev. June 1957.

Kane, Col. F. X. "Security Is Too Important to Be Left to Computers," *Fortune*, Vol. 69, No. 4, April 1964, pp. 146, 231+.

Kaplan, Abraham. *The Conduct of Inquiry*, Chandler Publ. Co., San Francisco, 1964.

Katzenbach, E. L., Jr. "The Horse Cavalry in the Twentieth Century: A Study in Policy Response," *Public Policy*, Graduate School of Public Administration, Harvard, 1958, pp. 120–149.

Kaufman, A. *Methods and Models of Operations Research*, Prentice-Hall, Englewood Cliffs, N. J., 1962.

Kaufmann, W. W. *The McNamara Strategy*, Harper & Row, New York, 1964.

Kecskemeti, Paul. *Utilization of Social Research in Shaping Policy Decisions*, The RAND Corporation, P-2289, April 24, 1961.

Kennedy, J. L. *A Display Technique for Planning*, The RAND Corporation, P-965 (DDC No. AD 605106), October 1956.

Kent, Maj. Gen. Glenn A. "On Analysis," *Air University Review*, Vol. XVIII, No. 4, May–June 1967, pp. 50–55.

Kermisch, J. J., and A. J. Tenzer. *On the Role of the Cost Analyst in a Weapon System Study*, The RAND Corporation, P-3360 (DDC No. AD 636497), May 1966.

Klein, B. H. *The Decision-Making Problem in Development*, The RAND Corporation, P-1916 (DDC No. AD 224166), February 19, 1960.

Klein, B. H., and W. H. Meckling. *Applications of Operations Research to Development Decisions*, The RAND Corporation, P-1054 (DDC No. AD 422570), March 3, 1958.

Koopman, B. O. "Fallacies in Operating Research," *Operations Research*, Vol. 4, No. 4, August 1956, pp. 422–426.

Levien, R. E. *An Appreciation of the Value of Continental Defense*, The RAND Corporation, RM-3987-PR (DDC No. AD 437699), March 1964.

Luce, R. D., and H. Raiffa. *Games and Decisions*, Wiley, New York, 1957.

McCloskey, J. F., and J. N. Coppinger (eds.). *Operations Research for Management*, Vol. II, Johns Hopkins Univ. Press. Baltimore, Md. 1956.

McCloskey, J. F., and F. N. Trefethen (eds.). *Operations Research for Management*, Vol. I, Johns Hopkins Univ. Press, Baltimore, Md., 1954.

McGarvey, D. C. *Problems of Force Posture Evaluation*, The RAND Corporation, P-2138, January 30, 1961.

McKean, R. N. *Efficiency in Government through Systems Analysis: With Emphasis on Water Resources Development*, A RAND Corporation Study, Wiley, New York, 1958.

McKinsey, J. C. C. *Introduction to the Theory of Games*, A RAND Corporation Study, McGraw-Hill, New York, 1952.

Machol, R. E., W. P. Tanner, Jr., and S. N. Alexander (eds.). *Systems Engineering Handbook*, McGraw-Hill, New York, 1965.

Markowitz, H. M., B. Hausner, and H. W. Karr. *SIMSCRIPT: A Simulation Programming Language*, The RAND Corporation, RM-3310-PR (DDC No. AD 291806), November 1962. (Published commercially by Prentice-Hall, Englewood Cliffs, N. J., 1963.)

Marshall, A. W. *Problems of Estimating Military Power*, The RAND Corporation, P-3417 (DDC No. AD 637779), August 1966.

Meckling, W. H. "Strategies for Development," in E. S. Quade (ed.), *Analysis for Military Decisions*, The RAND Corporation, R-387-PR (DDC No. AD 453887), November 1964, Chap. 12.

Micks, W. R. *Material Engineering in the Systems Context*, The RAND Corporation, P-2470, November 1961.

Mood, A. M. *War Gaming as a Technique of Analysis*, The RAND Corporation, P-899 (DDC No. AD 605063), September 3, 1954.

Morse, P. M., and G. E. Kimball. *Methods of Operations Research*, Wiley, New York, 1951.

Nelson, R. R. *The Economics of Parallel R&D-Efforts*, The RAND Corporation, P-1774 (DDC No. AD 224155), August 1959. (Also published in *The Review of Economics and Statistics*, Vol. 43, November 1961, pp. 351–364.)

Noah, J. W., and R. W. Smith. *Cost-Quantity Calculator*, The RAND Corporation, RM-2786-PR (DDC No. AD 279346), January 1962.

North, H. Q. *A Probe of TRW's Future: The Next 20 Years*, TRW Systems, Redondo Beach, California, July 1966.

Novick, David. *System and Total Force Cost Analysis*, The RAND Corporation, RM-2695-PR (DDC No. AD 257269), April 1961.

Novick, David. *Lead-Time in Modern Weapons*, The RAND Corporation, P-1240, December 26, 1957.

Novick, David (ed.). *Program Budgeting: Program Anaylsis and the Federal Budget*. 2nd ed. Harvard University Press, Cambridge, 1967.

Office of the Secretary of Defense. *Initiation of Engineering and Operational Systems Development*, DOD Directive 3200.9, July 1, 1965.

Paxson, E. W. *War Gaming*, The RAND Corporation, RM-3489-PR (DDC No. AD 297755), February 1963.

Peat, Marwick, Livingston & Company. *A Report on Contract Definition* (DDC No. AD 646240), January 2, 1967.

Peck, Merton J., and Frederic M. Scherer. *The Weapons Acquisition Process: An Economic Analysis*, Harvard Univ. Press, Cambridge, Mass., 1962.

Posvar, Col. Wesley W. "The Realm of Obscurity," in *American Defense Policy*, prepared by the Associates in Political Science, United States Air Force Academy, Johns Hopkins Univ. Press, Baltimore, Md., 1965, pp. 218-231.

Quade, E. S. *Some Problems Associated with Systems Analysis*, The RAND Corporation, P-3391 (DDC No. AD 634375), June 1966.

Quade, E. S. *Systems Analysis Techniques for Planning-Programming-Budgeting*, The RAND Corporation, P-3322 (DDC No. AD 629564), March 1966.

Quade, E. S. *The Limitations of a Cost-Effectiveness Approach to Military Decision-Making*, The RAND Corporation, P-2798 (DDC No. AD 425786), September 1963.

Quade, E. S. *Military Systems Analysis*, The RAND Corporation, RM-3452-PR (DDC No. AD 292026), January 1963.

Quade, E. S. *Pitfalls in Military Systems Analysis*, The RAND Corporation, P-2676 (DDC No. AD 291247), November 1962.

Quade, E. S. (ed.). *Analysis for Military Decisions*, The RAND Corporation, R-387-PR (DDC No. AD 453887), November 1964. (Published commercially by Rand McNally, Chicago, 1964.)

Quinn, James Brian. "Technological Forecasting," *Harvard Business Review*, Vol. 45, No. 2, March–April 1967, pp. 89–106.

Rauner, R. M., and W. A. Steger. *Game-Simulation and Long-Range Planning*, The RAND Corporation, P-2355, June 22, 1961.

Saaty, T. L. *Mathematical Methods of Operations Research*, McGraw-Hill, New York, 1959.

Sasieni, M., A. Yaspan, and L. Friedman. *Operations Research: Methods and Problems*, Wiley, New York, 1959.

Savage, L. J. *The Foundations of Statistics*, Wiley, New York, 1954.

Schamberg, R. "Technological Considerations," in E. S. Quade (ed.), *Analysis for Military Decisions*, The RAND Corporation, R-387-PR (DDC No. AD 453887), November 1964, Chap. 9.

Schelling, T. C. "Controlled Response and Strategic Warfare," Adelphi Papers No. 19, Institute for Strategic Studies, June 1965.

Schlesinger, J. R. *Systems Analysis and the Political Process*, The RAND Corporation, P-3464, June 1967.

Schlesinger, J. R. *On Relating Non-technical Elements to System Studies*, The RAND Corporation, P-3545 (DDC No. AD 650846), February 1967.

Schlesinger, J. R. "Quantitative Analysis and National Security," *World Politics*, Vol. XV, No. 2, January 1963, pp. 295–315.

Shubik, Martin. *Game Theory and Related Approaches to Social Behavior*, Wiley, New York, 1964.

Shubik, Martin. "A Game Theorist Looks at the Antitrust Laws and the Automobile Industry," *Stanford Law Review*, Vol. 8, 1956, pp. 594–630.

Smith, Bruce L. R. *The RAND Corporation: Case Study of a Nonprofit Advisory Corporation*, Harvard Univ. Press, Cambridge, Mass., 1966.

Smith, T. C. *SAMSOM: Support-Availability Multi-System Operations Model*, The RAND Corporation, RM-4077-PR (DDC No. AD 601813), June 1964.

Smithies, A. *A Conceptual Framework for the Program Budget*, The RAND Corporation, RM-4271-RC (DDC No. AD 453162), September 1964.

Specht, R. D. *War Games*, The RAND Corporation, P-1041 (DDC No. AD 224018), March 18, 1957.

Subcommittee on National Security and International Operations. *Planning-Programming-Budgeting: Selected Comments*, Committee on Government

Operations, U.S. Senate, 90th Congress, 1st Session, U.S. Government Printing Office, Washington, D.C., 1967.

Tenzer, A. J., O. Hansen, and E. M. Roque. *Relationships for Estimating USAF Administrative and Support Manpower Requirements*, The RAND Corporation, RM-4366-PR (DDC No. AD 611567), January 1965.

Thomas, C. J. "Some Past Applications of Game Theory in the United States Air Force," Paper presented at the NATO Conference on the Theory of Games and Its Military Applications, June 29–July 3, 1964, Toulon, France.

Thrall, R. M., C. H. Coombs, and R. L. Davis. *Decision Processes*, Wiley, New York, 1954.

Tucker, Samuel A. (ed.). *A Modern Design for Defense Decisions: A McNamara-Hitch-Enthoven Anthology*, Industrial College of the Armed Forces, Washington, D.C., 1966.

Von Neumann, John, and Oskar Morgenstern. *Theory of Games and Economic Behavior*, Princeton Univ. Press, Princeton, N. J., 1944.

Watts, A. F. *Aircraft Turbine Engines – Development and Procurement Cost*, The RAND Corporation, RM-4670-PR (Abr.) (DDC No. AD 624094), November 1965.

Weiner, M. G. "Gaming Methods and Applications," in E. S. Quade (ed.), *Analysis for Military Decisions*, The RAND Corporation, R-387-PR (DDC No. AD 453887), November 1964, Chap. 11.

Weiner, M. G. *An Introduction to War Games*, The RAND Corporation, P-1773 (DDC No. AD 221676), August 17, 1959.

Weiner, M. G. *War Gaming Methodology*, The RAND Corporation, RM-2413 (DDC No. AD 234505), July 1959.

Wiener, Norbert. *The Human Use of Human Beings*, Doubleday, Garden City, N. Y., 1956.

Williams, J. D. *The Compleat Strategyst: Being a Primer on the Theory of Games of Strategy*, A RAND Corporation Study, McGraw-Hill, New York, rev. ed., 1966.

Wilson, E. B., Jr. *An Introduction to Scientific Research*, McGraw-Hill, New York, 1952.

Wohlstetter, A. J. "Strategy and the Natural Sciences," in Robert Gilpin and Christopher Wright (eds.), *Scientists and National Policy Making*, Columbia Univ. Press, New York, 1964.

Wohlstetter, A. J., F. S. Hoffman, R. J. Lutz, and H. S. Rowen. *Selection and Use of Strategic Air Bases*, The RAND Corporation, R-266 (DDC No. AD 413271, April 1954.

Wolf, Charles, Jr. "Defense and Development in the Less Developed Countries," *Journal of the Operations Research Society of America*, Vol. 10, No. 6, November–December 1962, pp. 828–838.

Wolfers, Arnold. "'National Security' as an Ambiguous Symbol," *Political Science Quarterly*, Vol. 67, No. 4, December 1952, pp. 481–502.

BIBLIOGRAPHIC NOTE

Additional references to RAND's analytic studies can be found in the one-volume *Index of Selected Publications of The RAND Corporation*, which covers the years 1946–62, and in the follow-on series, *Selected RAND Abstracts*, published quarterly and collected in annual volumes beginning in 1963. These indexes are available in many public and academic libraries throughout the free world.

Of general interest are the following two publications, sponsored by the Operations Research Society of America:

A Comprehensive Bibliography on Operations Research, Wiley, New York, 1963.

International Abstracts in Operations Research, Operations Research Society of America, Baltimore, Md. (published quarterly).

Aside from studies that have appeared in commercial editions, the Reports (R's) and Memoranda (RM's) of The RAND Corporation listed in this bibliography may be obtained by writing directly to the RAND Reports Department, 1700 Main Street, Santa Monica, California 90406, U.S.A. Wherever a DDC document (AD or ATI) number is given in the citation, the publication is also available from the Defense Documentation Center, Cameron Station, Alexandria, Virginia 22314, U.S.A. or from the Clearinghouse for Federal Scientific and Technical Information, U.S. Department of Commerce, Springfield, Virginia 22151, U.S.A.

Papers (P's) are contributions of individual RAND authors to their professional fields. Copies of those cited in the bibliography are available from RAND and, if an AD number is provided, from the Defense Documentation Center or from the Clearinghouse for Federal Scientific and Technical Information.

Commercially published books by RAND authors are available *only through the publisher or through a bookseller.*

INDEX

A fortiori analysis, definition, 77
  use of, to handle uncertainty, 95–96
  see also Dominance
Agathon, on the improbable, 323
Aggregation, definition, 63
  level of, in the synthesis of component studies, 77–78
  role of, in systems analysis, 63–64
  see also Component studies
Air Battle Model, *see* STAGE
Aircraft, cost-sensitivity of alternatives for SLBM defense, Chapter 8 (138 ff.)
  future bomber, resource analysis of, Chapter 9 (153 ff.)
  simulation model of support requirements for, Chapter 13 (255 ff.)
  see also Interceptor aircraft
Aircraft engines, illustration of problems in selection of, 55–59
  "state-of-the-art" illustrated by reference to, 104–107
Air defense, difficulties of selecting the "best technical approach" illustrated by reference to, 108–118
  game-theoretic model of, 229–233
Airpower, trade-offs with groundpower, Chapter 2 (20 ff.), Chapter 21 (388 ff.)
Air warfare, tactical, game-theoretic model of, 233–239
Alternatives, danger of arbitrarily excluding, 80, 420
  definition, 12, 55
  identification of, 37–38, 423
  illustration of approaches to reduce the number considered, 46–48
  necessity of using indirect measures of effectiveness of, 361–362
  use of cost-effectiveness analysis in comparing, 4–5
  see also Criteria, Evaluation, Interpretation
Analysis, *see* Cost-effectiveness analysis, Cost-sensitivity analysis, Operations analysis, Operations research, Resource analysis, Sensitivity analysis, Systems analysis, Trade-off analysis, *etc.*
Analytic models, *see* Models
Applied research, relation of costs, criteria, and objectives in, 55–59
  technological considerations in the decision to proceed to Contract Definition, 98–101
Assured destruction, and the desire to avoid collateral damage, 367–368, 369
Ayres, L. P., cited, 2*n*l
  on a statistical problem from World War I, 346–347

B-52, resource analysis of a hypothetical successor to, Chapter 9 (153 ff.)
"Best estimate" analysis, definition, 77
"Best technical approach," problems in selecting, for weapon systems, 108–118
  see also RDT&E, Technological considerations
Bias, as a source of error in systems analysis, 350–352, 386–387, 420
  see also Fallacies
Biryuzov, S. S., cited, 319
Bombing, strategic, influence of cherished beliefs on studies of, 351–352
Branch-point technique, in gaming, 268
Brown, T. A., cited, 285*n*3
Budgetary constraints, influence of, on force posture planning, Chapter 7 (138 ff.) 280, 371–372
  see also PPBS, Resource analysis, Strategic forces
Burke, T. F.: model of hard point defense, 214 ff.

California, pioneering efforts to apply systems analysis to state problems, 10
Campaign models, *see* Models
Central war, *see* Thermonuclear war
Closed-play technique, in gaming, 268
Cold War game, purpose, 267
Committees, avoiding the disadvantages of, through the Delphi technique, 334
  weaknesses of, as an instrument for utilizing groups of experts, 32–33, 225, 328–329
Communication failure, among experts, as a source of error in systems analysis, 108, 353
  role of the model in helping to reduce, 417, 423
  see also Experts, Fallacies, Models, Nonquantitative considerations
Component studies, dependence of systems analysis on, 9, 41–42
  methods of estimating the accuracy of, 72
  necessity of viewing in a systems context, 420
  role in alleviating uncertainties, 71–72

synthesis of, 64, 67–68, 75–78, 80, 233
see also Aggregation
Computational modeling, *see* Computers, Models, Simulation
Computers, danger of overemphasizing the use of, 421
growing usefulness of, in systems analysis, 427–428
models designed for use on, 222–225
role of, in analysis, 15–17
use of, in simulations, 242–243, 245–246, 248–250, 252–254, Chapter 13 (255 ff.)
Concept Formulation, as a decision-point in the RDT&E process, 98–101
cost-effectiveness analyses in, 100, 118–123
definition, 99–100
selection of the "best technical approach" in, 108–118
Conflict, as a major consideration in the analysis of strategic forces, 282, 311–312
as an essential ingredient in some models, 8–10, 222
game-theoretic treatment of, 93–94, 228–229
necessity of accounting for, 219–220, 365–366, 421
Context, example of, in a trade-off analysis of groundpower and airpower, 393–394
inattention to, as a source of error in systems analysis, 357
influence of, in determining the scope of cost-effectiveness comparisons, 78–80, 118–120
influence of, in resource analysis, 127–130
influence of, in selecting the "best technical approach" among alternative weapon systems, 108–110
major changes in, for systems analysis over the last decade, 365–372
political, modeling of through scenarios, 305–309
Contextual map, as a tool for facilitating communication among experts, 329
Contingencies, as a technical term in systems analysis, defined, 71
basic importance of, in systems analysis, 78–80
strategic, model of, 91–93
techniques for limiting the number to consider, 77–78
see also Effectiveness
Contract Definition, as a decision-point in the RDT&E process, 98, 99, 101
selection of the "best technical approach" in, 118
Cost categories, for aircraft (Table), 166
Costs, as a fundamental variable in systems analysis, 78
definition, 12, 55
non-dollar, difficulty of estimating, 21
see also Cost-sensitivity analysis, Non-quantitative considerations, Resource analysis
Cost-effectiveness analysis, advantages of using, in early stages of a development program, 259
and the present composition of U.S. missile forces, 380–384
as an outgrowth of operations analysis, 3
choice of criteria for, 57–59
dangers of nonrelative comparisons in, 120–121
difficulty of comparing systems of varying state-of-the-art through, 119–120, 154
in Concept Formulation, 100, 118–123
relation to systems analysis, 3–5
relation to trade-off analysis, 20–21
some common misconceptions about, 15–17
use in comparing competitive weapon systems, 100, 104, 118–123
Cost-sensitivity analysis, basic types of, 192–193
definition, 138
example of, to compare aircraft systems for SLBM defense, 139–151
methods of, for future aircraft systems, 192–206
usefulness of, 151–152
see also Resource analysis, Sensitivity analysis
COW (Cold War game), as a game model, 267
Crisis exercises, as a type of model, 224
Criteria, choice of, in selection of strategic forces, 75–78, 373–380
confusion with "effectiveness scale," 55
danger of overspecifying, 57
dangers of using side issues as, 357
definition, 12, 55
illustration of, for evaluating mixes of groundpower and airpower, 390–400
influence of cost-effectiveness considerations on, 56–59, 424
necessity of allowing for uncertainty in the selection of, 75–80, 365–372, 384–387

need for care in selecting, 50–53, 218
selection of, through utility theory, 85–91
usefulness of scenarios in specifying, 300, 304–305
use of a maximized effectiveness-cost ratio as, 57–58
use with probabilistic models, 73–75
see also Alternatives, Dominance, Evaluation, Fallacies, Interpretation, Nonquantitative considerations, Uncertainty

Damage limitation, analysis of, in recent studies, 366–369
as a strategic objective, 65
example of the problem of allocating resources to achieve, 373–380
Data collection, difficulties of, in systems analysis, 41–42
see also Models, Nonquantitative considerations, Search, Verification
Decisionmaking, accuracy of game models of, 332
advantages and limitations of simulation models in, 250–252
alternatives to systems analysis as sources of advice for, 31–33
as a political process, 17, 51
contributions systems analysis can make to, 425–426
dangers of neglecting subjective elements in, 359–360
declining usefulness of traditional methods of, 6
Delphi technique as a source of inputs for, 333–342
example of strategies for, to determine composition of U.S. missile forces, 380–384
future of systems analysis in contributing to, 427–429
illustration of problems in allocating a strategic budget, 64–78
illustration of problems in selecting a new aircraft engine, 55–59
illustration of problems in selecting best control mode for interceptor forces, 59–64
in the interpretative stage of systems analysis, 50–53
levels of, 7–10
major changes in the environment for, over the last decade, 365–372
role of PPBS in, 6–7
role of quantitative factors in 4–5, 13–15
selection of appropriate criteria for, in risky or uncertain situations, 85–91
"technical feasibility" as a consideration in, 104–108
typical decisions involving technological considerations, 98–104
usefulness of cost-effectiveness analyses in, 4–5, 15–17
usefulness of game theory to, 239–240
usefulness of political analyses in, 311–313, 322–323
usefulness of resource analysis in, 129
usefulness of SAFE in providing inputs for, 295–297
use of the computer in, 17, 427–428
use of the multidimensional scenario to model, 327–328
see also Criteria, Interpretation, Intuition, Judgment, Nonquantitative considerations
Decision theory, usefulness of, in systems analysis, 384–385
see also Decisionmaking, Utility theory
Defense management, range of problems amenable to analysis, 7–10
see also Military planning
Delphi technique, advantages and limitations of, 342–343
as a source of inputs to verbal models, 225
as a technique for utilizing groups of experts, 45, 333–334
example of use of, to obtain policy advice, 335–342
example of use of, to select a number 334–335
likely future developments of, 343
Deployment, as a decision involving technological considerations, 102
see also RDT&E, Technological considerations
Deterrence, as a strategic objective, 51, 65
increased sophistication in the analysis of, 366–369
Development, of systems, as a decision-point in the RDT&E process, 98, 101
see also RDT&E, Technological considerations, Weapon systems
Dogmatism, as a source of error in systems analysis, 349–350
Dominance, as a criterion, 58–59, 95–96
see also A fortiori analysis
Downs, Anthony, cited, 225

Eaker, Ira C., on cost-effectiveness analysis, 16

Effectiveness, approximate character of measures of, 361–362
as a technological consideration in the RDT&E process, 98–99
difficulties in measuring, 20–21, 59–60, 61–62
examples of the use of gaming to measure, 273–277
measurement of, within contingencies, 72–75
measures of, in a game-theoretic sense, 230–231, 235–236
of strategic forces, influence of national goals on the measurement of, 64–66
role of simulation models in measuring, 247
scales of, in strategic studies, 66–68
use of an effectiveness scale to define, 54–55
use of campaign models in measuring, 61–62
use of utility theory to determine, 85–91
see also Alternatives, Cost-effectiveness analysis, Evaluation, Interpretation, Nonquantitative considerations, Uncertainty
Einstein, Albert, on the uniformity of nature, 311
Ellsberg, Daniel, on models and systems analysis, 226–227
End product orientation, in resource analysis, 131, 133–136
Engineering, relation of systems analysis to, 3–4, 30–31
Enthoven, A. C., on analysis at the national policy level, 4
on quantification as a guide to understanding, 48
on the role of judgment in systems analysis, 428
on the use of sensitivity analysis in the DOD, 356
Errors, *see* Fallacies, Uncertainty
Estimating relationships, use of, in resource analysis, 136–137
Evaluation, as a step in systems analysis, 33, 42–50
danger of selecting an alternative on too narrow a basis, 424
dangers of using side issues as criteria in, 357
disregard of initial assumptions as a source of error in, 355
failure to reappraise results as a source of error in, 360
necessity of preparing cost estimates in, 424
of trade-off analyses, by examining their basic structure, 28, 388
overconcentration on the model as a source of error in, 353
uncritical acceptance of model outputs as a source of error in, 354–355
use of models in, 42–50
see also Alternatives, Fallacies, Models, Nonquantitative considerations, Uncertainty, Verification
Expected value, as a criterion, 85–91
Expected-value models, contrast with Monte Carlo models, 216, 244–245, 355–356
definition, 216, 244
see also Models, Simulation
Experimentation, in systems analysis, through working with the model, 416–417
see also Verification
Experts, efficient use of, the essence of systems analysis, 41, 423
judgment of, in handling nonquantitative considerations, 324–325
methods of utilizing, when analytic models are lacking, 327–328, 328–343
objective measures of the qualifications of, 325–326
obtaining judgments of, through Delphi technique, 333–342
see also Decisionmaking, Judgment

Fallacies, adherence to cherished beliefs, 350–351, 386–387
arbitrarily establishing external system requirements, 121–123
attempting to build universal models, 357–359
communication failure, 108, 353
comparing systems of varying state-of-the-art, 119–120
disregarding initial assumptions in interpreting model outputs, 355
excessive attention to detail, 353–354
failure to consider alternatives across contingencies, 79–80
failure to reappraise the work, 360
inattention to uncertainties, 356–357
inflexibility in the face of evidence, 349–350
neglect of subjective elements, 359–360
neglect of the question, 354
nonrelative comparisons, 120–121

not-invented-here factor, 108, 352
overconcentration on statistical uncertainty, 355–356
overconcentration on the model, 353
overemphasis on quantitative factors, 13–15, 40, 359–360, 387, 421
overlooking discontinuities in the results, 112–113
overoptimism, 108
oversimplification of assumptions, 113
parochialism, 352
uncritical use of model outputs, 354–355
use of side issues as criteria, 357
see also Evaluation, Formulation, Interpretation, Nonquantitative considerations, Uncertainty
Flexibility, of systems, as a hedge against uncertainty, 40, 95, 371–372, 385–386
FLIOP, as a type of computer simulation model, 224, 248–249
Force structures, analysis of, to achieve damage limitation, 373–380
determinants of, 279–281
increased emphasis on highly specialized weapon systems in, 369–371
limitations of trade-off analysis in comparing, 413–417
necessity of flexibility in, 369–372, 385–386
relation of strategic and general purpose forces, 367, 368–369
tools for the analysis of, 281–286
usefulness of analysis in planning, 281–286
use of gaming to examine requirements for, 274–275, Chapter 15 (279 ff.)
see also Weapon systems
Formulation, central place of, as a step in systems analysis, 33–41, 419–420
dangers of overemphasizing statistical uncertainty in, 355–356
definition of, 33, 35–36
examples of, in game theory, 229–231, 233–235
excessive attention to detail as a source of error in, 353–354
fallacy of underemphasizing, 348–349
in gaming, 269–271
major changes in the character of, over the last decade, 365–372
problem of, in an air defense analysis, 110
subjectivity in, 36, 420
see also Intuition, Judgment, Models, Nonquantitative considerations
Fort, D. M., on balancing theoretical and empirical research, 42
on the explicit treatment of uncertainty, 46–47, 48

Game theory, applications of, 93–94, 239–240
extensive form of models of, 229
illustration of, in modeling air defense, 229–233
illustration of, in modeling tactical air warfare, 233–239
models used in, 222–223
normal form of models of, 229
role of, in systems analysis, 228–229
see also Gaming, Models, Verification
Gaming, analysis phase in, 272–273
as a research technique, when analytic models are lacking, 329–333
as a way of enhancing communication among experts, 332–333
definition, 265–266
essential characteristics of, 267
examples of 273–277, 330–332
inputs, major types of, 270
most fruitful applications of, 332
operational, as a technique of systems analysis, 44, 330–333
play phase in, 271–272
preparation phase in, 269–271
realism of results of, 272–273, 278, 332
techniques of, 268–269
types of games, 266, 285–286
usefulness of, in analyzing strategic force postures, 282–285
use in modeling limited war, 268–269
use of scenarios in, 301–302
see also Game theory, Models, RAND Corporation, SAFE, Verification
General purpose forces, relation to strategic forces, 367, 368–369
Golubev-Novozhilov, S., on the usefulness of simplified game-theoretic models, 240
Groundpower, trade-offs with airpower, Chapter 2 (20 ff.), Chapter 21 (388 ff.)

Helmer, Olaf, cited, 285*n*3
on methods of identifying experts, 325–326
on refining projections derived through operational gaming, 331–332
on the necessity of relying on expert judgment in systems analysis, 325
on the need for enhancing communication among experts, 329

on the usefulness of operational games, 330
on the usefulness of simulation models in aiding communication among experts, 330
Hitch, C. J., cited, 7
on cost-effectiveness analysis, 17
on fixed rules or models for decisionmaking, 358–359
on judgment in systems analysis, 325
on the limits of systems analysis, 213–214
Hoffman, F. S., cited, 291
Huxley, T. H., on attention to detail, 346

Initial investment, as a type of cost, 129, 165
method of estimating costs of, for future aircraft systems, 173–180
see also Resource analysis, Resource analysis categories
Interceptor aircraft, difficulties of selecting the "best technical approach" illustrated by reference to, 108–118
illustration of problems in selecting best method of control of, 59–64
Interdisciplinary research, in systems analysis, 11, 413
Interpretation, as a step in systems analysis, 33, 50–53
dangers of neglecting subjective elements in, 359–360
parochialism as a source of error in, 352
uncritical acceptance of model outputs as a source of error in, 354–355
see also Criteria, Decisionmaking, Evaluation, Judgment, Nonquantitative considerations
Intuition, central place of, in systems analysis, 220–221, 333, 344, 345, 363, 387, 420, 426, 427
defects of, as an alternative to systems analysis, 31
see also Decisionmaking, Judgment, Nonquantitative considerations

Jantsch, Erich, on the potential of the Delphi technique, 343
Johnson, Lyndon B., cited, 314, 320
on Planning-Programming-Budgeting System, 7
Jones, R. V., on the use of chaff, 352
Jones, W. M., on the importance of organizational considerations in systems analyses, 425
Judgment, analysis as a substitute for, 66
efficient use of, the essence of systems analysis, 423
errors arising through the neglect of, 359–360
expert, measures of, 325–326
fallacy of attempting to build models to supplant, 357–359
of a committee, as an alternative to systems analysis, 32–33
of a single expert, as an alternative to systems analysis, 31–32
of experts, methods for obtaining when analytic models are lacking, 327–343
role of, in building models, 220–221
role of, in problem formulation, 36, 420
sources of differences in, among experts, 108
subjective, key role of in systems analysis, 4–5, 11, 17, 31–32, 36, 78, 324–325, 333, 344, 345, 363, 426, 427
subjective, in allocating resources for damage limitation, 373–380
see also Decisionmaking, Intuition, Nonquantitative considerations

Kahn, Herman, on cherished beliefs as sources of error in systems analysis, 351
Katzenbach, E. L., Jr., on the persistence of horse cavalry, 350
Kennedy, J. F., cited, 314, 319
Kennedy, J. L., on the origin of the contextual map, 329
Kent, G. A., on the primary output of an analysis, 52
Khrushchev, Nikita, cited, 319

Laird, Melvin, cited, 298
on cost-effectiveness analysis, 16
Lavin, M. M., on judgment in systems analysis, 221
Logistics, SAMSOM model as a tool of analysis for, Chapter 13 (255 ff.)

MAGIC (Manual Assisted Gaming of Integrated Combat), as a game model, 267–268
Maintenance, aircraft, use of SAMSOM to estimate requirements for, Chapter 13 (255 ff.)
Malinovsky, R. Y., cited, 319
Management science, aim of, 7–8
as outgrowth of operations analysis, 3
relation to systems analysis, 3–4

Mann, I., on cherished beliefs as sources of error in systems analysis, 351
Map exercises, as a type of model, 246, 266, 267–268
Mathematical models, types of, 222–224
 see also Models
Mathematics, danger of overemphasizing, in systems analysis, 421
 central role of, in narrower analytic problems, 7–8
 see also Quantitative considerations
McNamara, Robert S., cited, 7, 382
 on DOD's reliance on cost-effectiveness analyses, 298
Measures of effectiveness, *see* Effectiveness
Methods and procedures, of systems analysis, *see* Cost-sensitivity analysis, Resource analysis, Sensitivity analysis, Trade-off analysis, *etc.*
Military assistance programs, examination of, through gaming, 275–277
Military planning, advantages of systems analysis over its competitors in, 31–33, 348, 418–419, 426–427
 as a political process, 17, 51
 basic conceptual approaches to, 43, 124
 for strategic forces, role of analysis in, 281–286
 future of systems analysis in contributing to, 427–429
 limitations of systems analysis for, 360–363
 major changes in the environment of, 365–372
 new approaches to, since 1961, 1
 ultimate dependence on human judgment, 4–5, 78
 usefulness of SAFE to, 295–297
 see also Decisionmaking, Uncertainty
Military space program, political analysis of requirements for, 314–322
Minimax theorem, use of, as a criterion, 93–94, 365–366
Missile defense, model of, illustrated, 214–220
 cost-sensitivity analysis of aircraft systems for, Chapter 8 (138 ff.)
Missile forces, example of decisionmaking strategies to determine composition of, 380–384
Models, and the selection of criteria, 218
 as the essence of systems analysis, 11, 31
 campaign type, 43, 61–62
 classification of, 221–226
 cost models, 43
 cost-sensitivity, illustration of types of output from, 141–151
 dangers of failing to let the question determine the design of, 354, 420–421
 dangers of overemphasizing statistical uncertainty in building, 355–356
 dangers of uncritical acceptance of outputs from, 354–355
 definition, 12, 211–213
 difficulty of structuring, 43–44
 excessive attention to detail as a source of error in constructing, 353–354
 expected-value type, 216, 244–245, 355–356
 families of, 253–254, 357*n*10
 game-theoretic type, 228–229
 game-theoretic, examples of, 229–239
 game type, 225
 mathematical type, 222–223, Chapter 11 (228 ff.), 242, 282–285
 Monte Carlo type, 216, 244–245, 355–356
 nature of, in operations research, 4
 necessity of accounting for nonquantifiable considerations in using, 49, 232–233
 nonquantitative, and expert judgment, 324–325
 nonquantitative or verbal type, 222, 225–226
 overconcentration on, as a source of error in systems analysis, 353
 predictive accuracy as the test of, 50
 probabilistic type, use of, 73–75
 refinement of, 49–50, 416–417
 role of judgment and intuition in constructing, 220–221, 415
 simulation type, 224, 242
 simulation type, methods of handling time in, 243–244
 simulation type, use of, 250–252
 universal, fallacy of attempting to design, 245, 357–359
 use in evaluating alternatives, 42–50
 usefulness to decisionmakers, 213–214
 verification of, 212, 220
 see also Evaluation, Interpretation, Intuition, Judgment, Nonquantitative considerations, RAND Corporation, Verification
Modification, of weapon systems, as a decision involving technological considerations, 102–103, 116–117

see also RDT&E, Technological considerations
Monte Carlo models, contrast with expected-value models, 216, 244-245, 355–356
definition, 216, 244–245
SAMSOM as an example of, 224, Chapter 13 (255 ff.)
see also Models, Simulation
Multidimensional scenarios, as a technique for modeling organizational behavior, 327–328
see also Scenarios

Nelson, Richard, cited, 283*n*l
Nonquantitative considerations, example of importance of, in determining strategic forces for damage limitation, 373–380
limitations of using proximate quantifiable variables to express, 85–91
methods of handling, when analytic models are lacking, 49–50, 222, 225–226, 327–328, 328–343
necessity of considering, in systems analysis, 13–15, 49, 219, 232–233, 387, 415, 427–428
neglect of, as a source of error in systems analysis, 359–360
role of expert judgment in handling, 324–325
treatment of, in game theory, 232–233
treatment of, through political analyses, 311–313, 322–323
treatment of, through scenarios, 304, 305–309
treatment of, through simultation models, 250–252
use of the Delphi technique to handle, 333–342
see also Decisionmaking, Experts, Interpretation, Intuition, Judgment, Uncertainty, Verification
Novick, David, cited, 7
Nuclear war, *see* Thermonuclear war

Objectives, consistency of, with higher goals, 39
definition, 12, 55
discovery of, through systems analysis, 39
essentially nonquantitative character of, on national level, 312–313
identification of, as a key task of systems analysis, 4–5, 36
importance of selecting carefully, 423
impossibility of measuring attainment of directly, 361–362
proximate expressions of, in uncertain or risky situations, 85–91
strategic, nature of, 64–66
see also Formulation, Models, Nonquantitative considerations
Objective probability distributions, defined, 81–82
role in decisionmaking, 82–85
see also Decision theory, Judgment, Uncertainty, Utility theory
Open-play technique, in gaming, 268
Operating costs, definition, 129, 165
method of estimating, for future aircraft systems, 180–188
see also Resource analysis, Resource analysis categories
Operational gaming, *see* Gaming
Operations analysis, development and applications in World War II, 2–3
Operations research, as an outgrowth of operations analysis, 3
problems handled by, 7–8
relation to systems analysis, 3–4
use of the techniques of, in early systems analyses, 13–14
Optimization, comparison with insensitivity as the aim of systems analysis, 15
see also Criteria, Effectiveness, Objectives, Suboptimization, Technological considerations

Parametric analysis, relation of trade-off analysis in Concept Formulation and Contract Definition, 113–115
use in establishing design requirements for weapon systems, 110–113
see also Sensitivity analysis
Parochialism, as a source of error in systems analysis, 352–353
Pastore, John O., on cost-effectiveness analysis, 16
Personnel requirements, method of estimating, for future aircraft systems, 156–165
Phase-out, of systems, as a decision involving technological considerations, 103–104
Pitfalls, *see* Fallacies
Planning - Programming - Budgeting System, *see* PPBS
Policy planning, *see* Military planning, Systems analysis

Political analysis, as a method for obtaining the judgment of an individual expert, 327
as a technique of systems analysis, 311–313, 322–323, 327
example of, involving U.S. space policy, 314–322
Political factors, credibility of, in scenarios, 305–309
impact of, in selecting alternatives for analysis, 122–123
see also Nonquantitative considerations
Posvar, W. W., on subjective judgments in systems analysis, 32
PPBS (Planning-Programming-Budgeting System), adoption throughout the Federal government, 7
as a framework for obtaining quantitative estimates, 15
as a management tool, 6–7
distinguishing characteristics, 7
role of systems analysis in, 6–7
Postattack recovery, use of the Delphi technique to identify policies for, after thermonuclear war, 335–342
Probabilistic models, use of, 73–75
see also Models
Probability distributions, objective, defined, 81–82
subjective, defined, 82
Probability theory, *see* Decision theory, Risk, Uncertainty Utility theory
Program budgeting, *see* PPBS

Quantitative considerations, dangers of overemphasizing, 13–15, 40, 359–360, 387, 421
role in decisionmaking, 4–5
usefulness in systems analysis, 48–49
see also Decisionmaking, Judgment, Mathematics, Models, Nonquantitative considerations

RAND Corporation, models devised at, examples, 214–220, 224, 248–249, 253–254, Chapter 13 (255 ff.), 267–268, 273–277, Chapter 15 (279 ff.), 300–301, 403–409
pioneering work on systems analysis, v, 1
role in developing modern form of PPBS, 7
studies of strategic bombing, 351–352
study of strategic air bases, 36–37
RDT&E (Research, Development, Test and Evaluation), as a major consideration in the analysis of strategic forces, 282
costs of, defined, 165
dangers of emphasizing optimization in the analysis of questions of, 423–424
decision-points in the process of, 98–101
design of programs of, to ensure flexible force structures, 369–372, 385–386
method of estimating costs of, for future aircraft systems, 167–173
modeling of, in SAFE, 285–286, 289–290
need to develop highly specialized weapon systems, 369–371
resource analysis for, 129
rising costs of, 371–372
strategies for, to hedge against uncertainty, 371–372
see also Military planning, Technological considerations, Weapon systems
REDWOOD, as a game model, 267–268
Replacement, of systems, as a decision involving technological considerations, 103–104
Research, as a step in systems analysis, *see* Search
Research and development, *see* RDT&E
Resource analysis, definition, 125
essential role of, in the choice among alternatives, 424
example of, involving a hypothetic bomber system, Chapter 9 (153 ff.)
incremental resource requirements as the focus of, 133
influence of technological considerations in, 98–104
relation to systems analysis, 124–125
scope of problems considered in, 129–130
static display of results in, 190–191
time horizon of, 127–129
time-phased display of results in, 136, 191–192
types of decisions informed by, 129
use of functional categories in, 131–132
use of resource categories in, 131–132
see also Cost-sensitivity analysis.
Resource analysis categories, for aircraft (Table), 166
Risk, contrasted with uncertainty, 82, 385
treatment of, 82–89
Rivers, L. Mendel, on cost-effectiveness analysis 16

SAFE (Strategy-And-Force-Evaluation game), advantages of, 296–297

as a game model, 267
comparison with other analytic tools, 285–286
definition, 285
example of plays in, 291–295
range of weapon systems that can be studied in, 286–289
routine of play, 290–291
limitations of, 295–296
SAMSOM (Support-Availability Multi-System Operations Model), as a type of logistics simulation model, 224
example of outputs from, 258–264
key features of, 256
origin and purpose, 255–258
Scale of effectiveness, *see* Effectiveness
Scenarios, as an input to larger studies, 43, 300–301
as a method for obtaining the judgment of an individual expert, 327–328
as a type of model, 225
causes of controversy about, 307–309
content of, on various decision levels, 302–305
credibility of assumptions in, 305–309
definition, 299–300
form and content of, determined by research task, 301–302
multidimensional, as a technique for modeling organizational behavior, 327–328
usefulness of, in analyzing strategic force postures, 282–285
Scientific method, compared with systems analysis, 30–31, 363, 416, 427
Search, as a step in systems analysis, 33, 41–42
definition, 33, 41
role of component studies in, 41–42
see also Models, Uncertainty
Seminar technique, in gaming, 268
Sensitivity analysis, importance of, in systems analysis, 423
types of, in resource analysis, 192–193
use of, to handle uncertainty, 72, 96, 116, 117, 118, 119–120, 133, 356–357, 416
see also Cost-sensitivity analysis, Uncertainty
Series and variation technique, in gaming, 269
SIERRA, as a game model, 267–268
Simulation, advantages of, 250–251
as an input to systems analyses, 241, 247
as a substitute for traditional analytic models, 44
computer languages for, 253
customary elements of, 243
definition, 44, 224, 241
event technique for handling time in, 243–244
examples of, 248–250
initial procedures, 247–248
interval technique for handling time in, 243
limitations of, 251–252
new developments in, 252–254
reasons for use of, 246–247
relation to mathematical models, 242
use of computers in, 242–243, 245–246, 248–250, 252–254, Chapter 13 (255 ff.)
usefulness of, in analyzing strategic force postures, 282–285
see also Gaming, Models
SLBM (Submarine-Launched Ballistic Missile), cost-sensitivity analysis of aircraft systems to defend against, Chapter 8 (138 ff.)
interception, methods of, 139–140
Soviet Union, *see* U.S.S.R.
Space, political analysis of military role in, 314–322
Space technology, as a major consideration in formulating U.S. space policy, 315–317
Staff studies, relation to systems analyses, 5
STAGE, as a type of model, 224, 267
operation and outputs of, 248
use of scenarios in, 301
State-of-the-art, and technical feasibility, 104–108
as a consideration in the RDT&E process, 98–99
see also Technological considerations
Strategic Air Command, RAND study of SAC bases, 36–37
Strategic Air War game, purpose, 267
Strategic forces, analytic tools for the study of, 281–286
illustration of problems in allocating budget for, 64–78, 373–380
interactions of policy and posture for, 279–281
relation to general purpose forces, 367, 368–369
Strategic Operations Model, *see* STAGE
Strategic studies, increased sophistication of, over the last decade, 366–369
Strategy, optimal, in a game-theoretic sense, 229–230, 231, 236–238
Strategy-And-Force-Evaluation game, *see* SAFE

Subjective probability distributions, definition, 82
role in decisionmaking, 82–85
see also Decision theory, Judgment, Uncertainty, Utility theory
Subjectivity, *see* Decisionmaking, Intuition, Judgment, Nonquantitative considerations, Uncertainty
Submarine-launched ballistic missiles, *see* SLBM
Suboptimization, dangers of carelessness in, 421
definition, 63
necessity of, as a limitation of systems analysis, 361–362
parametric analysis as a technique for, 110–112
role of, in systems analysis, 38, 63
"rubber", 71
see also Criteria, Effectiveness, Objectives, Optimization, Technological considerations
Systems analysis, advantages of, over its alternatives, 31–33, 348, 418–419, 426–427
applications of, 7–11
as an outgrowth of operations analysis, 3
basic conceptual approaches to, 43, 124
causes of changes in the conduct of, over the last decade, 365–372
central role of models in, 11, 31, 212–214, 232–233, 418
comparison with engineering, 3–4, 30–31
comparison with scientific method, 3, 30–31, 363, 427
considerations in evaluating an analysis, 28–29
definition, 1–2
differing views of analysts about the uses of, 385
early preoccupation with mathematical tools, 13–14
elements of, 12–13, 54–55
factors that complicate the task of, 345
future of, 427–429
impossibility of predicting the future through, 363
incomplete studies typical, 52–53, 361
initial Pentagon resistance to, 3
limitations of, 18, 360–363, 426–427
methods of handling nonquantitative considerations in, Chapter 18 (324 ff.)
nonmilitary applications of, v, 10
process of, 12–13, 33–53
purpose of, 2, 3, 40, 97, 232–233
reasons for its acceptance at the national policy level, 5–7
resource analysis as an input to, 124–125
role in PPBS, 6–7
role of political analyses in, 311–313 322–323
sources of error in, 13–15, 40, 79–80, 108, 112–113, 119–120, 120–121, 121–123, 348–360, 386–387, 421
summary of principles for conduct of, 419–425
technological considerations in, Chapter 6 (97 ff.)
use of gaming in, Chapter 14 (265 ff.) 282–285
use of scenarios in, 282–285, Chapter 16 (298 ff.)
see also Cost-effectiveness analysis, Cost-sensitivity analysis, Criteria, Decisionmaking, Judgment, Military planning, Resource analysis, Sensitivity analysis, Trade-off analysis, *etc.*
Systems design, comparison with system study, 364–365
misleading use of the term, 9*n*7
Systems engineering, as an outgrowth of operations analysis, 3
misleading use of the term, 9*n*7
relation to systems analysis, 3–4
Systems research, misleading use of the term, 9*n*7

TAGS (Theater Air-Ground Study model), as a type of model, 224
TAGS-II, design and outputs of, 403–409
Technical feasibility, *see* State-of-the-art, Technological considerations
Techniques, of systems analysis, *see* Cost-sensitivity analysis, Resource analysis, Sensitivity analysis, Systems analysis, Trade-off analysis, *etc.*
Technological considerations, analysis of, in studies of weapon systems, Chapter 6 (97 ff.)
as an influence on posture planning for strategic forces, 280
dangers of seeking optimal solutions, 423–424
increased emphasis on highly specialized weapon systems, 369–371
increased emphasis on improving compatibility among systems, 368
role of, in formulating U.S. space policy 315–317, 322

rising costs of research and development, 371–372
see also RDT&E, State-of-the-art, Uncertainty
Theater Air-Ground Study model: *see* TAGS, TAGS-II
Thermonuclear war, increased sophistication in the analysis of, 366–369
initiation of, 368–369
trade-offs between strategic offensive and strategic defensive forces to achieve damage limitation, 373–380
use of the Delphi technique to identify policies for recovery after, 335–342
Thomas, C. J., on the applications of game theory, 239, 239–240
Time horizon, effects of, on resource analysis, 127–129
Time-phasing, illustration of, 151, 191–192
of resource requirements, in resource analysis, 129–130, 136
Titov, G. S., cited, 319
Trade-off analysis, between strategic offensive and strategic defensive capabilities, 373–380
contrast with cost-effectiveness analysis, 20–21
definition, 21
example of, for comparing airpower and groundpower, Chapter 2 (20 ff.), Chapter 21 (388 ff.)
example of approaches toward, 389–390
in Concept Formulation, 100
levels of analysis and criteria for evaluating groundpower and airpower, 390–400
limitations of, for comparing alternative force mixes, 413–417
through SAFE, 296–297
use of, to improve compatibility among systems, 368
various meanings of, 113–115

Uncertainty, contrasted with risk, 82, 385
dangers of overconcentration on just one type of, 355–356
essential difficulty of handling in systems analysis, 39–41, 311–312, 424
explicit treatment of, 46–49, 133, 421
general guidelines for handling, in systems analysis, 94–96
hedges against, in formulating an R&D program, 371–372
inattention to, as a source of error in systems analysis, 356–357
in SAFE, 286, 295–296
major types of, 9, 45–46, 69–71, 91–94, 115–118, 138, 311–312, 389
models of, 91–94
treatment of, in resource analysis, 133, 136
treatment of, in strategic analyses, 69–72, 91–94, 282, 384–385
treatment of, through political analyses, 311–313, 322–323
treatment of, through scenarios, 305–309
treatment of, through sensitivity analysis, 151–152, 356–357
use of the Delphi technique to reduce, 333–344
use of gaming to reduce, 329–333
use of models to reduce, 91–94, 226–227, 247
see also Intuition, Judgment, Models, Verification
United States, and Soviet attitudes toward thermonuclear war, 366–369
decisions leading to present missile forces, 380–384
development of systems analysis in, 3, 13–14, 364–372, 384–387, 427–429
space policy of, 314–322
Universal models, *see* Models
U.S.S.R., changed U.S. perceptions of, in the last decade, 365–366
intentions in space, and U.S. space policy, 314–322
see also United States
Utility theory, use in selecting criteria, 85–91

Verification, as a step in systems analysis, 33–34
difficulty of, 212, 220, 226–227, 363
in SAFE, 293–297
of game results, 272–273, 278, 332
of political analyses, 322–323
of scenario assumptions, 305–309
of simulation models, 251–252
see also Decisionmaking, Intuition, Judgment, Models, Uncertainty

War games, as a type of model, 224, Chapter 14 (265 ff.)
War gaming, *see* Gaming
Watson-Watt, Robert, cited 352
Weapon systems, changing concept of, 368
design and development of, through analysis, 8–11

future aircraft, resource analysis of, Chapter 9 (153 ff.)
increased emphasis on high degree of specialization of, 369–371
increased emphasis on improving compatibility among, 368
necessity of flexibility in, 369–372, 385–386
rising costs of research and development of, 371–372
technological considerations in the analysis of, Chapter 6 (97 ff.)
types of decisions made during the lifespan of, 98
see also Cost-effectiveness analysis, Cost-sensitivity analysis, Force structures, Resource analysis, SAMSOM
Williams, J. D., on the applications of game theory, 239
Wilson, E. B., Jr., on selecting manageable problems for analysis, 38
"Worst case" analysis, definition, 77
and game theory, 94
Worth, *see* Effectiveness

SELECTED RAND BOOKS

Bellman, Richard, and Stuart E. Dreyfus. *Applied Dynamic Programming.* Princeton, N. J.: Princeton University Press, 1962.

Bellman, Richard, R. E. Kalaba, and M. C. Prestrud. *Invariant Imbedding and Radiative Transfer in Slabs of Finite Thickness.* New York: American Elsevier Publishing Company, Inc., 1963.

Bellman, Richard, H. H. Kagiwada, R. E. Kalaba, and M. C. Prestrud. *Invariant Imbedding and Time-dependent Transport Processes.* New York: American Elsevier Publishing Company, 1964.

Bellman, Richard, and R. E. Kalaba. *Quasilinearization and Nonlinear Boundary-Value Problems.* New York: American Elsevier Publishing Company, 1965.

Bellman, Richard, R. E. Kalaba, and J. Lockett. *Numerical Inversion of the Laplace Transform.* New York: American Elsevier Publishing Company, 1966.

Brodie, Bernard. *Strategy in the Missile Age.* Princeton, N. J.: Princeton University Press, 1959.

Buchheim, Robert W., and the Staff of The RAND Corporation. *The New Space Handbook: Astronautics and its Applications.* New York: Vintage books, A Division of Random House, Inc., 1963.

Dantzig, G. B. *Linear Programming and Extensions.* Princeton, N. J.: Princeton University Press, 1963.

Dorfman, Robert, Paul A. Samuelson, and Robert M. Solow. *Linear Programming and Economic Analysis.* New York: McGraw-Hill Book Company, Inc., 1958.

Downs, Anthony. *Inside Bureaucracy.* Boston, Mass.: Little, Brown and Company, 1967.

Dresher, Melvin. *Games of Strategy: Theory and Applications.* Englewood Cliffs, N. J.: Prentice-Hall, Inc., 1961.

Ford, L. R., Jr., and D. R. Fulkerson. *Flows in Networks.* Princeton, N. J.: Princeton University Press, 1962.

Harris, Theodore E. *The Theory of Branching Processes.* Berlin, Germany: Springer-Verlag, 1963.

Hitch, Charles J., and Roland McKean. *The Economics of Defense in the Nuclear Age.* Cambridge, Mass.: Harvard University Press, 1960.

Judd, William R. (ed.). *State of Stress in the Earth's Crust.* New York: American Elsevier Publishing Company, Inc., 1964.

SELECTED RAND BOOKS

McCall, J. J., D. C. Jorgenson, and R. Radner. *Optimal Replacement Policy*. Chicago, Illinois: North-Holland Publishing Company and Rand McNally, 1967.

McKean, Roland N. *Efficiency in Government Through Systems Analysis: With Emphasis on Water Resource Development*. New York: John Wiley & Sons, Inc., 1958.

Nelson, Richard R., Merton J. Peck, and Edward D. Kalachek. *Technology, Economic Growth and Public Policy*. Washington, D.C.: The Brookings Institution, 1967.

Novick, David (ed.). *Program Budgeting: Program Analysis and the Federal Budget*. 2nd ed. Harvard University Press, 1967.

Sheppard, J. J. *Human Color Perception*. New York: American Elsevier Publishing Company, Inc., 1968.

Williams, J. D. *The Compleat Strategyst: Being a Primer on the Theory of Games of Strategy*. New York. McGraw-Hill Book Company, Inc., Revised ed., 1966.